NEW ZEALAND'S
ISLANDS

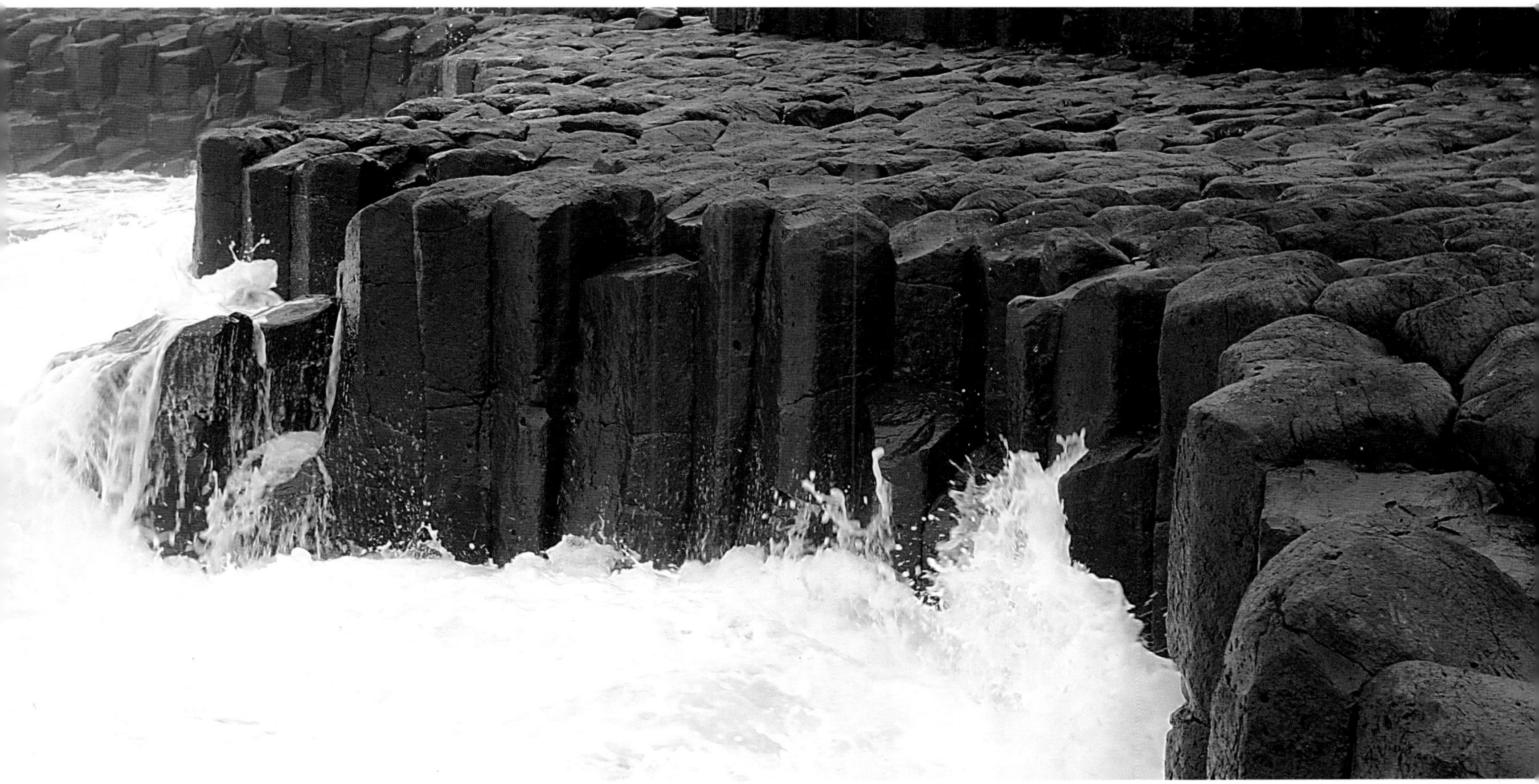

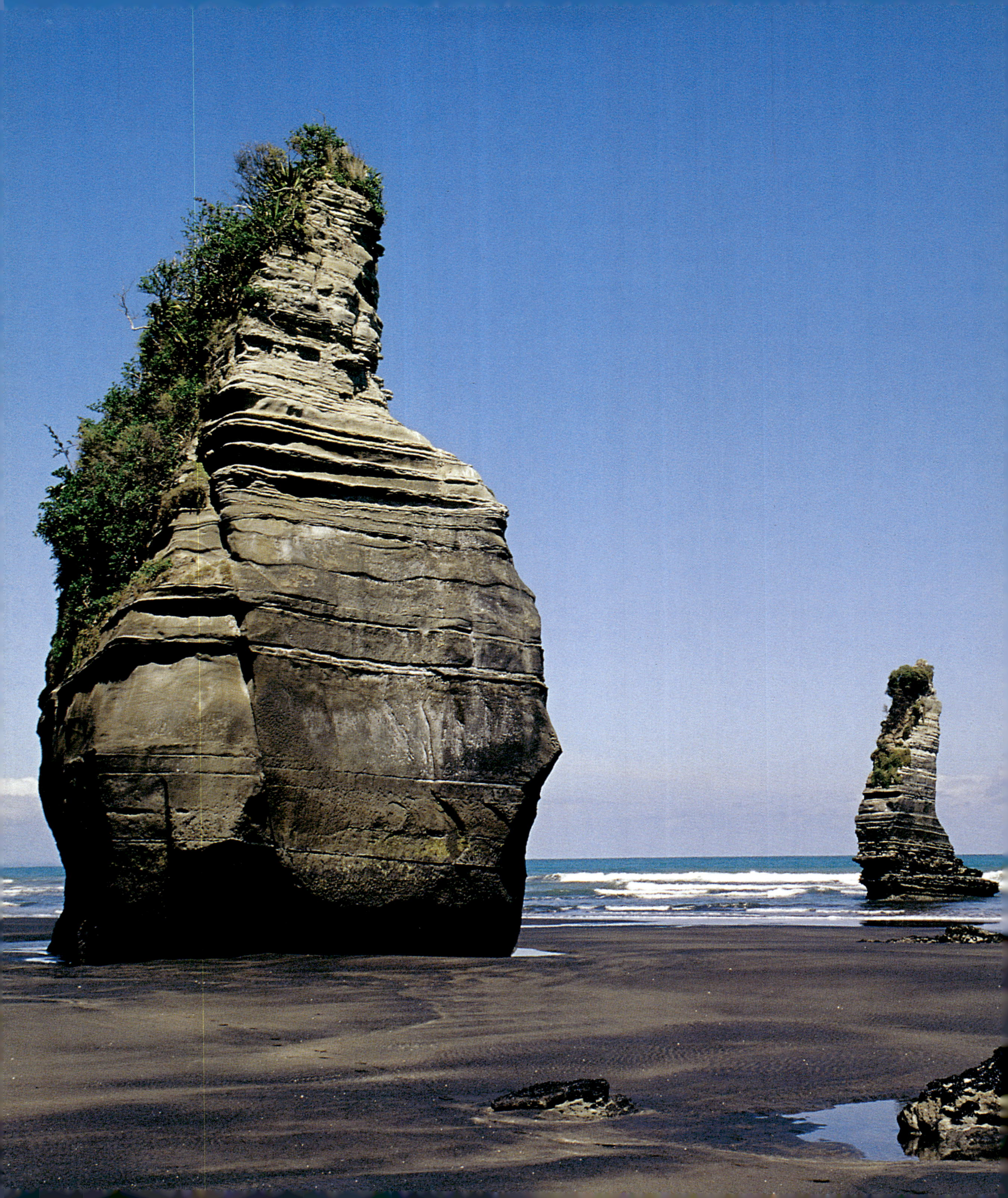

NEW ZEALAND'S ISLANDS

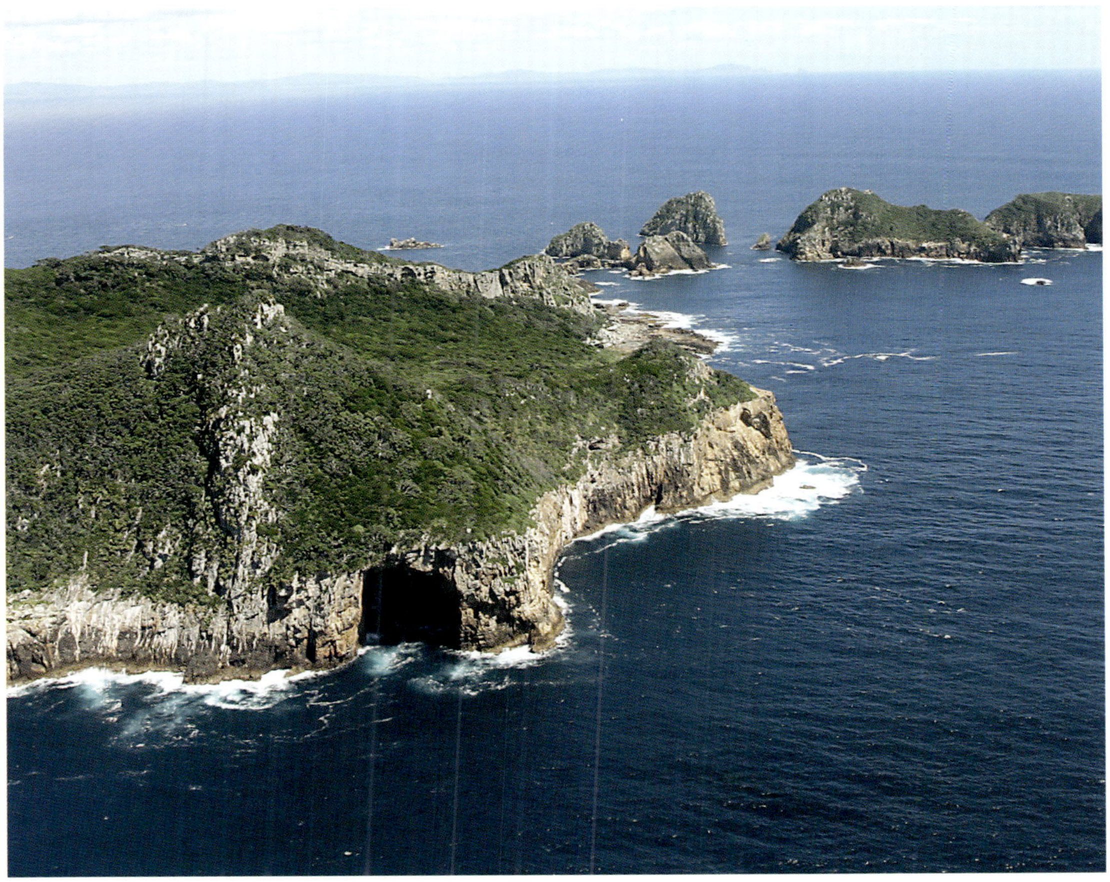

Pamela McGeorge
Photographs by Russell McGeorge

David Bateman

Acknowledgements

Researching the history of the islands, talking to people who live on them, and above all visiting the off-shore islands and photographing them has been a tremendous experience. This is their story.

To the pilots – both the old and the bold – who flew at just the right angle; to the boat skippers who took us where we wanted to go; to the islanders who told us their stories; to our family who housed and fed us on our travels, thank you. To Tracey Borgfeldt, our editor at David Bateman, a heartfelt thank you for suggesting the project and encouraging us along the way.

You can visit many of the islands that feature in the book. Most of them are criss-crossed by roads or tracks which make them accessible on foot, on bike or by car. They form a fantastic addition to our mainland islands. A whole world of adventure lies around the perimeter of the coast. We urge you to get out there and explore it!

Page 1: The typical formation of these basalt columns at Chatham Island occurs as the volcanic rock cools and ages.
Page 2: Rock stacks at Tongaporutu, Taranaki.
Page 3: Poor Knights Islands.
Page 4–5: Surfers enjoy the wild ocean beaches, like Kaitoke, that fringe the eastern side of Great Barrier.
Page 6: Rocky, precipitous and surrounded by sea that is teeming with a fascinating variety of under-water life, the Poor Knights Islands attract visitors from all over the world.
Page 7 Woodhen Cove between Resolution Island and Five Finger Peninsula, at right.

First published 2004 by David Bateman Ltd,
30 Tarndale Grove, Albany, Auckland, New Zealand

ISBN 1-86953-586-3

Maps by Max Oulton
Book design by Red Design, Auckland
Printed in China through Colorcraft, H.K.

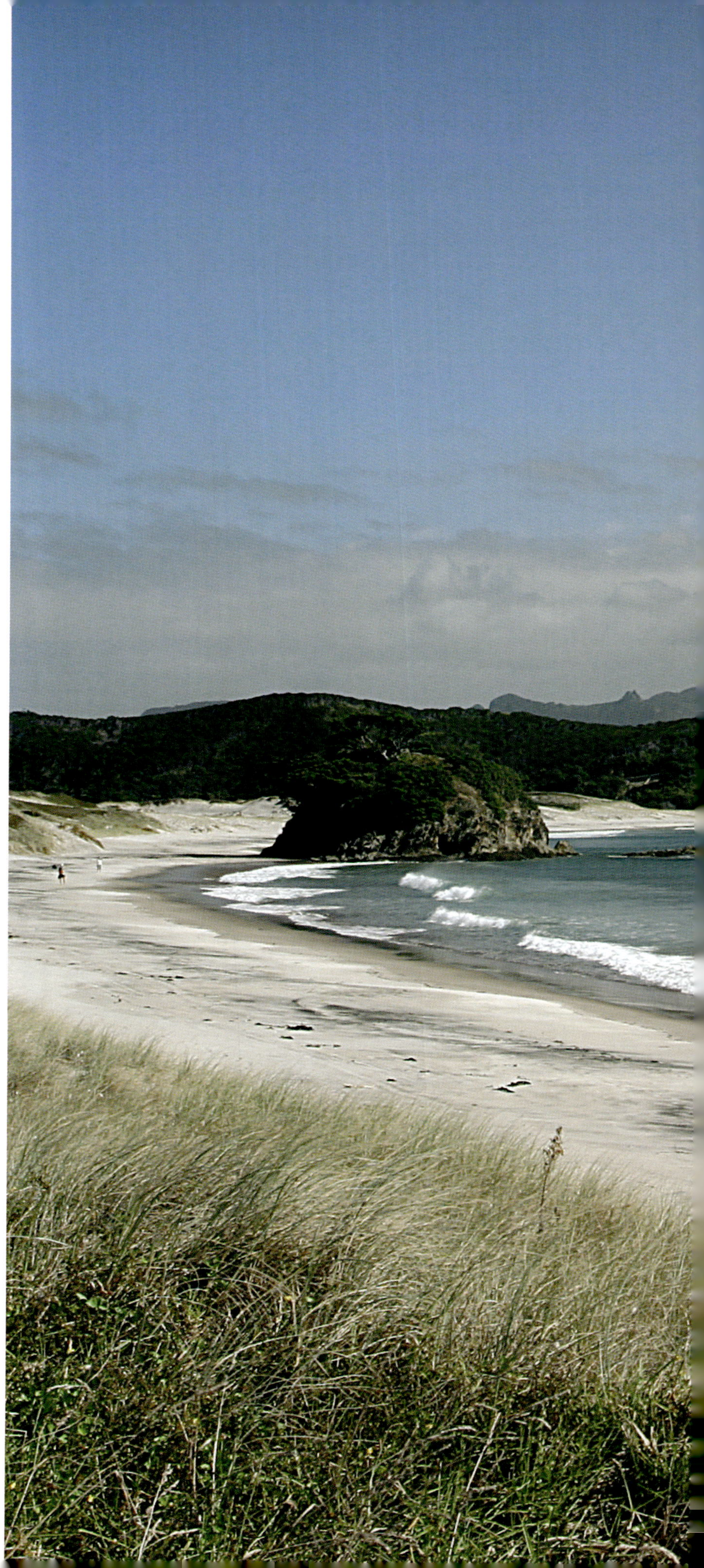

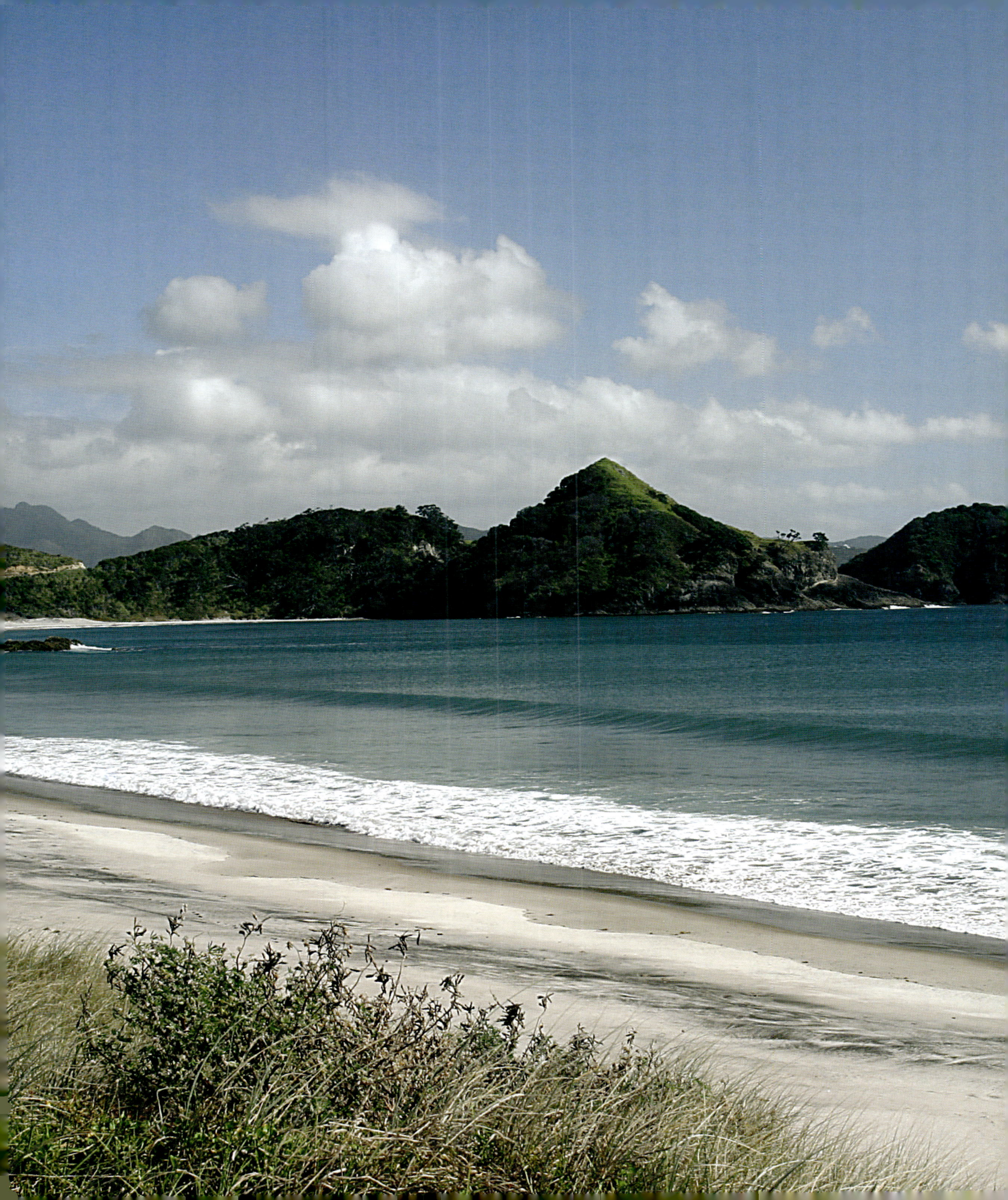

Contents

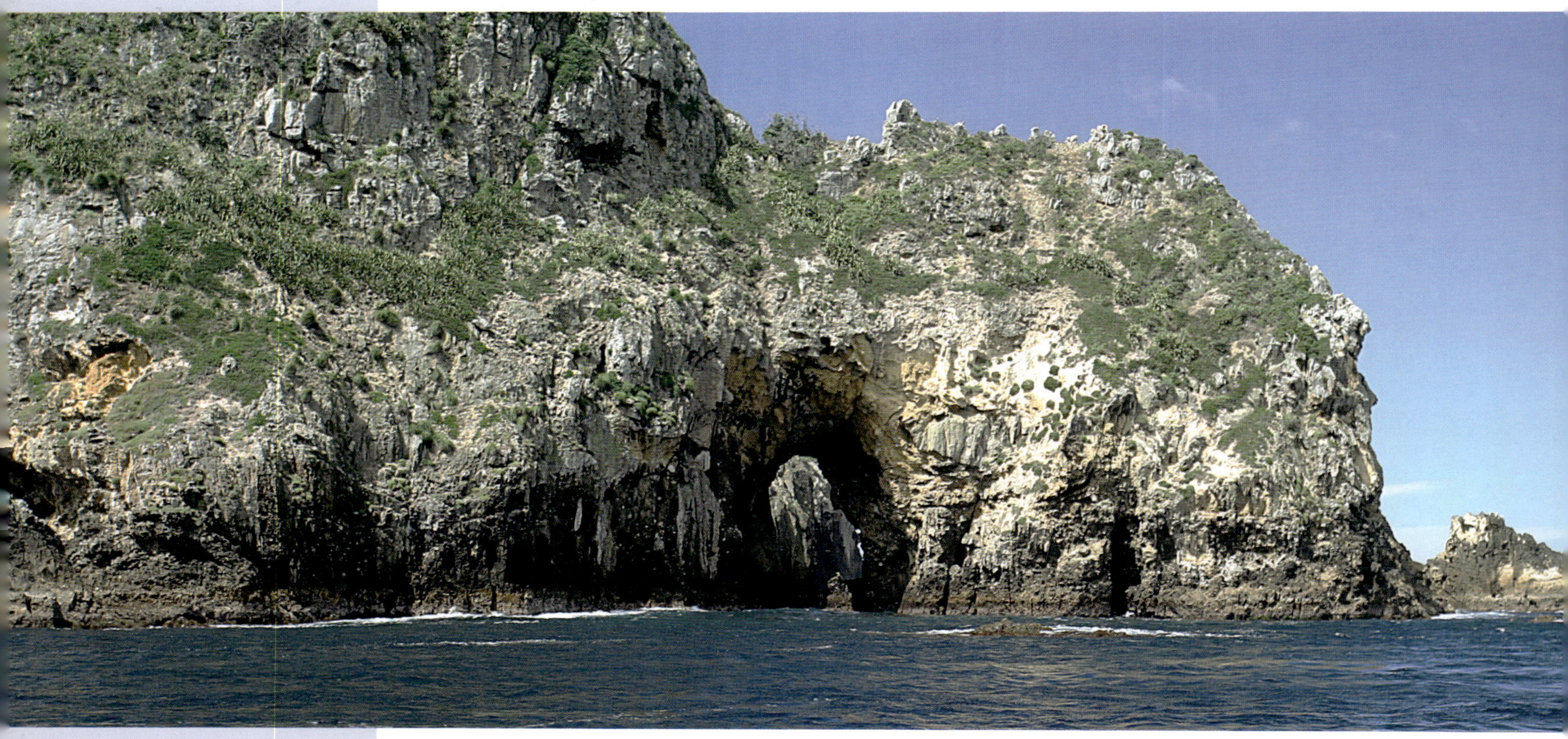

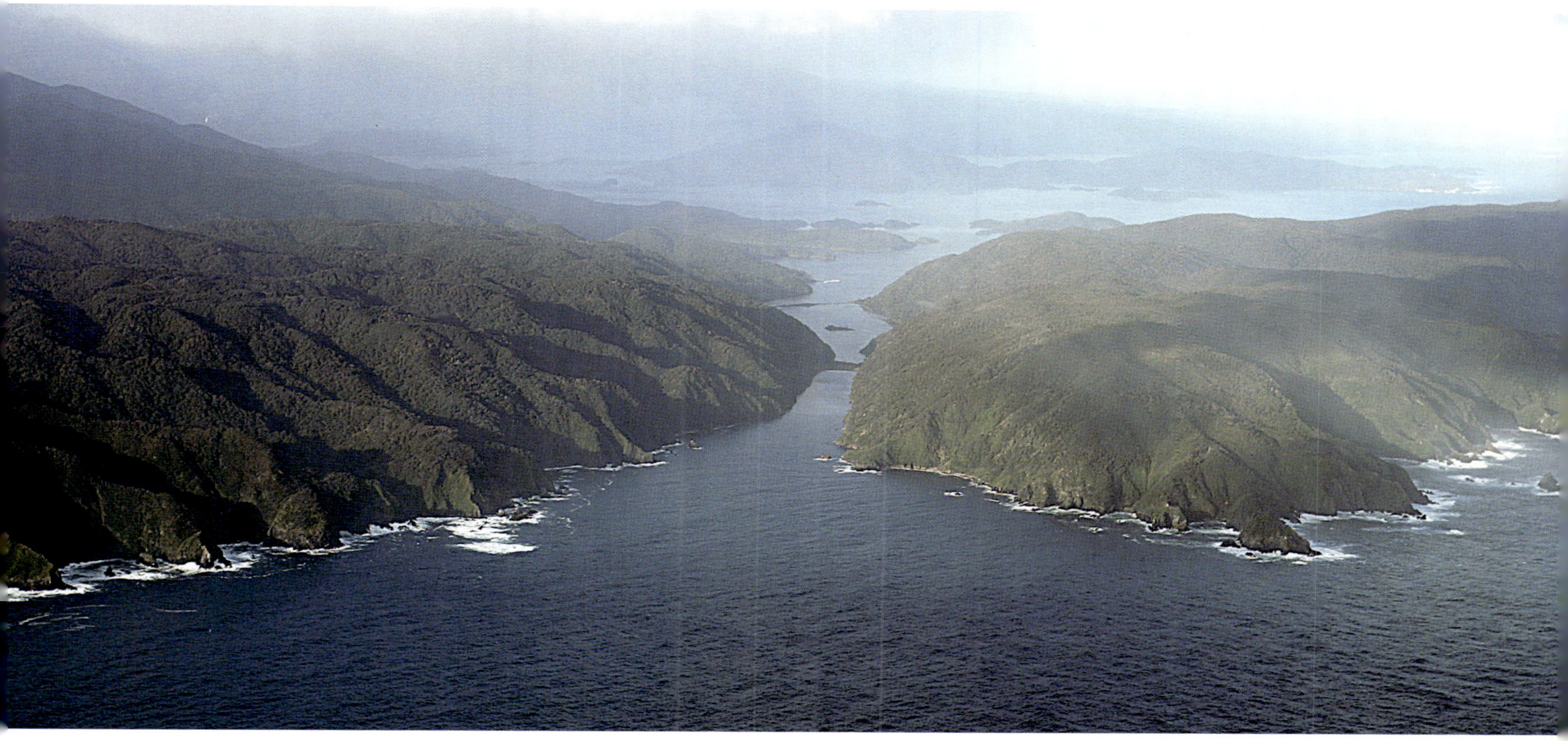

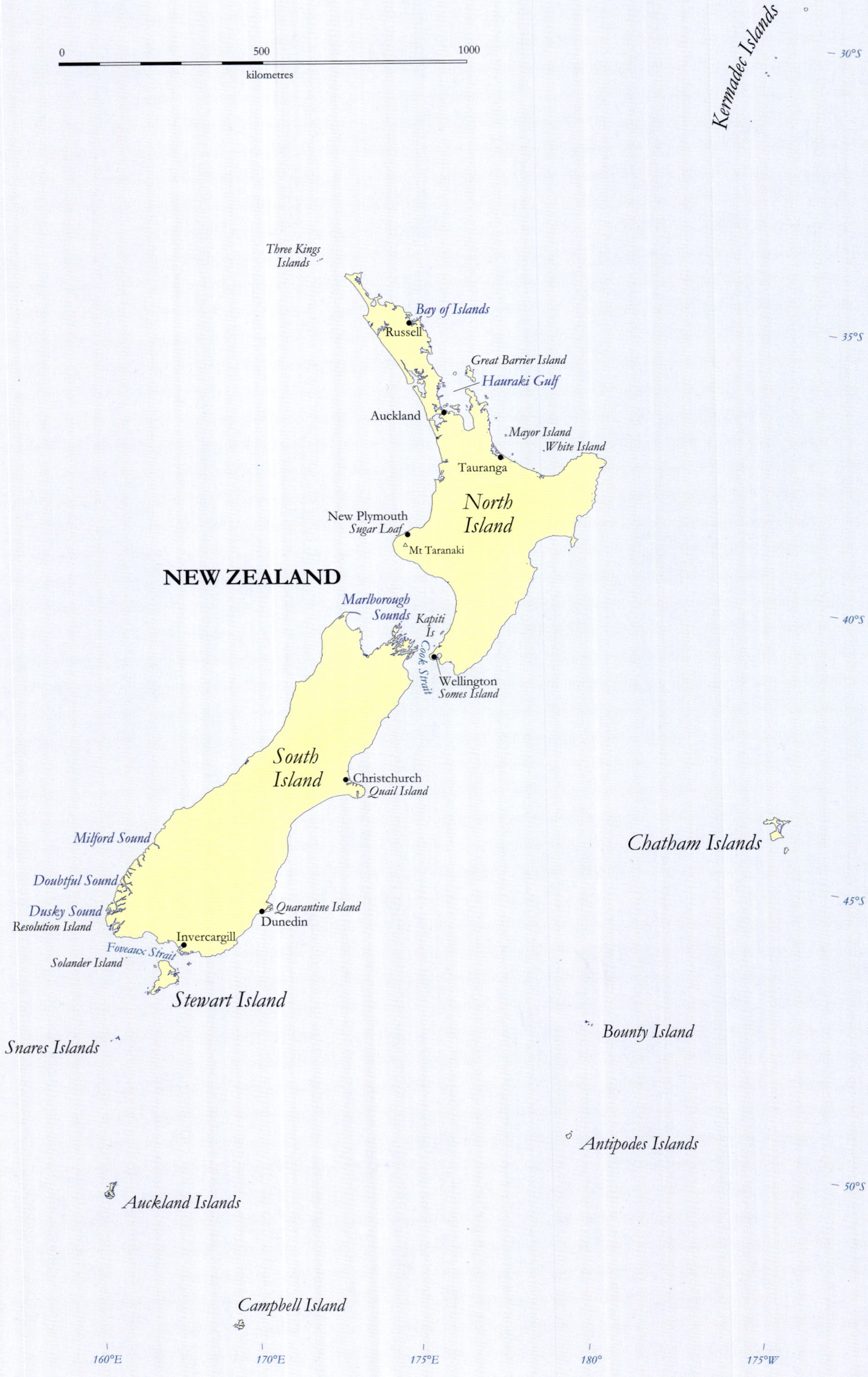

0
500
1000
kilometres
Kermadec Islands
30°S
Three Kings Islands
Bay of Islands
Russell
35°S
Great Barrier Island
Hauraki Gulf
Auckland
Mayor Island
White Island
Tauranga
North Island
New Plymouth
Sugar Loaf
Mt Taranaki
NEW ZEALAND
Marlborough Sounds
Kapiti Is
40°S
Cook Strait
Wellington
Somes Island
South Island
Christchurch
Quail Island
Chatham Islands
Milford Sound
Doubtful Sound
Dusky Sound
Resolution Island
Quarantine Island
Dunedin
45°S
Invercargill
Foveaux Strait
Solander Island
Stewart Island
Snares Islands
Bounty Island
Antipodes Islands
50°S
Auckland Islands
Max Oulton
Campbell Island
160°E
170°E
175°E
180°
175°W

Introduction

THE STORY OF NEW ZEALAND IS THE STORY OF ISLANDS: THEIR IMPRINT SHAPES OUR PSYCHE. THE SOUND OF THE SEA ECHOES IN THE CORNERS OF OUR MINDS; WE LIVE WITH EVER-RESTLESS WINDS; WE HAVE INHERITED THE DREAMS THAT BROUGHT THE FIRST AND SUBSEQUENT VOYAGERS HERE. WE ARE ISLAND DWELLERS WITH ALL THE PRIDE AND STUBBORNNESS AND FAITH IN OUR OWN SOLUTIONS THAT ISOLATION BREEDS.

In the view of most people, New Zealand consists of only two large islands characterized by snow-capped alps and boiling mud pools, dense rain forest, new world cities and the seething ocean. Yet the New Zealand poet A.R.D Fairburn viewed our country differently – and more holistically. These islands:

> the remnant peaks of a lost continent,
> roof of an old world, molten droppings
> from the earth's bowels, gone cold —
> is his description.

His words embrace not only the main islands but also the small chunks of land that sit off the coast of both the North and South Islands – survivors of the last ice-age melt-down or later arrivals, thrown up by angry under-water eruptions.

No ruined Celtic castles record the legends of these islands; no timeless whitewashed houses glitter against a brilliant Mediterranean sky. But prehistoric lizards crawl across their rocks, unique plants cling to their cliffs and when the Celts

Anchor Island, almost bisected by a lake at its 'waist' and surrounded by a flotilla of smaller islands, sits at the entrance to Dusky Sound. Captain Cook explored this area in 1773 and charted the waterways and islands.

were building their castles, Polynesians were establishing fragile footholds on a few of these tiny islands.

Make no mistake, the story of our islands is a long one.

Much has been written about the North and South Islands. This book is not about them. It is a celebration of the *other* islands that surround the shores of the 'mainland', the assemblage of incredibly numerous islands scattered over a broad expanse of ocean that make up the whole country of New Zealand. Depending on your definition of an island there are hundreds, ranging in size from Stewart Island to rock stacks that are little more than sites for gannet colonies or hazards to navigation.

Our government has jurisdiction over islands that extend from the Kermadecs, 1000 kilometres north-east of New Zealand, to the island groups of the New Zealand subantarctic zone 400 kilometres and more south of Stewart Island to the Chatham Islands 800 kilometres east of Christchurch.

Given the spread of latitude encompassed between these specks of land, they experience enormous changes in climate, much greater than those of the 'mainland'. Their topography and geology are likewise extremely varied – from flat tablelands of green fields to raw rock stacks to the smoking marine volcano of White Island. The plants they nurture and the birds that live on them reflect these wide variations of climate and habitat and include some rare species. In the case of *Pennantia baylisiana*, a single tree growing on one of the Three Kings Islands, it is the only wild specimen in the world.

Surprisingly, given their remoteness and frequently wild terrain, people have lived on most of these islands, no matter how small or how bleak, at some stage in their history. Motumaire, a tiny island at almost wading distance from the town of Paihia in the Bay of Islands, is said to have been a centre of learning in pre-European times. Stephen Gerard, a sea captain in the 1930s, described Stephens Island in Cook Strait as 'a gaunt, lofty crag of solid rock towering above tide-troubled waves, a bleak and isolated isle ... The endless winds pour over its bald brow'.

Yet Maori lived there, and later lighthouse keepers, even enjoying the life it offered.

In the past, as now, the inhabited smaller islands of New Zealand have always represented a microcosm of national life. Their histories reflect the history of the whole country but it's like the view seen through a telescope, set in sharp focus, delineated by uncompromising boundaries, honed by separation. So too are the characteristics of the people. Self-sufficiency, resourcefulness and a certain disregard for authority that New Zealanders pride themselves on are more clearly defined, more resolutely guarded among those who have spent their lives on the islands dotted around the coastline of the North and South Islands. The appeal of the last frontier or the sense of escape from mainstream problems is still a potent force for some.

The islands attract international visitors because of their varying, and in some cases unique, qualities. Vulcanologists from around the world regularly come to monitor the activity on White Island. Botanists arrive to study the plant life on islands such as the Three Kings and Poor Knights; bird watchers and biologists visit to learn more about the native birds that have been transferred to protected outposts; and marine biologists come to explore the sea life in the marine reserves that surround several of our islands

The New Zealand subantarctic islands are so remarkable for their high level of biodiversity that they were awarded World Heritage status in 1998 and a marine reserve has been established around the Auckland Islands. Much of their bird and plant life – notably the endemic albatrosses, shags, land birds and megaherbs – is unique. The subantarctic islands also are the haunt of scientists.

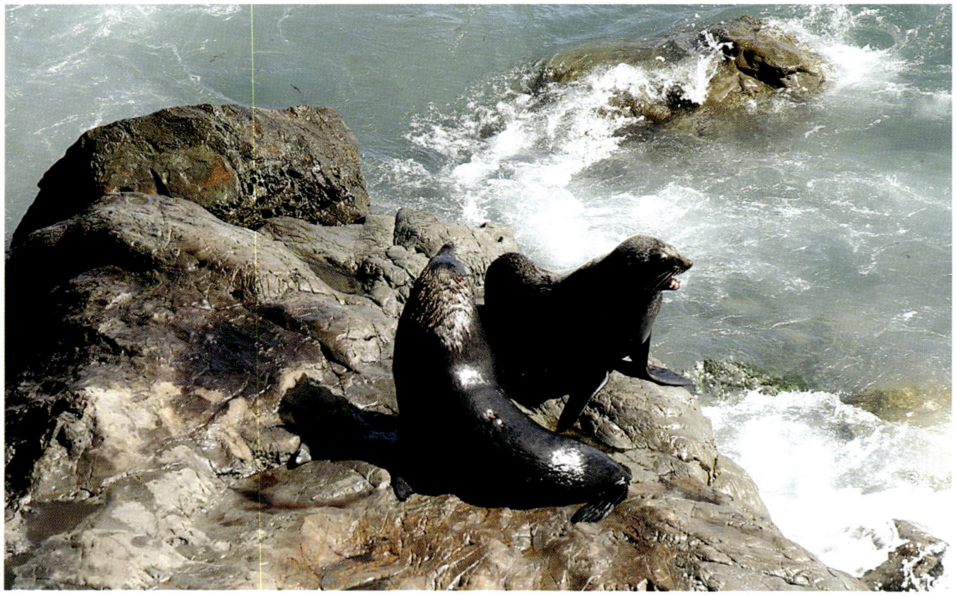

The New Zealand fur seal, formerly prolific on many of the southern islands around New Zealand, is gradually making a comeback after the depredations of sealing in the 19th century.

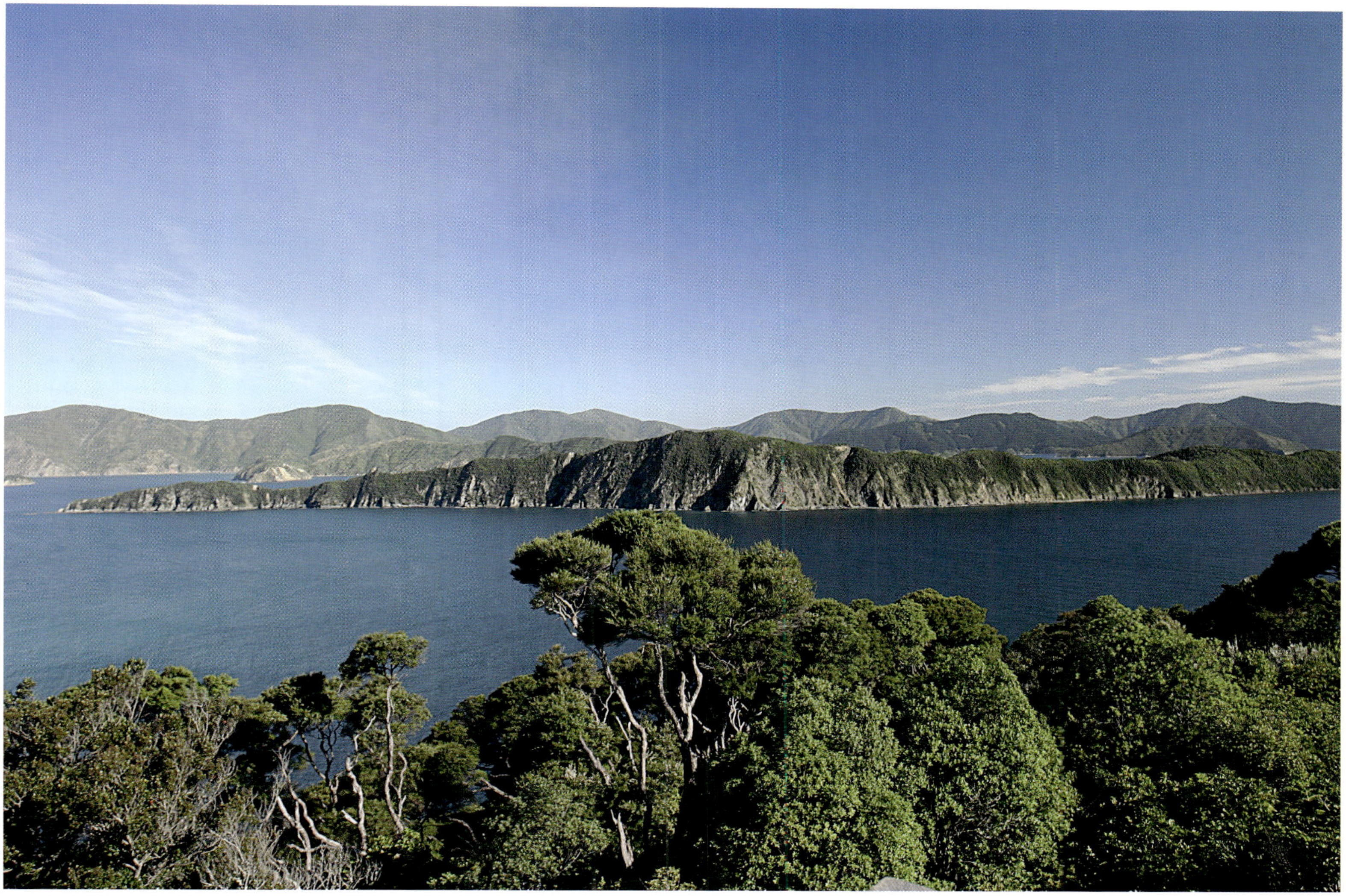

In 1993 a marine reserve was established at Long Island, seen here from the Cook Memorial on Motuara Island.

Why did people settle on these often remote islands originally? Romance, certainly, was far from the minds of the early Polynesians who arrived first. Their visits were prompted by the need for food – seasonal delicacies that could be exploited. The next group to arrive sought profit from minerals, whales, seals, timber and fish. All these natural resources fed the greed or need of successive colonisers, along with food cultivation that required large-scale clearance of bush cover.

Even the subantarctic islands with their freezing winds and inhospitable terrain have suffered their share of misguided human endeavour. An attempt in the mid-19th century to establish a settlement on the Auckland Islands, combining agriculture and whaling, failed miserably. The only legacy, 100 years later, of a subsequent attempt to set up farming on Enderby Island in 1894 was a breed of cattle that somehow survived on a diet of scrub, southern rata and seaweed, long after the farming enterprise was abandoned.

Life on the smaller islands continued to be a pioneering existence well into the 20th century and long after most other parts of New Zealand had progressed beyond the demands of such a self-reliant way of life. Danger was a constant companion. Access to education was – and often still is – a problem. In many cases the need for mothers to teach their own children by correspondence imposed a heavy burden. Success depended heavily on the role women played, for it is usually they who have had to bear the brunt of the inconveniences of island life. Lack of electricity reticulation caused daily hardship in the home until recently on many islands. Arapawa Island in the Marlborough Sounds, for example, did not link into the national power grid until the mid-1980s and Great Barrier Island residents still rely on their own individual systems

– be it windmill, solar or generator – to fill their energy needs.

People were attracted not just to the islands' resources; they were often exploited too for their strategic importance. A rocky outpost, surrounded by seething ocean, was a much easier site to defend than a mainland hilltop, and archaeological evidence of defensive pa sites on remote islands emphasises this reality. Imagination quails, however, at the thought of escape from a sea-girt fortress should the defence fail. And certainly the fate of the resident Maori tribe on the Poor Knights, all slaughtered or enslaved when they were ambushed by invaders in the 19th century, makes me glad I was not one of these island dwellers of the past.

Europeans, settling in New Zealand in the 19th century, were quick to make use of the strategic isolation of islands. When the spread of life-threatening diseases caused panic in their new settlements, island quarantine stations, with rigidly enforced restrictions, were the only effective places to contain such infections. At the same time, increased shipping traffic in New Zealand waters, with the attendant need for safe navigation, saw lighthouses sprout on many island outposts. Lighthouses had to be manned and the keepers and their families suffered years of loneliness, occasional deprivation and often mental depression, maintaining the life-preserving lights. Not until the early 1990s were all the lights around New Zealand automated.

The threat of war in the 20th century re-iterated the strategic importance of the outlying islands. For as New Zealand poet Allen Curnow has written:

> Always to islanders danger
> Is what comes over the sea …

Gun emplacements were built on several islands in the Hauraki Gulf and on Somes Island in Wellington Harbour, but these installations did not stop a German submarine from entering New Zealand waters. Even on far away Campbell Island, a coast watching station was set up in 1941 and manned for the remaining years of the Second World War. Later a weather station was established in its place and meteorologists lived there continuously until 1995 when it was automated.

Living in such isolation makes extraordinary demands on those who choose that lifestyle or have it forced upon them.

No story of islands is complete without mention of the boats that service them. In the early days of New Zealand's European history

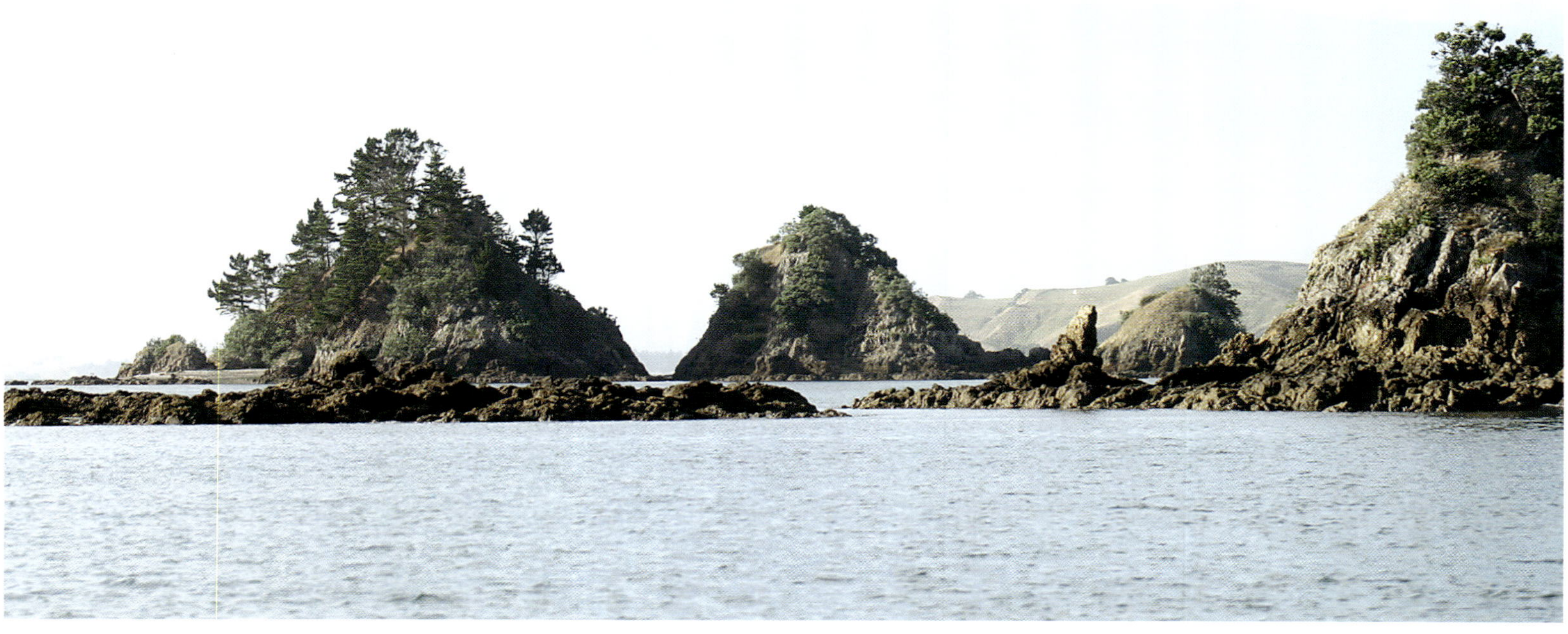

Rocky outcrops like these islets on the western side of Rakino Island add to the beauty and interest of the Hauraki Gulf, right on Auckland's doorstep.

roads throughout the country were few and far between. With a long coastline and settlements at widely spaced intervals, the early commerce of the country was dependent on boats. Shipping goods to or from a little island offshore was no more complicated than shipping around the coast of the main islands.

The real workhorse boats of New Zealand waters in those early years and up until the Second World War were shallow-hulled scows, often built of kauri, powered first by sail, later by motor. They plied their way between islands and the 'mainland' carrying sand, shingle, timber, building materials, livestock and all the other necessities of island living.

Passenger services were provided initially by small sailing boats but by the last two decades of the 19th century steam had all but replaced sail. In the early years of the 20th century steamboats also led to a vast increase in excursion travel, especially out of Auckland where the islands of the Hauraki Gulf provided excellent venues for day picnics. In 1903, a total of 10 steamers ran 15 trips from Auckland to Motutapu for an Oddfellows' annual picnic, carrying between 12,000 and 14,000 day-trippers!

As recently as 2003 a 24-metre general-purpose, ocean-going landing barge was launched in Auckland. It was designed by Graeme Lidgard, member of a well-known boat-building family with a long connection to Kawau Island. Its purpose is to serve the smaller island communities on New Zealand's north coast.

The story of New Zealand's islands continues to evolve as world economies change. Transport

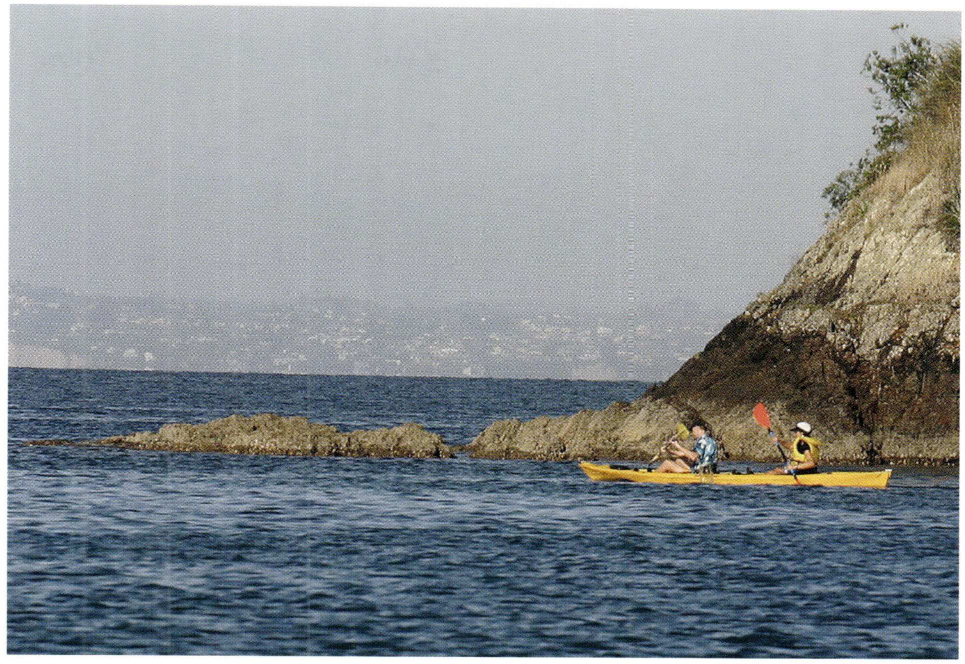

Kayakers enjoy a day out in the sheltered waters of the Hauraki Gulf.

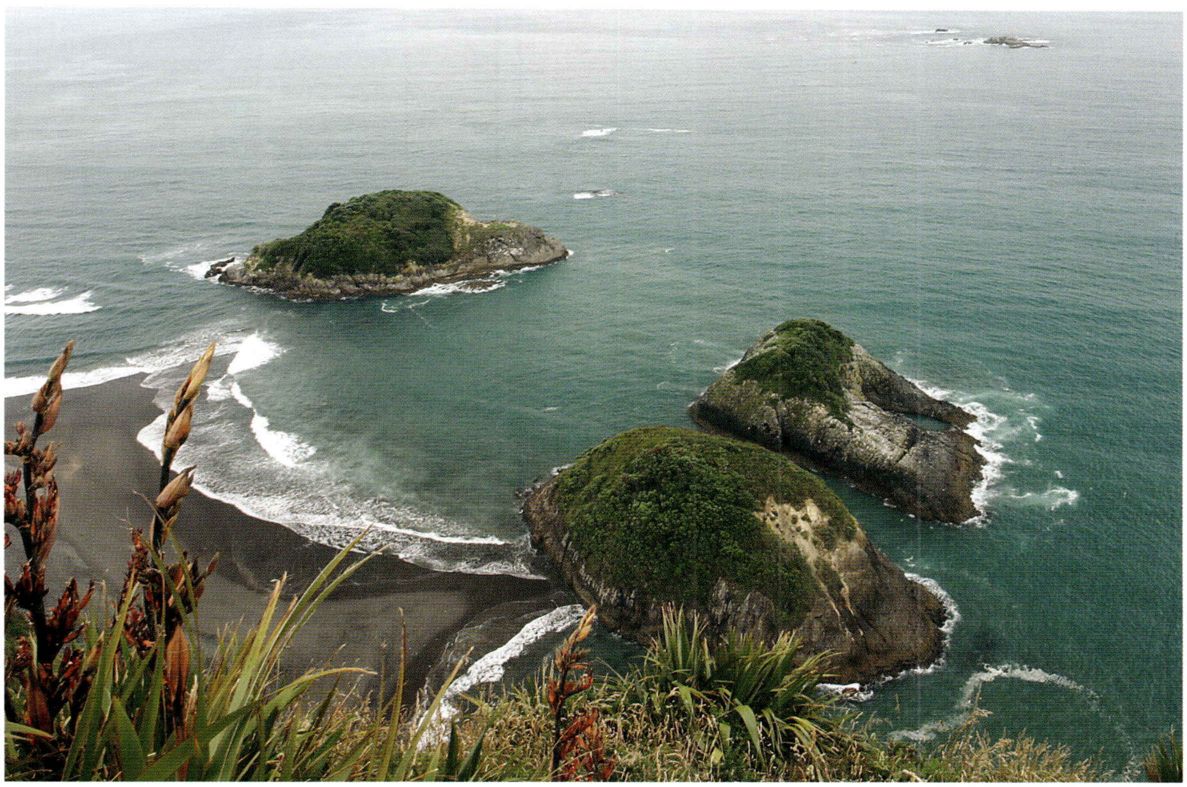

These three islands nestle in the sea below Paritutu, beside the city of New Plymouth.

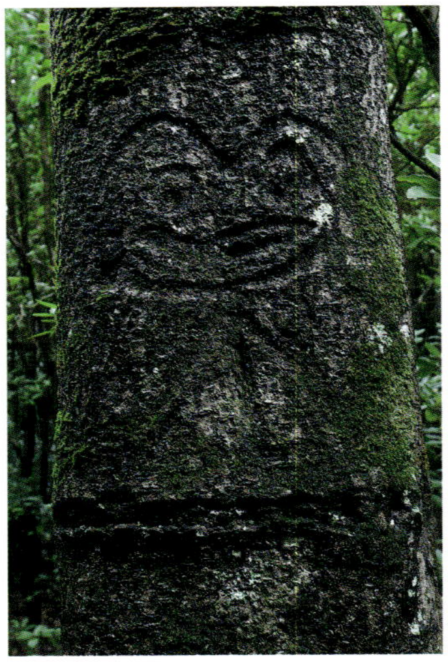

Dendroglyphs – or tree carvings – at the historic reserve of Hapupu on Chatham Island are among the few relics still in existence of Moriori life before the arrival of Europeans and Maori.

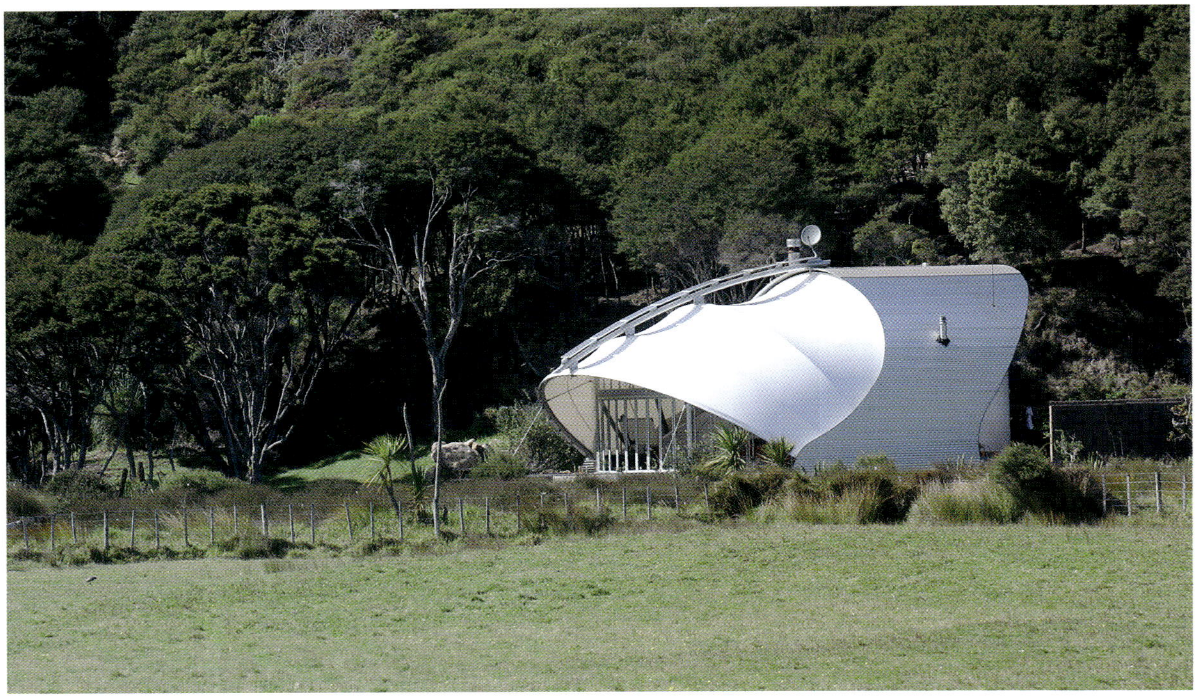

Right: Modern architecture on Great Barrier Island.

costs now make farming uneconomic on some islands but horticulture flourishes where the climate is suitable. Bird sanctuaries have long since replaced Maori strongholds, and beaches that once ran red with the blood of seals and whales are now restful holiday destinations. Tourism is replacing fishing. On some islands, life has come full circle and cleared farmland is reverting to native bush, a homeland for threatened species.

The concept of setting islands aside as wildlife reserves came surprisingly early in the colony's history. The idea was first suggested in 1870, but by the time Resolution Island, Kapiti and Little Barrier were granted reserve status in the 1890s, the huia had all but disappeared and it looked very much as if the hihi or stitchbird would go the same way. In fact a pair of huia had been captured in 1893, intended for the newly declared sanctuary of Kapiti Island. They never arrived. Somehow the naturalist Sir Walter Buller managed to appropriate them and take them back to England with him – stuffed, of course.

The idea of creating sanctuaries was good in theory but the difficulty in practice was actually acquiring the land. Most islands were still in Maori ownership at the end of the 19th century and the tribal owners were generally reluctant by then to part with more land. The answer, as always, was legislation!

From early, stumbling beginnings the concept of island conservation has burgeoned and several bird species have probably been saved from extinction due to the foresight of a small number of individuals – lighthouse keepers, private landowners or members of the New Zealand Wildlife Service – who recognised the special role islands could play as sanctuaries. The Department of Conservation (DOC) now manages or has an interest in more than 220 islands. The first step in restoration of threatened species – plant or animal – is pest eradication and New Zealand is viewed internationally as a world leader in island restoration. The kakapo breeding on

Originally built by the American ex-whaler Lewis Acker, around 1840, this stone house at Harrolds Bay is the oldest remaining building on Stewart Island.

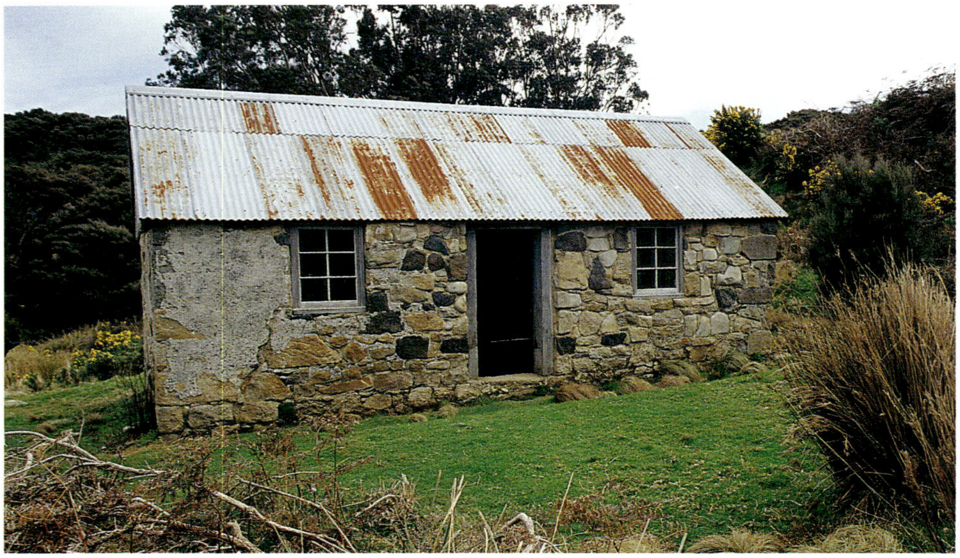

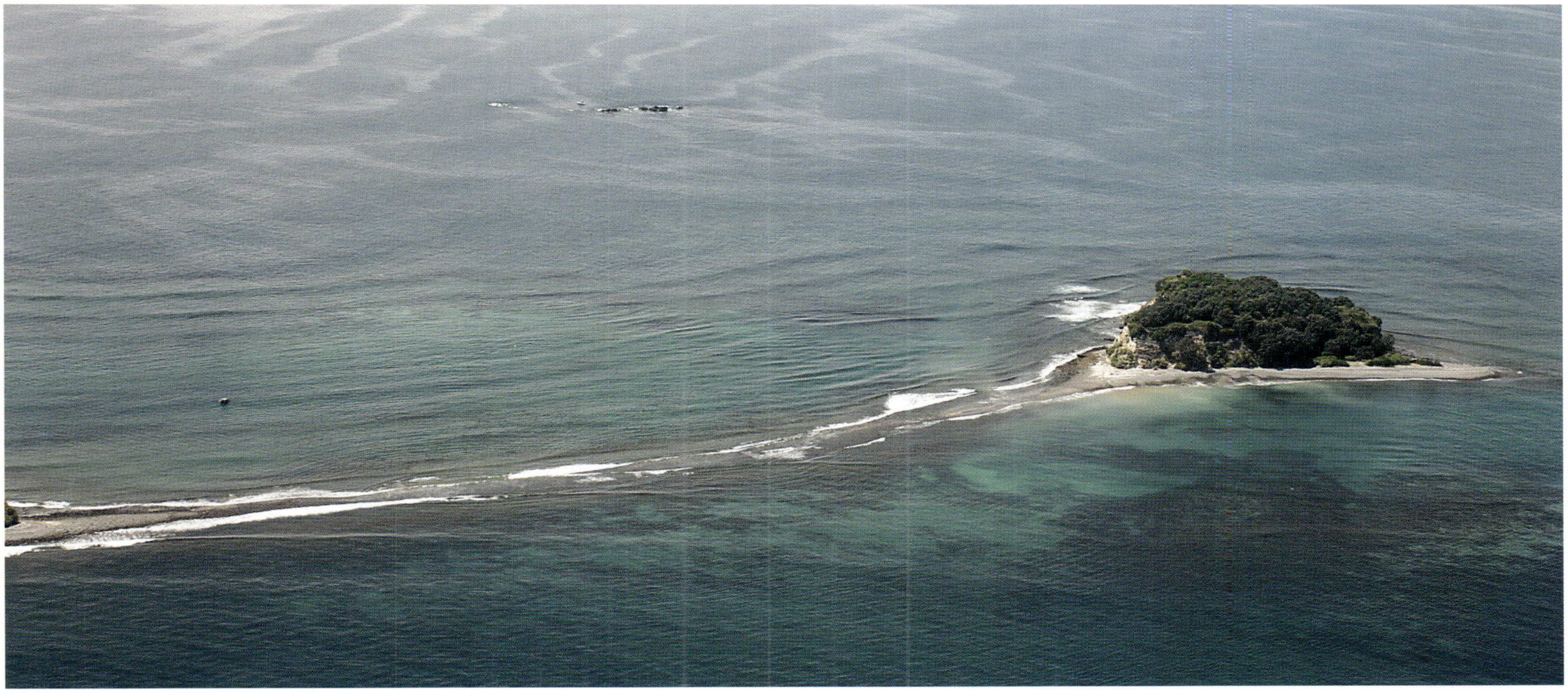

Tiny Taumaihi Island just south of Motiti Island, off Papamoa Beach, Tauranga.

Codfish Island, the saddlebacks or tieke we saw feeding unconcernedly on Ulva, Mokoia and Kapiti Islands and the blue penguins breeding on Motuara Island in Marlborough are all testament to the success of pest eradication.

Not only birds are relocated. Other threatened species include plants, invertebrates or reptiles such as tuatara and lizards. Birds tend to attract more public enthusiasm than creepie crawlies but all have a part to play in a balanced ecosystem.

Special interest groups, such as the Forest and Bird Society, have also worked long and hard towards island restoration and increasingly community groups too are working alongside DOC. It's that romance thing again – the concept of caring for an island has enormous appeal. In many cases, projects such as the transfer of birds are promoted by DOC and supported by local Maori, or local businesses, who enjoy the positive advertising such activities bring.

At Matakohe-Limestone Island, situated in Whangarei Harbour, Youth Hostel Association members have been involved in plantings of native species. At Moturemo, a tiny island in the Kaipara Harbour, the mid-North branch of the Forest and Bird Society paid for rat poison and helped with the laying of it. The Motuora Restoration Society, DOC, and a branch of Forest and Bird have joined forces to restore the vegetation and create nursery facilities for young zoo-bred kiwi on Motuora Island near Kawau.

Moturata – Rata Island – located off the mouth of the Taieri River, south of Dunedin, has been the focus of a revegetation project initiated by the Moturata whanau (of the Ngai Tahu tribe). Archaeological sites on the island had been exposed by erosion. Now, by encouraging topsoil to build up in which natives can be planted, it is hoped to restore the island to its former state. Maybe rata trees will once again bloom there.

Island dwellers all of us, we have developed a protective attitude towards these scraps of land with tantalising links to the world beyond our collective past.

Numerous people have written books about specific islands off the coast of the 'mainland', which include a detailed history of each island and the people who have been central to their

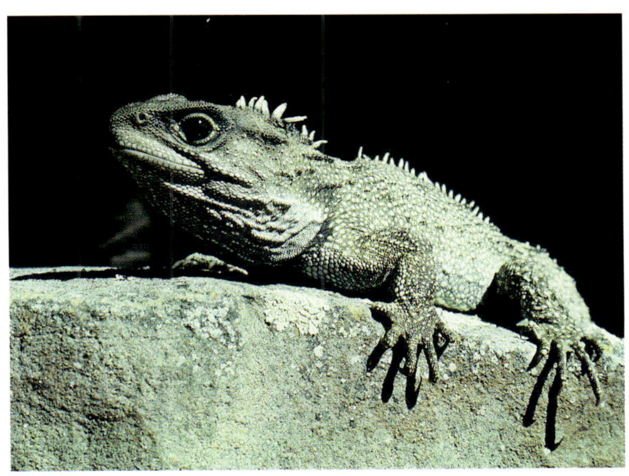

It is estimated that there may be more than 50,000 tuatara, along with other endangered native species, living on Stephens Island in Cook Strait.

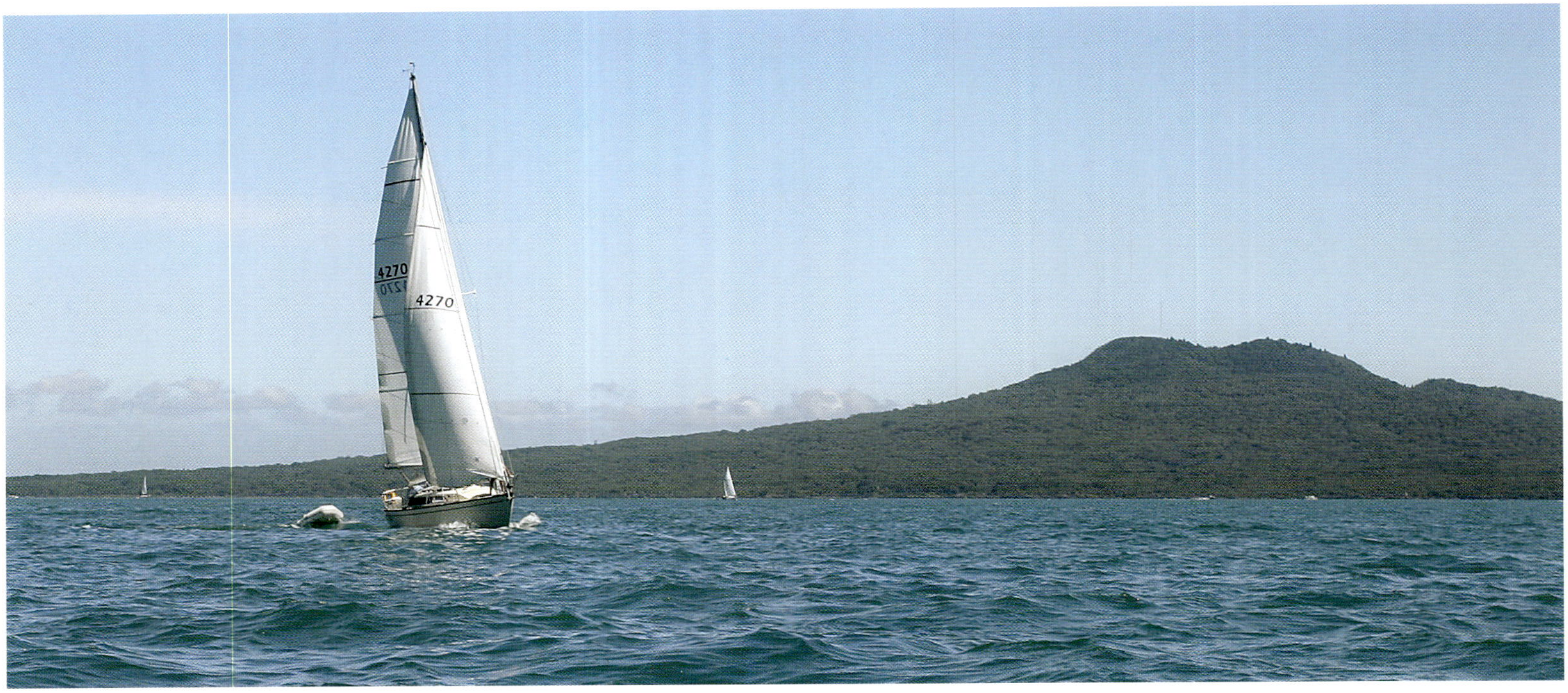

Billowing sails near Rangitoto Island.

Right: Boats moored at Rakino Island. The Hauraki Gulf, venue for the 2000 and 2003 America's Cup, is a mecca for sailors from all over the world.

development and commerce. Several of these are listed in the bibliography at the back of this book.

Our purpose has been more general. We have approached this book as a 'painting in words and pictures' to acquaint the reader with some of New Zealand's offshore and lake islands in the hope that you too will want to visit as many of them as possible and enjoy, as we have, their beauty and amazing diversity. Of necessity we could not include them all. We have had to make selections and decisions on what to include and how to treat our topic. The general inaccessibility and enormous expense of travelling to the subantarctic islands, to Bounty Island, the Snares Islands and the Antipodes or the Kermadecs meant that we reluctantly excluded them.

Not a rigorous geographical or historical record of the islands, this is a collection of essays that in our view tells the story of particular islands – be it their history, geology, topography or land use that is of interest.

But with all the islands the story of their human habitation is probably the most fascinating. We have spoken to people who have lived on islands all their lives and to others who have arrived on their chosen islands more recently. We have read about those who lived there in the past. There are stories we did not have space to include. Others, I'm sure, we did not uncover.

We visited as many of the islands as possible by boat. Then we flew over them and experienced them in a totally new dimension. Flying low over the land and sea in a small plane is an exhilarating way to gain an understanding of the geography of the country and the layout and texture of its islands.

Sometimes we explored new territory, other times we revisited favourite haunts. In the past we spent many holidays and weekends getting to know the islands of the Hauraki Gulf, the Bay of Islands and further afield. We have sailed around them, moored in their bays, clambered over them, snorkelled in their waters and gloried in the wonderful swimming opportunities around their shores. Going back to these places has reinforced for us what a spectacular country we live in.

Recently we sat on Palm Beach at Waiheke Island. It was late January, darkness was gathering. A few boats lay quietly at anchor. A family by the water were the only people in sight. They wandered up the beach and the children squatted beside us, eager for a chat.

'What are you doing here?' asked a bright-eyed, knowledgeable five year old.

'We're writing a book,' we answered.

'Are you an artist?' he asked the photographer.

'I guess so,' Russell replied.

'I'm an artist too,' he replied. 'I paint.'

A summer night. Moonlight on the sea. All things are possible.

This is the magic of islands.

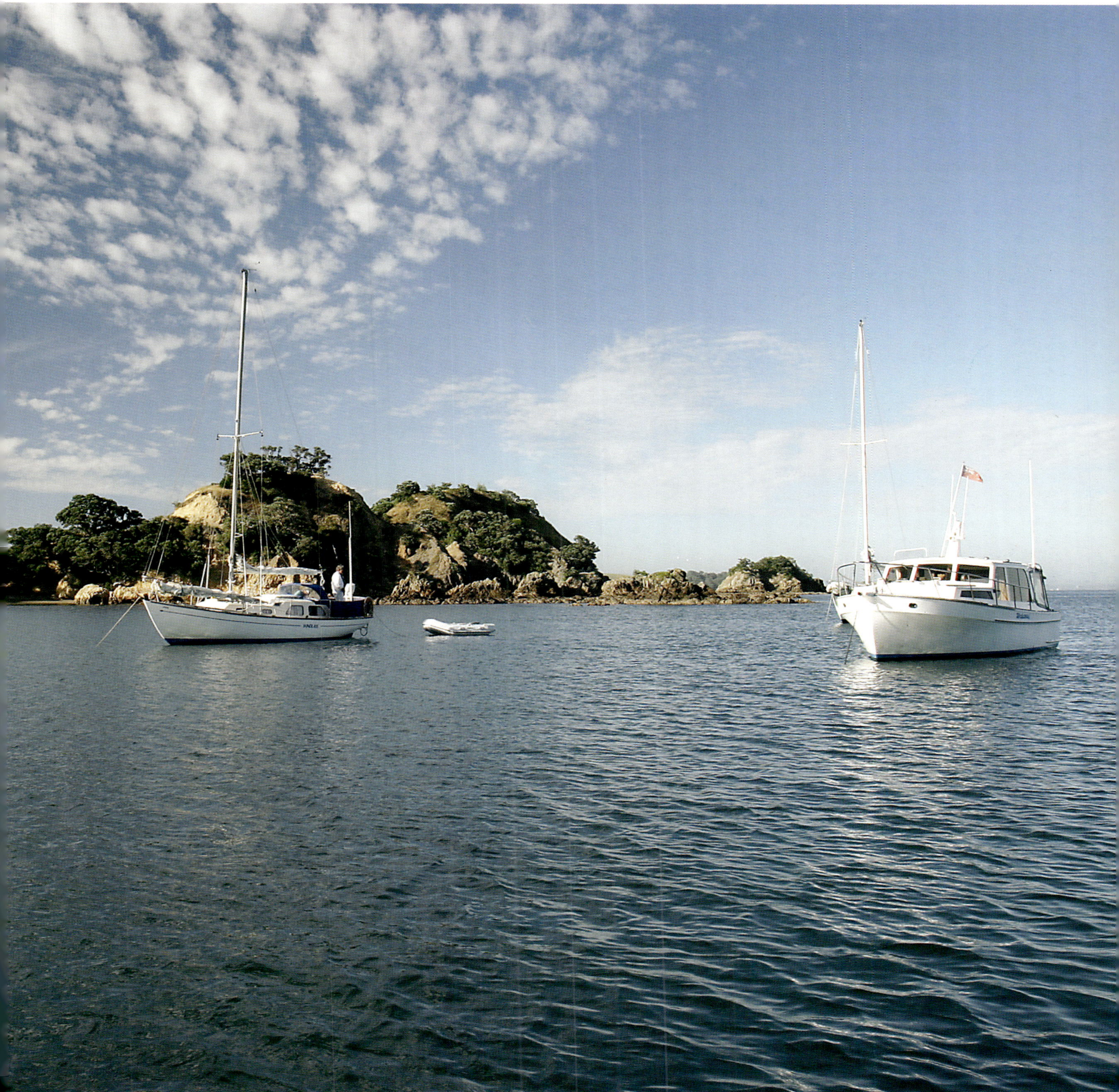

Islands of the North

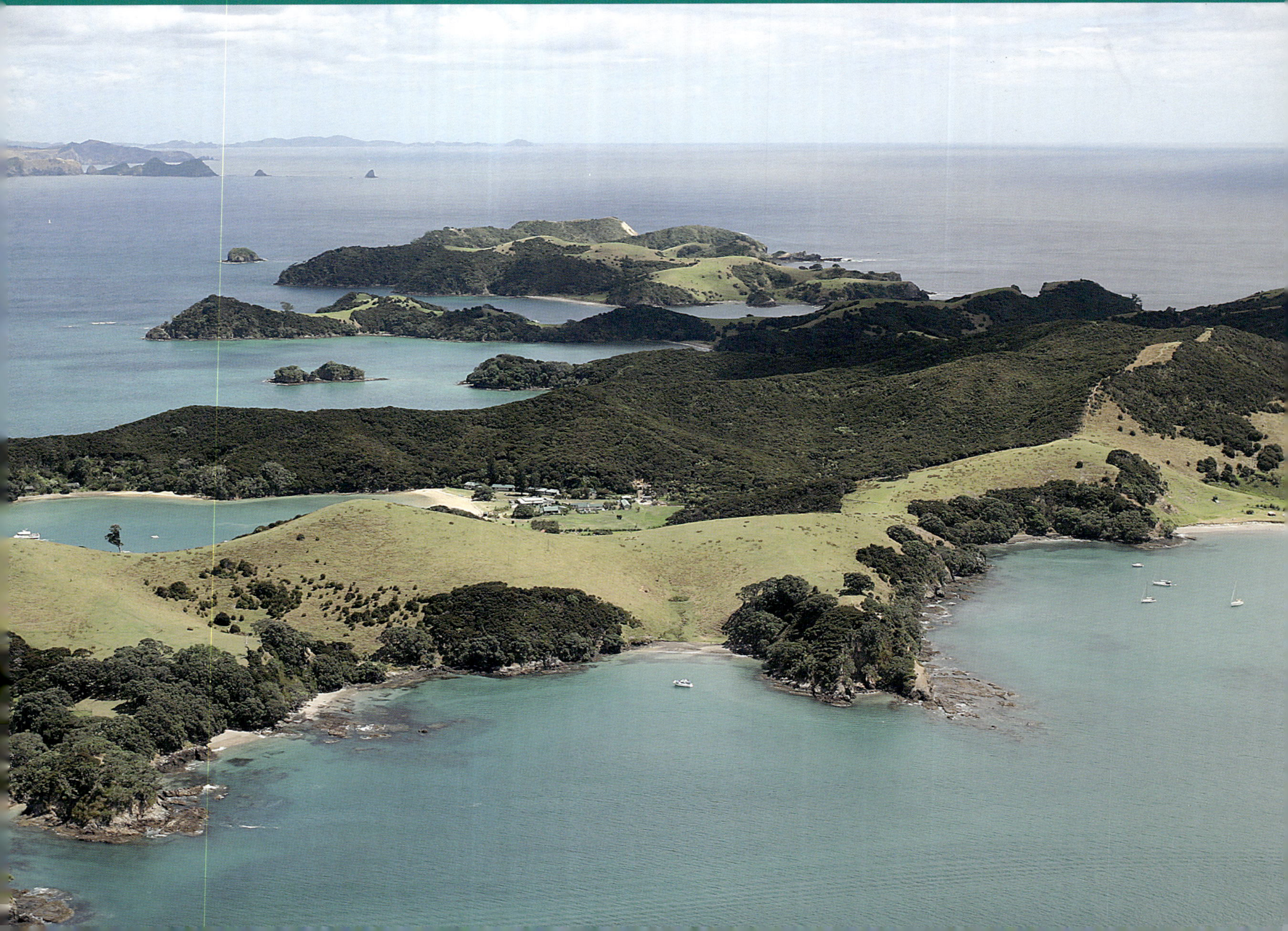

There is a profusion of islands off the coast of the mainland in the 'winterless north'. Once exploited for their food supplies by Maori or valued as easy-to-defend pa sites, some are still privately owned, many are tourist destinations and yet others are set aside as nature reserves. Visitors come to the islands to enjoy the magnificent fishing, diving and boating opportunities they offer or merely to relax on their beaches and revel in a largely unspoiled environment.

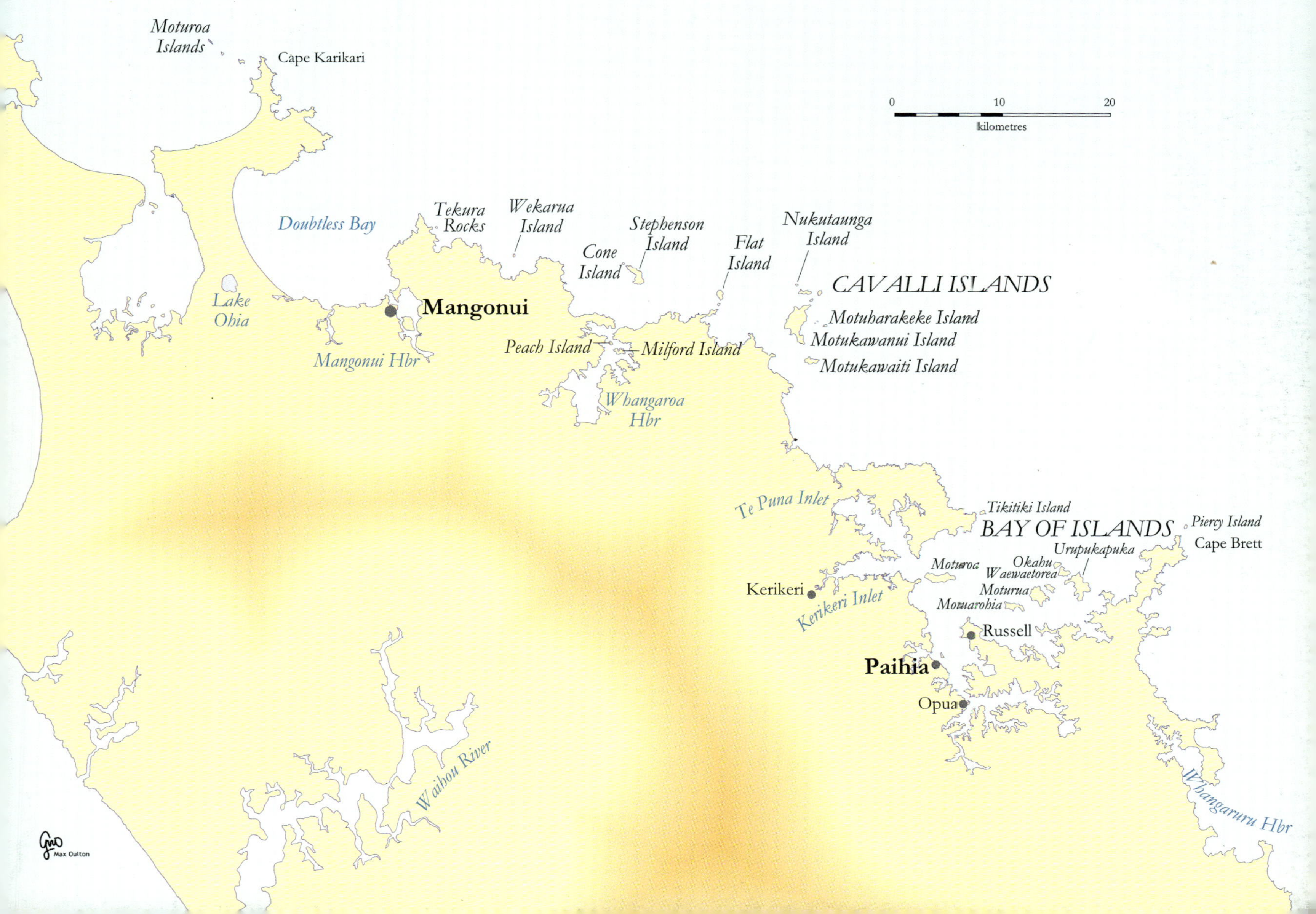

Three Kings Islands

WINDSWEPT AND FOG BOUND, THE THREE KINGS ISLANDS, NEARLY 60 KILOMETRES NORTH-WEST OF CAPE REINGA, ARE THE LAST GASP OF LAND THAT TRAVELLERS SEE AS THEY LEAVE NEW ZEALAND AND HEAD NORTH. THEY OFFER SPARSE COMFORT TO HUMANS.

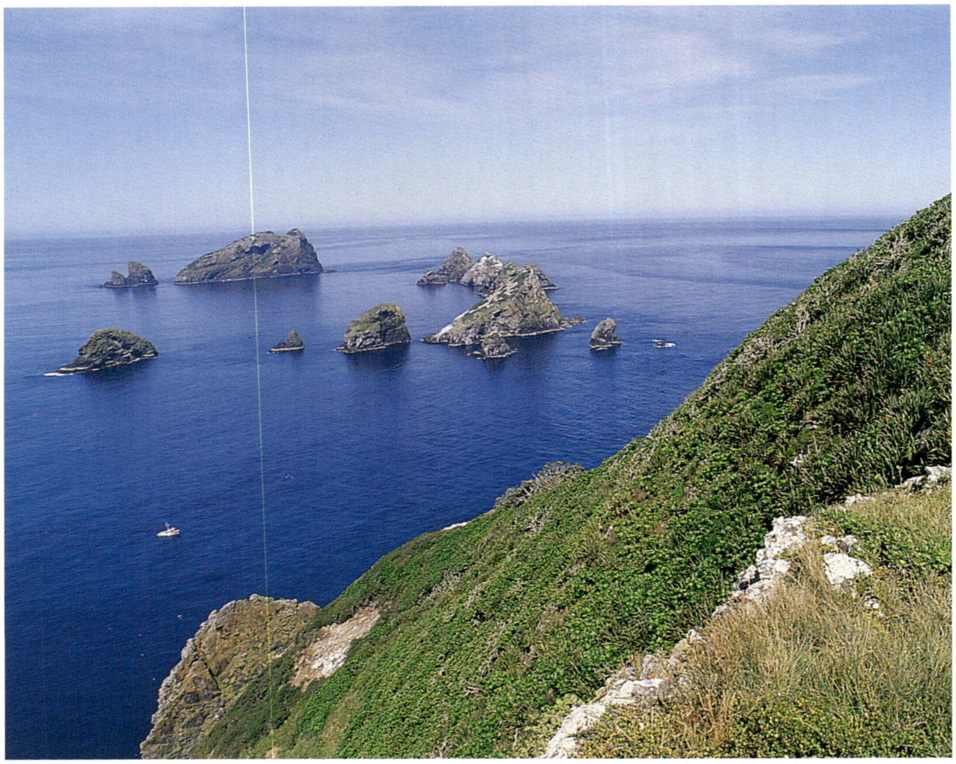

Three Kings Islands drift out to sea at the top of the North Island. Great Island, as its name suggests, is the largest. South West Island is much smaller with tiny West Island as the third 'king'. Shown here are the rocky pinnacles called the Princes Islands, with West Island at the back.

Previous Page: The idyllic Bay of Islands, looking north from Urupukapuka Island with Zane Grey's fishing camp in the foreground.

So steep that visitors have difficulty in finding a landing space, the bigger islands were nevertheless home to a limited number of Maori when Abel Tasman first saw the group in early January 1643. He described Maori gardens in a broad valley on Great Island and he attempted to send crewmen ashore to replenish his ship's water barrels from a waterfall that was visible from their anchorage. Heavy surf prevented them from landing on the first occasion. Thereafter, the sight of a group of Maori standing on the 'highest mountain' of the island, armed with clubs, dissuaded the Dutchmen from trying again. Tasman contented himself with naming the islands in commemoration of the day of Epiphany then sailed away, waiting until they reached Tonga to get fresh water.

The Three Kings consist of 13 islands surrounded by more islets and rocks. North East Island, where remains of Maori gardens have also been discovered, South West Island and West Island are all named in relation to Great Island, which sits in the middle of the group. Long abandoned by any human inhabitants, they were bought by the Crown from their Maori owners in 1908 and declared a sanctuary in 1930. Now the only legitimate visitors are Department of Conservation staff or the occasional scientist. The Three Kings Islands are among the island strongholds in the chain of wild islands around the coastline of New Zealand that shelter many of our native species – birds, reptiles, insects and plants.

These lonely islands are known in the botanical world for several originally rare native species that have become sought-after garden plants in recent years. With leathery leaves like large paddles, some of these trees have trunks that are 60 centimetres through. When pukanui (*Meryta sinclairii*) were discovered growing on the islands in 1889 by the botanist Thomas Cheeseman, they were rare plants. Still not common in the wild they are, however, widely available in garden centres and make handsome garden specimens. Growing in association with the pukanui is a kawakawa, recently described as the new species *Macropiper melchior*, a plant becoming popular in northern gardens.

In 1945 an exciting discovery was made by Dunedin botanist Geoff Bayliss when he found a single specimen of a tree subsequently named in his honour *Pennantia baylisiana*. The only tree of its kind existing in the wild anywhere in the world, it has been propagated by cuttings but is still extremely rare. Another plant discovered by Bayliss on a rock in Tasman Stream on Great Island was *Tecomanthe speciosa*, an attractive vine with large, tubular creamy flowers that sprout directly from the woody stems. It was the only specimen ever found and the many plants now in cultivation all derive from this one plant.

After reading descriptions of modern day visitors struggling ashore at Great Island onto a meagre rock platform drenched in salt spray, it is interesting to learn that Cheeseman was accompanied by his new bride on his second trip to Three Kings. It is difficult to imagine a Victorian lady in corset and long skirts making such a hazardous landing!

Even more difficult to picture is the terror felt by the survivors of the wreck of the *Elingamite* when they staggered ashore on the islands on an early November morning in 1902 after their ship struck rocks in thick fog and sank. The *Elingamite*, out of Sydney bound for Auckland, was carrying large amounts of coin destined for local banks, as well as a ton and a half of bullion. Forty-five of the 189 passengers and crew perished and the thought of sunken treasure tantalised divers for decades. Although some bullion was recovered in the few years following the disaster, it was not until the 1960s that well-known New Zealand divers Wade Doak and Kelly Tarlton secured salvage rights to the vessel. After several expeditions they recovered about 10,000 silver coins from the wreck and proceeds from the sale of this bounty helped finance Kelly Tarlton's Underwater World in Auckland. Yet another rare plant first found on the Three Kings in 1951 is named *Elingamita johnsonii* after the shipwreck and the botanist who discovered it.

More sinister than the wreck of the *Elingamite* was the sinking of SS *Wimmera* near the Three Kings in 1918, victim of one of numerous mines laid in New Zealand waters in 1917 by the German armed merchant ship *Wolf*.

Inhospitable as the islands are for any who want to go ashore, the waters nearby attract divers and anglers. Ocean currents, separated for hundreds of kilometres, swirl together in a cauldron of fascinating marine life and the *Elingamite* wreck makes for intriguing underwater exploration. Marlin, tuna, sharks and broad-billed swordfish abound and whale sharks, which are becoming more common in New Zealand waters, have been seen off Three Kings Islands.

Small in stature, the islands have contributed in unexpected ways to the tapestry of New Zealand life.

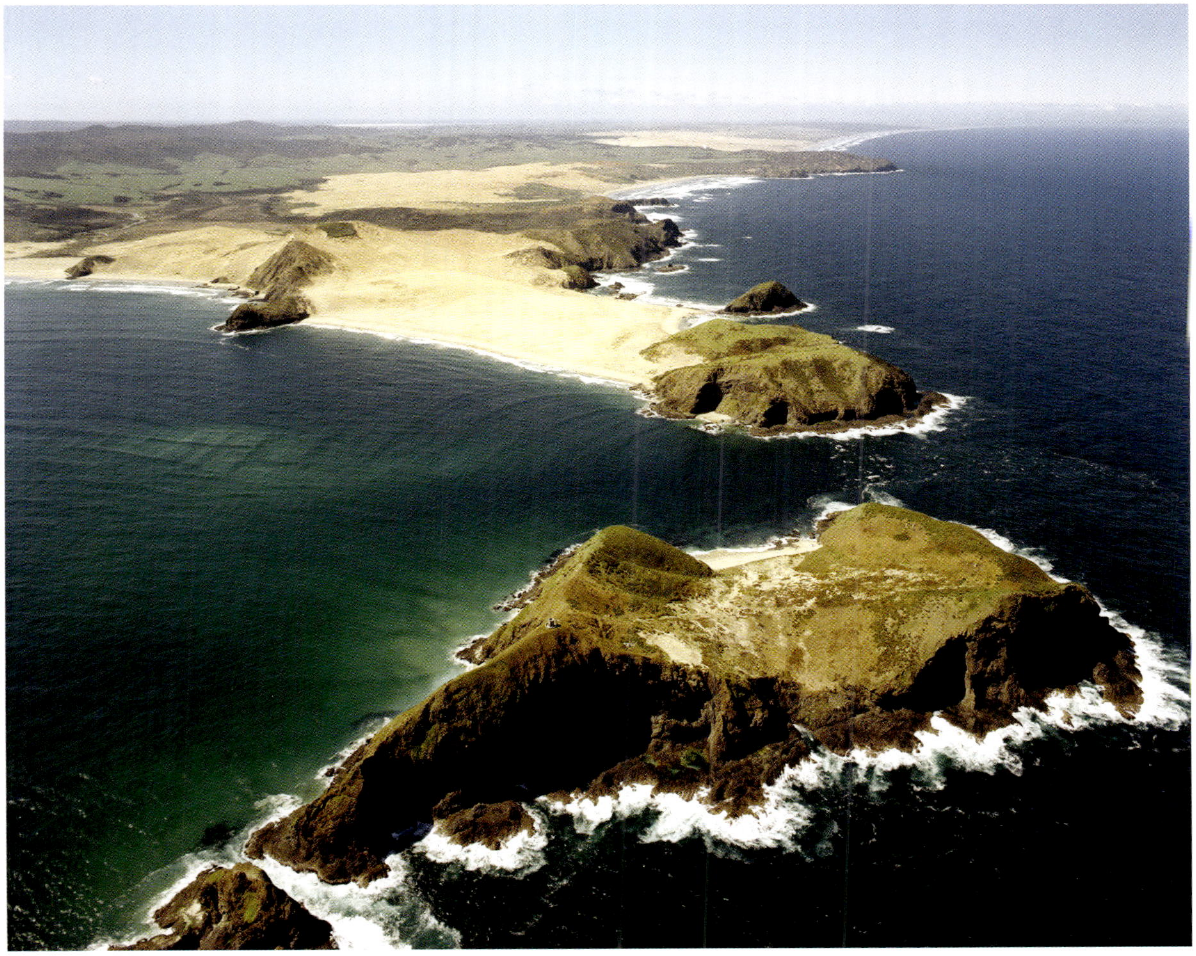

Motuopao

Sitting to the north of Cape Maria Van Diemen, Motuopao Island is probably the most western island off the coast of the North Island. It was home to the first lighthouse in the area, built in 1879, which helped ships navigate around the north of New Zealand until 1941 when it was moved to Cape Reinga. Now Motuopao is a nature reserve and valued for its breeding colonies of fairy prions, white-faced storm petrels and black-winged petrels. Rare skinks and geckos also live on the island.

Cavalli Islands and their neighbours

The Cavalli Islands are a special place. Comprising one large and four smaller islands plus a family of even smaller offspring, they lie north of the Bay of Islands and about five kilometres offshore from Matauri Bay. Whether you visit them by boat or view them from the mainland, they are intriguing.

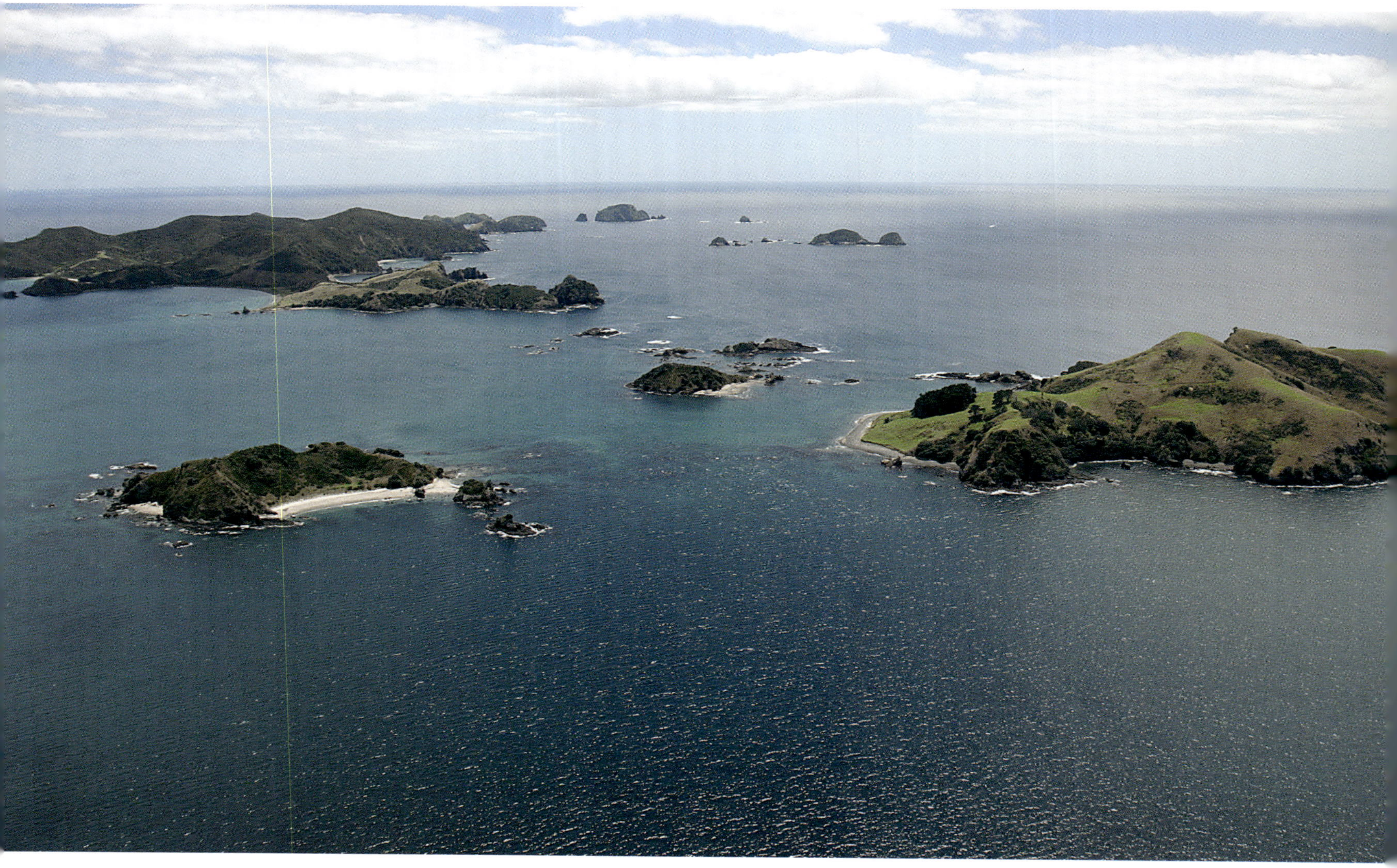

In the 19th century the Cavalli Islands were farmed. Now uninhabited, they are a magnet for boaties and holiday makers in the Far North.

Motukawanui, the largest island in the group and in the original Bay of Islands Maritime and Historic Park, has the feeling of an escape-it-all, island hideaway. When Captain Cook visited the group in 1769, Maori canoes came out to the ships and sold them fish which Cook called 'Cavelles', hence the name he gave to the islands. But it is probable that the fish in fact were trevally. Traces of pits, terrace sites, middens and cultivation areas suggest that there was a substantial population on the islands when Cook arrived in New Zealand.

A couple of decades after his visit in 1793, two young Maori were kidnapped at the Cavallis by a British Government ship. They were then taken to Norfolk Island where, it was hoped, they could give help to the authorities on the preparing of flax. At home, however, this was women's work and the two men were unable to help. They were later returned safely to New Zealand.

Like many islands around the country, Motukawanui was farmed during the 19th century and up until 1974. Now the island is uninhabited and native plant species are re-establishing, although much is still open pastureland. A three-kilometre walk along the island gives spectacular views of the general area and facilities exist for overnight stays.

As a group the Cavallis offer many beautiful secluded beaches where visitors can spend a day swimming, snorkelling, fishing or relaxing in what is essentially a subtropical paradise. The islands are a favoured destination for sailors, although numerous rocks below the surface are an eternal menace, while divers go there for the spectacular underwater scenery, which some regard as equal to that of the Poor Knights.

In 1987, two years after it was bombed and sunk in Auckland Harbour by French secret service agents, the Greenpeace flagship *Rainbow Warrior* was laid to rest as a dive wreck off Matauri Bay among the Cavalli Islands in about 25 metres of water.

In the intervening years, the skeleton of the ship has become almost totally encrusted with marine life. Moray eels keep watch from crevices and lazy goatfish shimmy along the bottom. The exterior is covered in coral and jewel sea anemones in pink, yellow and purple. Also of specific interest to divers are Tokananohia Reef and Taheke Reef, which have abundant fish with drop-offs to 53 metres and a reputation for sharks and barracouta.

Approximately 15 minutes by helicopter from Kerikeri is Flat Island, also known as Motueka Island – a privately owned, veritable paradise for fishing and diving. It is located on Northland's East Coast directly offshore from Mahinepua Bay, between Whangaroa Bay and the Cavalli Islands.

Clustered around the mouth of Whangaroa Bay, at both the north and south sides, there is a collection of tiny islands including Stephenson and Cone Islands and Cone Rock that offer good snorkelling and diving with many reef and pelagic fish, plus crayfish and shellfish.

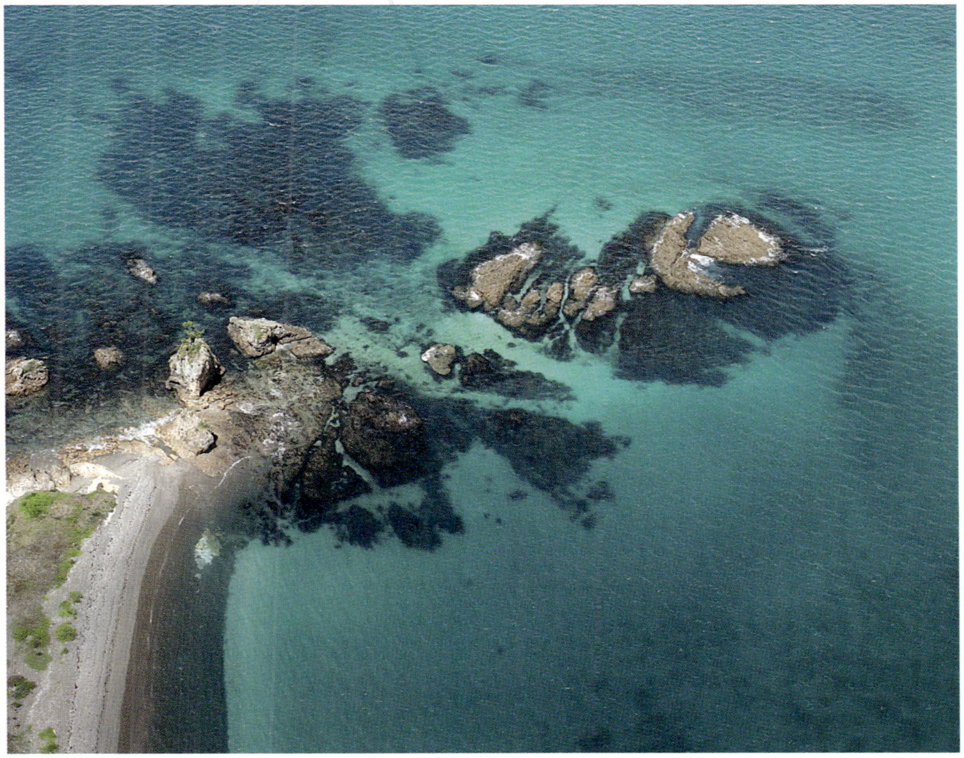

Alluring sea shades of turquoise and jade tint the crystal clear waters which surround the Cavalli Islands.

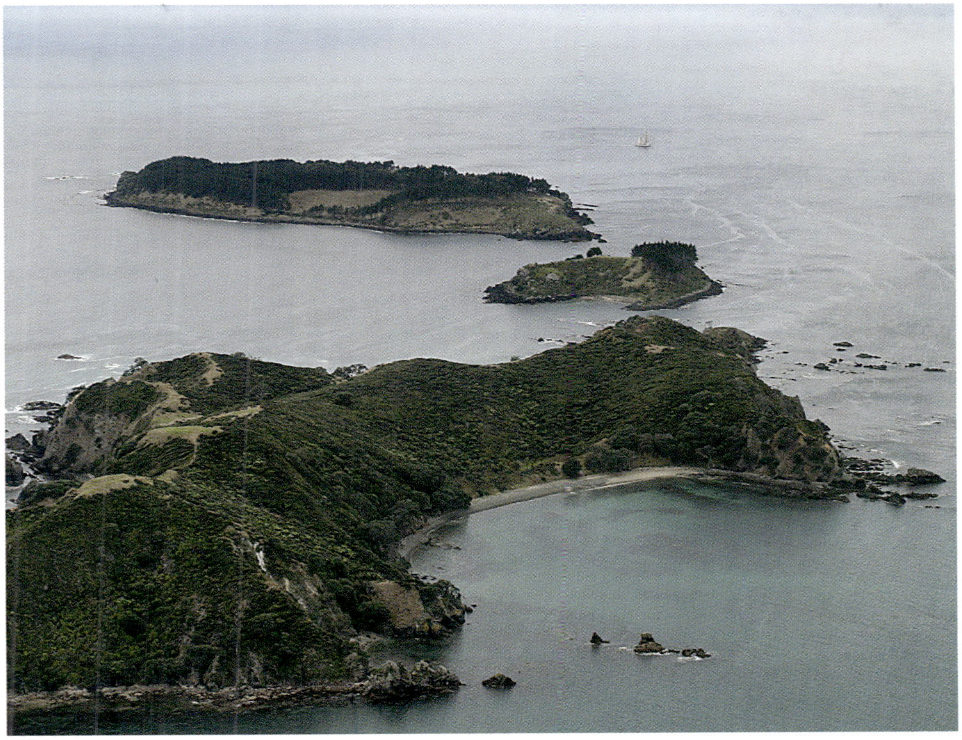

Flat Island (behind) and Motuekaiti Island, situated between the Cavalli Islands and Stephenson Island.

Bay of Islands

To explore the Bay of Islands by boat is to get to know one of the most beautiful areas in New Zealand – and this from a writer who is addicted to mountains and cold winters! There are islands of every shape and size.

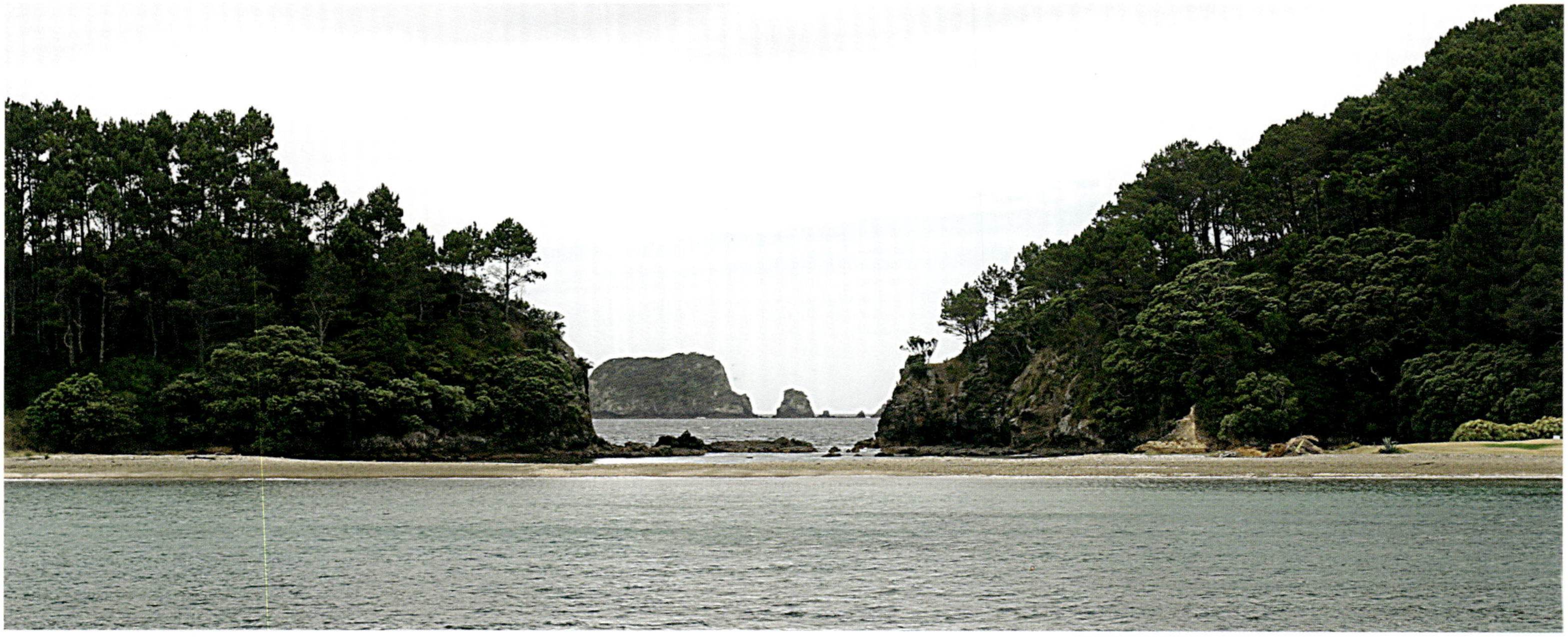

Sun, sand, sea and islands that go on forever – this is the magic of the Bay of Islands

Right: Cruising in the sub-tropical climate of the Bay of Islands is an unforgettable experience.

Some are mere rocks jutting out of the ocean, others consist of rolling hills of green pasture edged by rugged cliffs, and yet others are steeply rising mounds covered in native bush. There are countless sandy beaches and secluded, bush-lined bays. The water is so clear the seabed is visible in places 10 metres below the surface. In summer the crimson blooms of ancient pohutukawa contrast with the sea in shades of green, jade and turquoise. I have sailed near Cape Brett when the waves were so high it was impossible to see land, yet even in stormy weather when thunderclouds boil up in the sky the immense, island-studded bay is majestic.

It is little wonder that the early Polynesians, arriving in New Zealand from other islands in the Pacific, made this area their home. The climate is warm by New Zealand standards – suitable for growing kumara. Seafood was varied and abundant and the vegetation was there to supply raw materials for their domestic needs. As seafarers the proliferation of islands, with their numerous sheltered anchorages, must have appealed to the first settlers, and the topography of the area provided high places to build defensive pa in later times.

When Europeans first arrived in the Bay, the Maori population was dense and concentrated along the coastline, particularly in the south-east, and in some fertile inland areas. But historical records tell that Maori would move with the seasons, alternating fishing at the coast with cultivating their crops inland. As in all other regions of New Zealand, no island was too small on which to establish living space. Tiny Poroporo, a mere sliver of land at the foot of Urupukapuka Island, was the site of a pa at some time in the past though it is probable that these smallest islands served for temporary settlements. Seabirds abound on Poroporo now. Perhaps Maori in the past lived there during the season to harvest chicks.

The botanist Joseph Banks, aboard the *Endeavour* on Captain Cook's first voyage to New Zealand in 1769, described his impressions of the Bay of Islands where Cook made his first landfall:

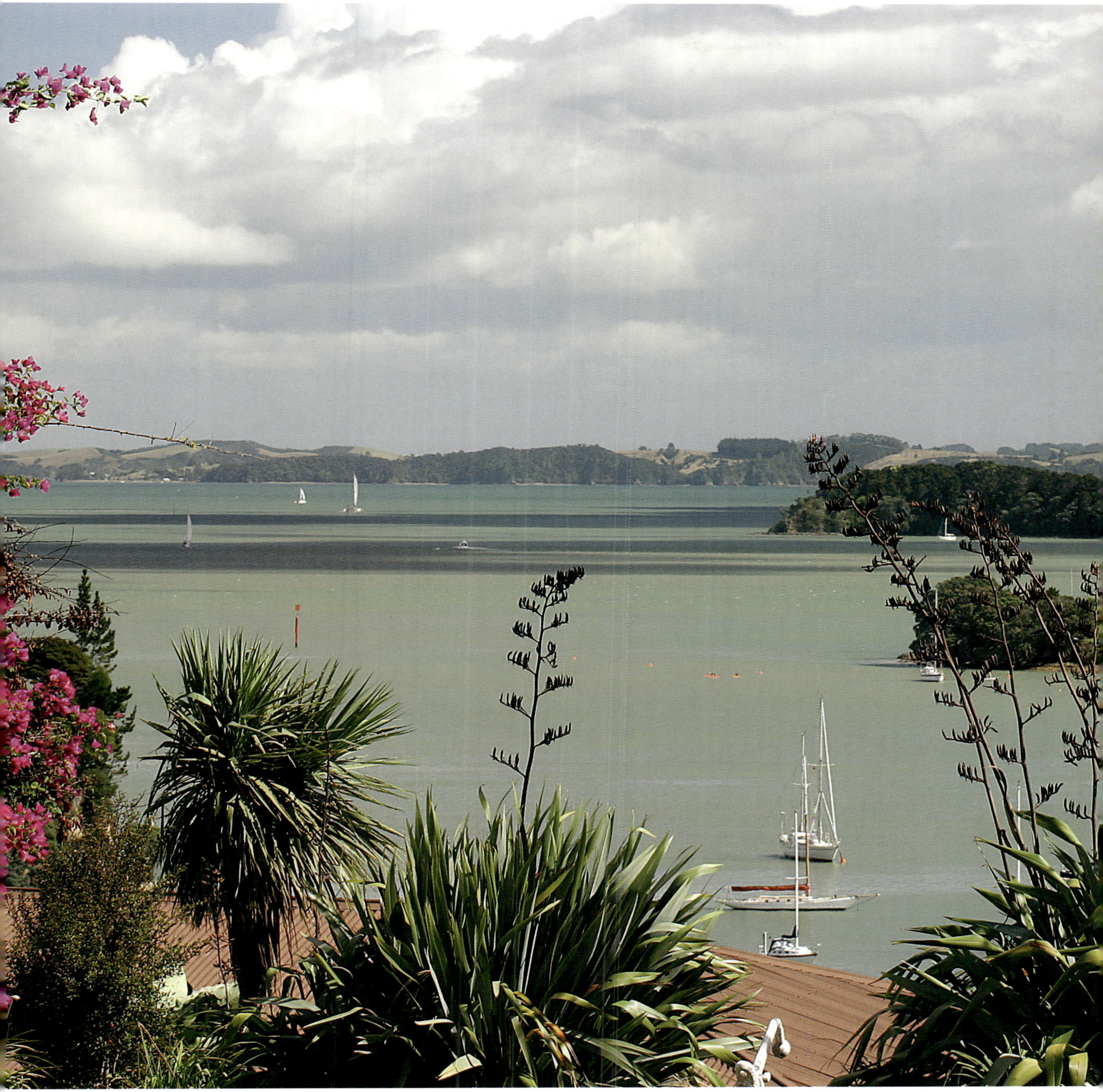

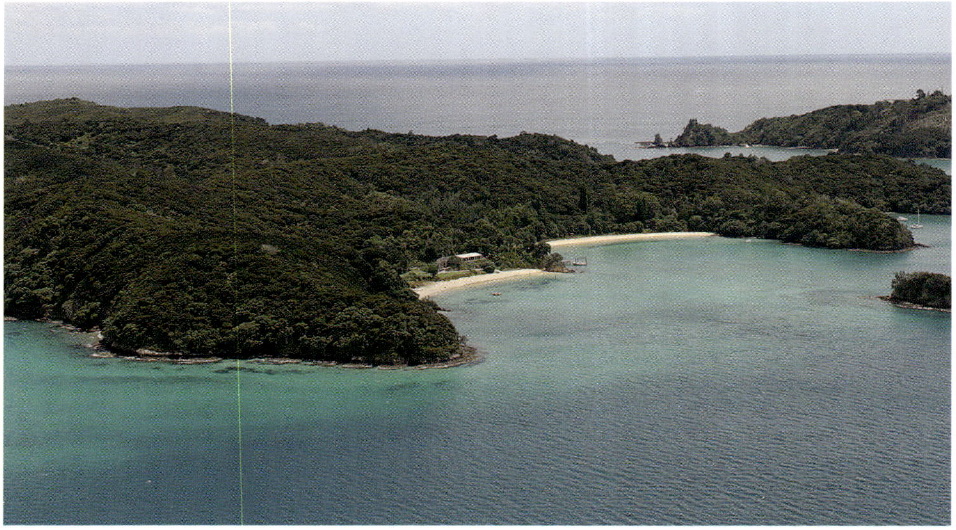

Romantic hideaway bays on Moturua Island.

'It was more thickly populated than those parts further south, the people more elaborately tattooed, some of their canoes more elaborately carved; the bay itself beautiful, with many good anchorages, the hills and valleys round it, forests and cultivations beautiful also. Cook called it the Bay of Islands.'

The Frenchman Marion du Fresne also made landfall in the Bay of Islands, two and a half years after Cook had been there. In need of fresh supplies and repairs to his ships, he anchored near the island of Moturua. When the local Maori at first seemed friendly, he established a hospital camp for his sick crewmen at Waipao, a bay on the south-west side of Moturua, with his contingent staying in the area for three months. This was the first protracted contact that Maori and Europeans had sustained. Unfortunately, relations deteriorated between the two races and perhaps as a result of du Fresne breaching a Maori tapu in ignorance, he and several fellow crewmen were murdered. His men wreaked vengeance on two Maori communities and, after declaring New Zealand a possession of France, they hastily departed.

When Europeans arrived in the decades following Cook's sojourn in New Zealand waters their assessment on where to live was based on criteria totally different from those of the Maori. Whaling initially brought these later seafarers to the Bay of Islands, and in their wake came traders, missionaries and soldiers. The whalers needed a protected landfall and a reliable anchorage where they could reprovision their boats. The village of Kororareka provided this and the Maori inhabitants were very ready to trade fresh vegetables, pork, ships' spars, fish and the favours of their women for European goods. Kororareka, or Russell as it was later named, became the capital of New Zealand, temporarily, and for 20 years from the 1820s, a rip-roaring hub of activity.

When Governor Hobson moved the capital to Auckland in 1840 after the signing of the Treaty of Waitangi, trade in the Bay of Islands suffered. Much unrest followed with Maori uprisings led by Hone Heke, and when peace finally came to the region the Bay of Islands no longer featured as a major arena for the events that went on to shape the country.

As the 20th century rolled on, however, numerous local people and public servants in the Department of Lands and Survey recognised the historical, recreational and scenic importance of the area. As some of the islands were bought by private individuals the realisation dawned that some form of protection was necessary to preserve the right of public access to a unique part of the country. Little by little the Crown purchased islands and key coastal areas when they became available. By 1978 sufficient islands and mainland reserves had been acquired to create an entity linking the individual parts under one umbrella. The Bay of Islands Maritime and Historic Park was formed.

A quarter of a century on, the concept of the park continues to evolve. No longer called a maritime park, the principles of conservation and preservation remain. Private landowners are increasingly prepared to co-operate with DOC, often allowing predator trapping on their land and participating in reforestation programmes. The people of the Bay of Islands are involved in the consultation process and the setting up of several marine reserves is high on the priority list at the time of writing.

Nowadays Russell is a village that wears its age with grace and an atmosphere of timelessness. Its roistering past is long gone, though during the height of the holiday season many boats anchor in the harbour and restaurants and cafés there do a fine trade.

Across the Bay from Russell are the towns of Waitangi, which attracts visitors who want to see where the historic document had its birth, and Paihia, which in recent times has developed from a sleepy seaside hideaway to a thriving centre for the tourism industry.

Two tiny islands, Motumaire and Motuarahi, sit at the entrance to Paihia Harbour, a short

Exploring the Bay of Islands

Paihia, Russell and Opua are the main towns with most tourist activities centred in Paihia.
A wide range of tourist accommodation available on the mainland.
Activities include fishing, diving, kayaking, swimming with dolphins, sailing, cruising.
Most available on a daily basis. Both chartered and scheduled boat trips available. Chartered dive trips to the Cavalli Islands leave from Paihia.
Chartered scenic flights also available by arrangement.
For information and bookings contact the Paihia Information Centre.

rowing distance from the waterfront. Few visitors realise that these were once inhabited. Traces of terraces and ditch defences on them indicate that, even sitting this close to the mainland, they were at some time occupied by Maori. Motumaire was reputed to be used in pre-European times as a centre of learning and is still held in respect by local hapu or sub-tribes. On both these islands, native forest is regenerating. Water around them is shallow at low tide, a trap for unwary yachties. I have vivid memories of a warm summer's day, the cockpit of our yacht laden with buckets of laundry, heading for the laundromat at Paihia, when suddenly the yacht ahead of us stopped short, caught fast on the sand. Too late the skipper yelled a warning. Another moment and we too were caught until a friendly launch came to tow us free.

In summer the Bay of Islands is swarming with visitors from all over the world. They come to dive, to indulge in big-game fishing, to sail, to swim with dolphins or just to relax in the midst of wonderful scenery and picture-book beaches.

This part of the Northland coast has been famous worldwide for deep-sea fishing since the American adventure writer, Zane Grey, spent time holidaying and fishing in the area in the 1920s and wrote about it. Black and Pacific blue marlin, mako and hammerhead sharks as well as smaller fish are all considered fair game. Numerous charter boats, fully equipped and surveyed for carrying passengers, operate out of Russell and Paihia. Nowadays the emphasis is frequently on the sport, rather than the catch, and many of the fish that are hooked get tagged and returned to the sea.

For nearly 90 years, the Cream Trip has been a popular and well-known activity. In 1919, Edmund Lane operated the launch *Dairymaid* for the Hikurangi Dairy Company. Three days a week he made the rounds of the outlying islands in the Bay, collecting the cream from all the small farms and delivering mail. When he gave up the run six years later to concentrate on farming, the Cream Trip was taken over by Fullers ferry company and it continues into the present. The boat departs from Paihia, meandering in and out of many of the smaller bays, delivering stores and mail as it goes and passing points of historical interest relating to both early European and Maori settlement. An ideal way to visit many of the islands, the Cream trip often also provides an opportunity for visitors to see seals, dolphins and sometimes whales, as well as bird and fish life. It is one of the few official Royal Mail runs by boat remaining in New Zealand.

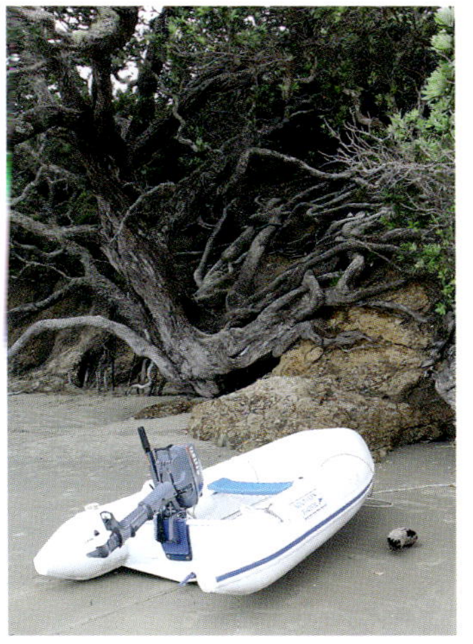

The historic islands of the north are ideal for exploring.

Lunchtime at Paradise Bay, Urupukapuka Island.

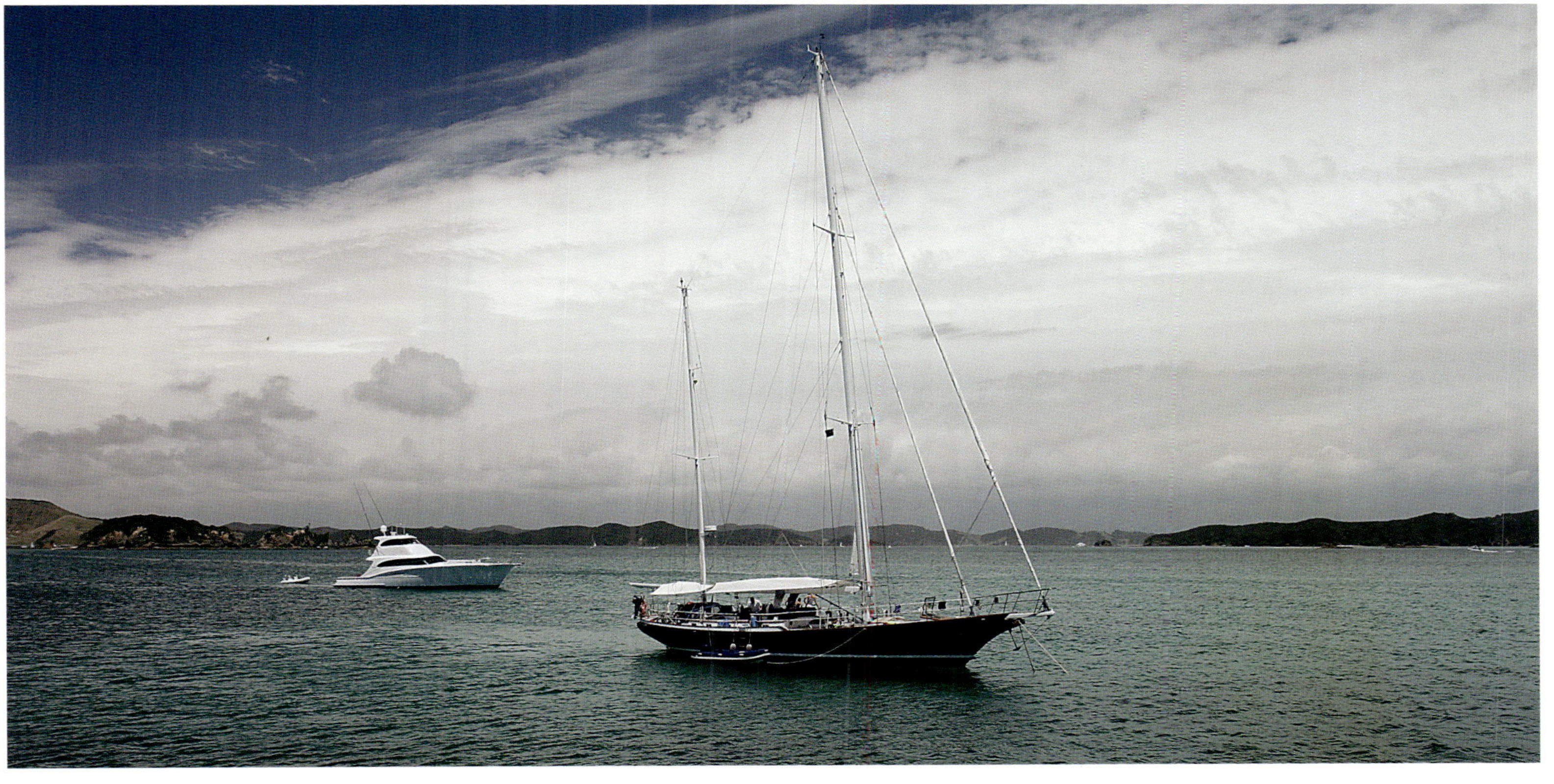

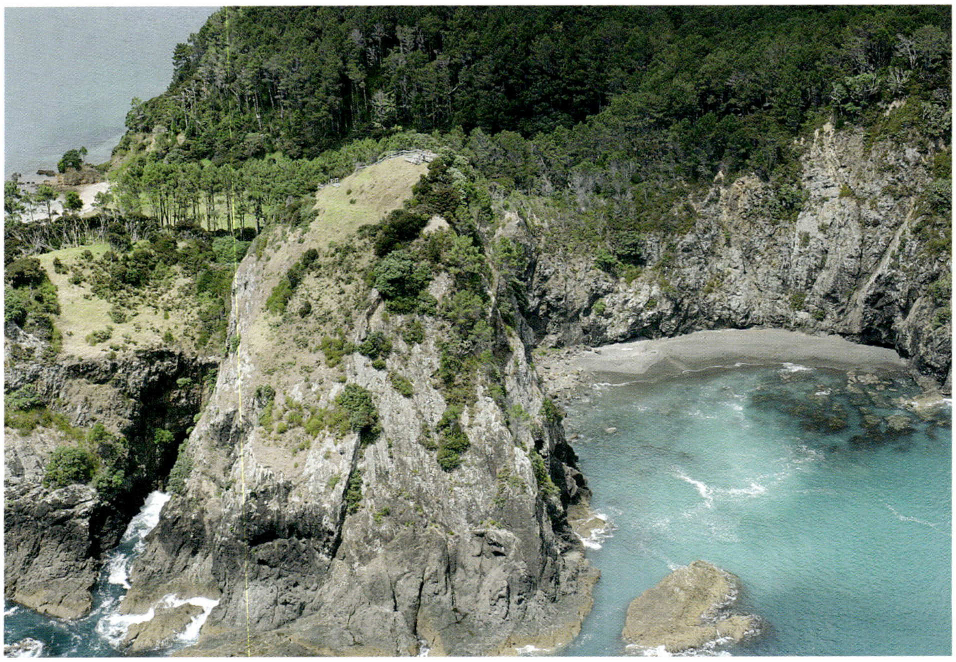

Not for boaties, this tiny, enclosed bay on Roberton Island is a good place for snorkellers to explore.

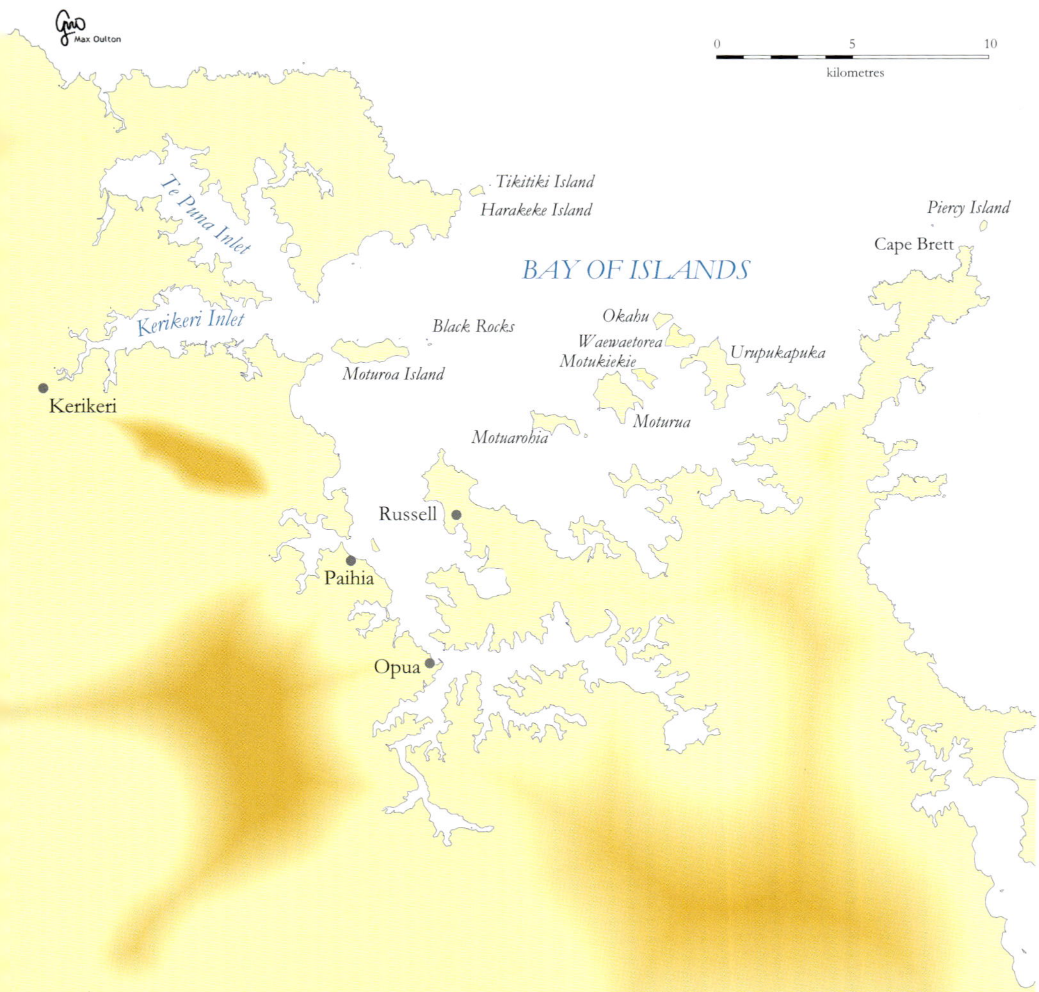

Roberton Island/ Motuarohia

Closest to the mainland of the larger islands is Motuarohia ('Island to be Desired'), also known as Roberton Island. A more evocative name than Motuarohia would be hard to imagine but it is an idyllic island with a haunted past, rich in both Maori and European history. It was off Roberton Island that Cook first anchored in the *Endeavour*. At the time there were 200 to 300 Maori on the island and although the first meeting of the two races developed into a minor skirmish, later relations between the newcomers and the residents were amicable. A commemorative plaque marks Cook's first anchorage.

Shortly after the signing of the Treaty, tragedy struck the island when relationships between Maori and Europeans went sadly awry. A quarrel arose between two employees of Mrs Roberton, a widow who continued farming on the island after her husband had drowned. Maketu, a young Maori, was reputedly ill-treated by Thomas Bull, a Pakeha. Maketu murdered Bull – justified in Maori eyes because of the abuse he had suffered – and when threatened with English justice by Mrs Roberton he killed her and her children, as well as a child in her care who was the granddaughter of a local chief. When justice ran its course, Maketu was hanged in Auckland, the first Maori to be tried and executed under the new law of the land.

In the centre of the island, a high jagged peak, once a pa site, is flanked by twin shallow lagoons where an interpreted underwater trail has been set up for snorkellers. The island is wasp-waisted and in a high sea waves occasionally crash over the land.

Partly in private ownership, with a few holiday houses along the coast and on the high points, the central area of Motuarohia is a scenic reserve. In summer, boats moor in the bay and visitors often walk to the high point of the island from where there is a spectacular outlook over the Bay – and a superb viewpoint for the annual tall ships race that takes place in January. It is from this high point that Maketu reputedly hurled the young Roberton boy to his death after chasing him to the top of the island. Every time I climb to the top of the island, it is difficult to look down over the cliffs to the sparkling water beneath without imagining this horror of long ago.

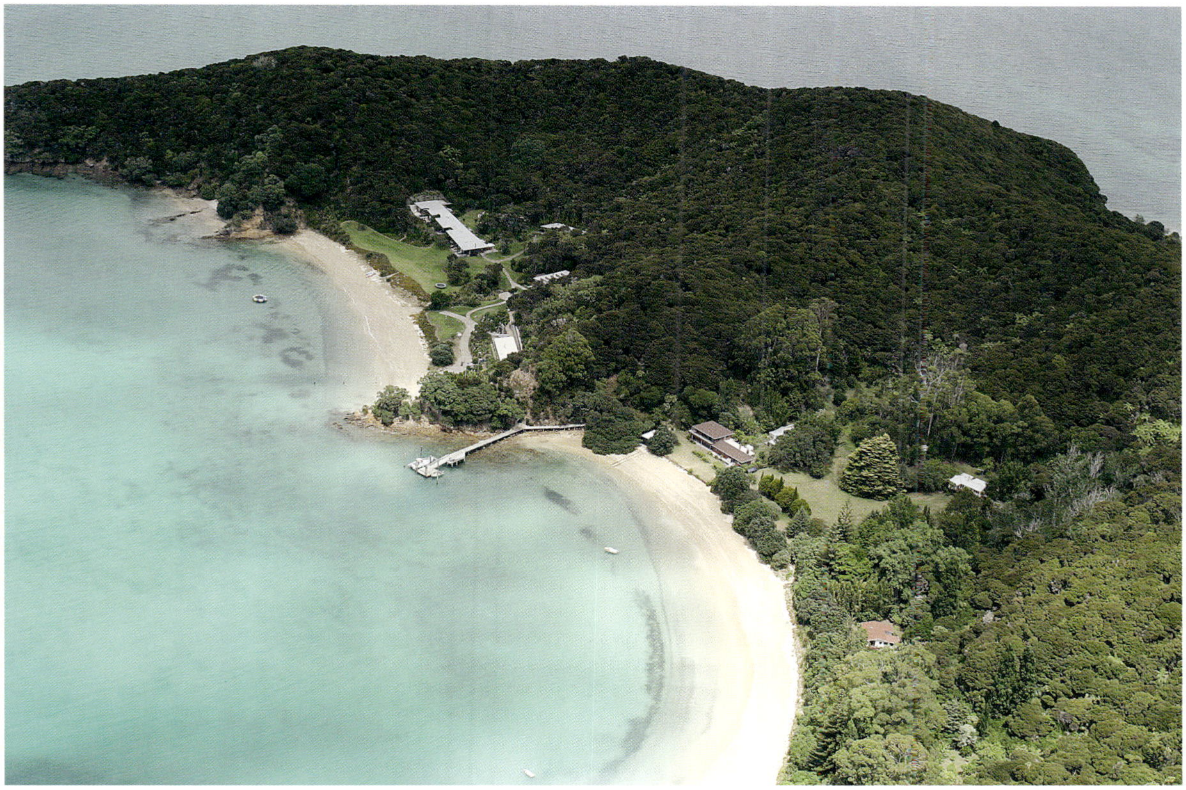

Secluded and private, these homes on beautiful Moturua Island are holiday destinations that most people can only dream about.

Moturua and Motukiekie

These two islands are close neighbours of Roberton Island, both offering great anchorages for visiting boats. I have spent many nights in Waiwhapuku Bay on the eastern side of Moturua Island and in the morning set off from the beach on the walking track that heads up over the hill and continues right around the island. Views from the high points are spectacular and there's good swimming at several of the beaches on the way.

Moturua is the second largest island in the Bay and originally home to the Ngare Raumati people who had been living there for generations when Marion du Fresne arrived in 1772 and set up his hospital facilities at Waipao Bay. When the French avenged du Fresne's murder by attacking one of the pa on the island, it was the first time a traditional Maori village had been attacked by a military force armed with muskets. Accustomed to hand-to-hand combat, the Maori were no match for the French. Over 200 defenders died, another estimated 200 escaped and the pa was destroyed. Subsequent tribal battles left Ngare Raumati defeated and Ngapuhi in control of the island, but through intermarriage, descendants of the original tribe lived on Moturua at Otupoho Bay, farming the land until 1968 when most of the island was finally bought by the Crown and designated as a scenic reserve.

The island was then left for native forest to regenerate. Manuka and kanuka have sprung up, providing an ideal nursery for young shrubs and tree-ferns. Pohutukawa thrive along the coast and particularly on the track that leads up to the high point of Hikurangi Pa, an ancient defence observation post that was also used during the Second World War as a lookout for spotting mines. North Island robins, extinct in the region on the mainland, and North Island kiwi have been reintroduced to the island.

Long, narrow and steep-sided, Motukiekie is one of the prettiest islands in the Bay. Tall, architectural Norfolk pines grow along its spine, contrasting with the pohutukawa and other natives, giving Motukiekie its distinctive skyline. I anchored in the bay on its eastern side once, in the off-season, with American friends on board. We were the only boat to be seen. As we prepared to jump overboard and swim ashore another yacht came into the bay, anchoring nearby. 'I guess you call this crowded,' our friend drawled.

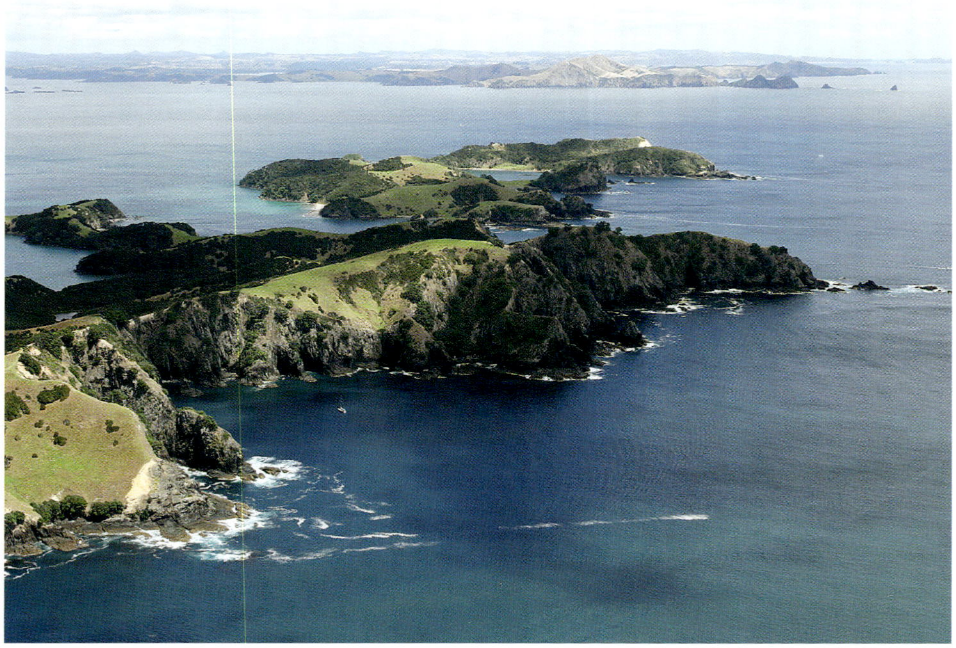

Urupukapuka Island in the foreground was inhabited by Maori as long as 800 years ago. Now it is one of the few islands in the Bay of Islands where camping is permitted.

Urupukapuka and its neighbours

Urupukapuka Island is a paradise for sailors and campers who like unsophisticated outdoor living. It is the biggest island in the Bay and the numerous beaches are idyllic for swimming and sunbathing. Many of the bays offer sheltered anchorages and diving is good, particularly on the eastern coast where there is plenty of reef life.

The first settlers on Urupukapuka, ancestors of modern Maori, may have arrived there around 800 years ago. Two centuries before the crew of the *Endeavour* visited the Bay of Islands, a sub-tribe of Ngare Raumati lived on the island. However, little is known about pre-European life on Urupukapuka although numerous archaeological sites give clues to their diet. A supposed sale of the island to a sea captain fell through in 1839 when the Maori chief Rewa claimed the purchase price was in fact only a deposit. Later the island was leased by two European settlers who began clearing the land and grazing it.

In 1971 most of the island was bought by the Crown and all but a small area around Otehei Bay, which remained in private hands, was designated a recreation reserve and farm park.

In 1927 Zane Grey set up his deep-sea fishing base at Otehei Bay, attracted in the first instance to this part of the world by the fishing camp at Deep Water Cove on the mainland. Soon after a well-known fishing resort, called the Otehei Bay Company, was established there with shareholders from Wellington and the Bay of Islands. First a wharf was built, followed by cabins with a dining room, bar and a house for the manager. Until the main hotel building burnt down in 1973, accommodation and refreshments were always available.

Nowadays the island amenities are much simpler. Outdoor showers and fresh water cater to the needs of boaties and campers. There is an established track around the island that takes about five hours to complete and winds past ancient pa sites, villages and complexes of kumara storage pits with signs that explain the various archaeological features. Paradise Bay is one of my favourite haunts. It is well-named. A bay of turquoise-blue water to swim in, golden sand to laze on, and a yacht to dream on under the stars – this is truly paradise.

Across a narrow channel from Urupukapuka is the smaller island of Waewaetorea, also with beautiful beaches. It was designated as a recreational reserve when it was bought by the Crown in 1980 from its American owner. The island is grazed to make it accessible to walkers and to help prevent fires.

Historical records from du Fresne's expedition in 1772 report that two of his ships anchored beside Waewaetorea Island before penetrating closer to the mainland and that they were surrounded by more than 100 canoes – indicating a large population of Maori in the vicinity. Two headland pa sites are visible, the final battle on the island taking place at the northern one, when Ngapuhi defeated Ngare Raumati near the Ororua Blowhole about 1829.

Okahu Island to the north-west of Waewaetorea and Poroporo Island to the south of Urupukapuka are both small islands with clear, safe water for diving. Pohutukawa trees overhang the cliffs of Poroporo; one of them at the southern tip is reputed to be more than 1000 years old. Once inhabited by Maori, the island now is home to many seabirds.

Okahu is the outermost island in this part of the Bay and its cliffs show signs of oxidized iron, giving the rock a reddish hue and the island its local name of Red Head.

Little blue penguins inhabit Red Head and they come ashore in August and September to nest in rocky caves and burrows.

Islands near the Kerikeri Inlet

Near the entrance to the Kerikeri Inlet is the privately owned island of Moturoa. Its original shareholders gained the right to build one house each on the island and the remainder of the land was farmed. In recent years native trees have been planted on Moturoa and bellbirds and North Island robins have been released.

Off the eastern end of Moturoa lie the Black Rocks, scarcely islands as such, but definitely a landmark feature of the Bay. They consist of a succession of flat-topped smooth rocks, with vertical sides, sitting in very deep water. It's an eerie feeling to anchor in their dark shadow. Clearly volcanic in origin, they are fine examples of columnar jointing in basalt lava flows. In any but smooth seas, the waves crash against the rocks, drenching them in spray and only the most hardy of plants survive. But seabirds love the environment and they are a breeding ground for black-backed and red-billed gulls as well as the white-fronted tern.

The waters around the Black Rocks are renowned for their fishing while underwater caves provide endless opportunities for divers to explore.

The island of Motupapa, also known as Cocked Hat Island, sits at the mouth of the Kerikeri Inlet and was bought by the Department of Conservation in 1989. Only four hectares in size, it was formerly overrun with kikuyu grass but in the early 1990s, the local branch of Forest and Bird initiated an ecological restoration. Over a period of two or three years, more than 5000 trees, raised from local seed, were planted on the island. Now, 10 years on, the trees are doing well and the island has taken on a new personality.

The Te Pahi Islands are a group of small islands with historical importance for local Maori situated on the northern side of the Bay of Islands in Wairoa Bay on the Purerua Peninsula. Te Pahi was a powerful Bay of Islands chief who lived at the end of the 18th and the beginning of the 19th centuries. One of the first chiefs to establish trade with the whaling ships, he lived on the westernmost of these islands. His meeting with Samuel Marsden in Sydney in 1805 led to the establishment of the first New Zealand mission 10 years later. However, Te Pahi, rightly or wrongly, was implicated in the burning of the vessel *Boyd* at Whangaroa in 1809 in which as many as 70 Europeans may have been killed. His island was subsequently attacked by whalers in reprisal for the deaths. About 60 Maori were killed and their village was destroyed. Te Pahi died later from his wounds.

The islands have been set aside as a Maori reservation, a permanent memorial to a chief who was much respected in his time.

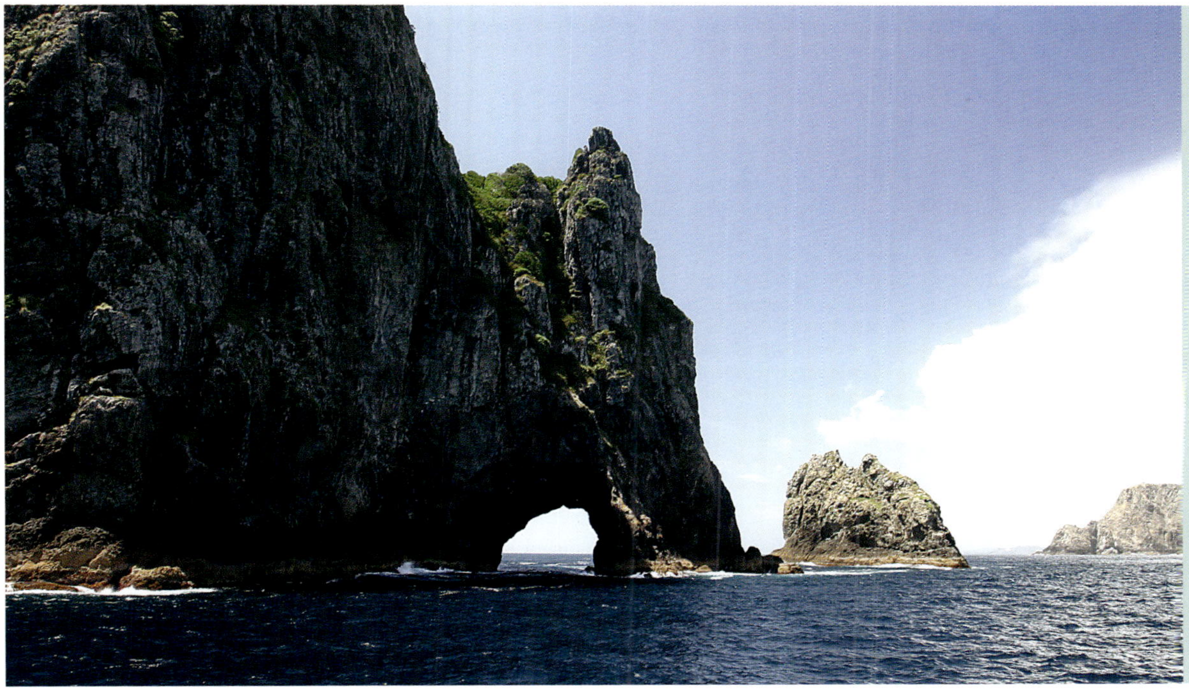

Piercy Island

Piercy Island or Motukokako, otherwise known as the Hole-in-the-Rock, was given its English name by Captain Cook and is situated at the southern entrance to the Bay of Islands, just off Cape Brett. The name is a clever pun, naming the island in honour of Rear Admiral Sir Piercy Brett while also recognising the notable feature of the island. On days when the sea is calm, tourist boats motor right through the hole, a rather eerie experience.

Poor Knights Islands

An awesome playground, is how divers describe them. Like breakfast pudding, spread with raspberry jam, is how Captain Cook saw them. A rich fishing ground, was the view of fishermen before the advent of a marine reserve. However they are seen, the Poor Knights Islands are spectacular.

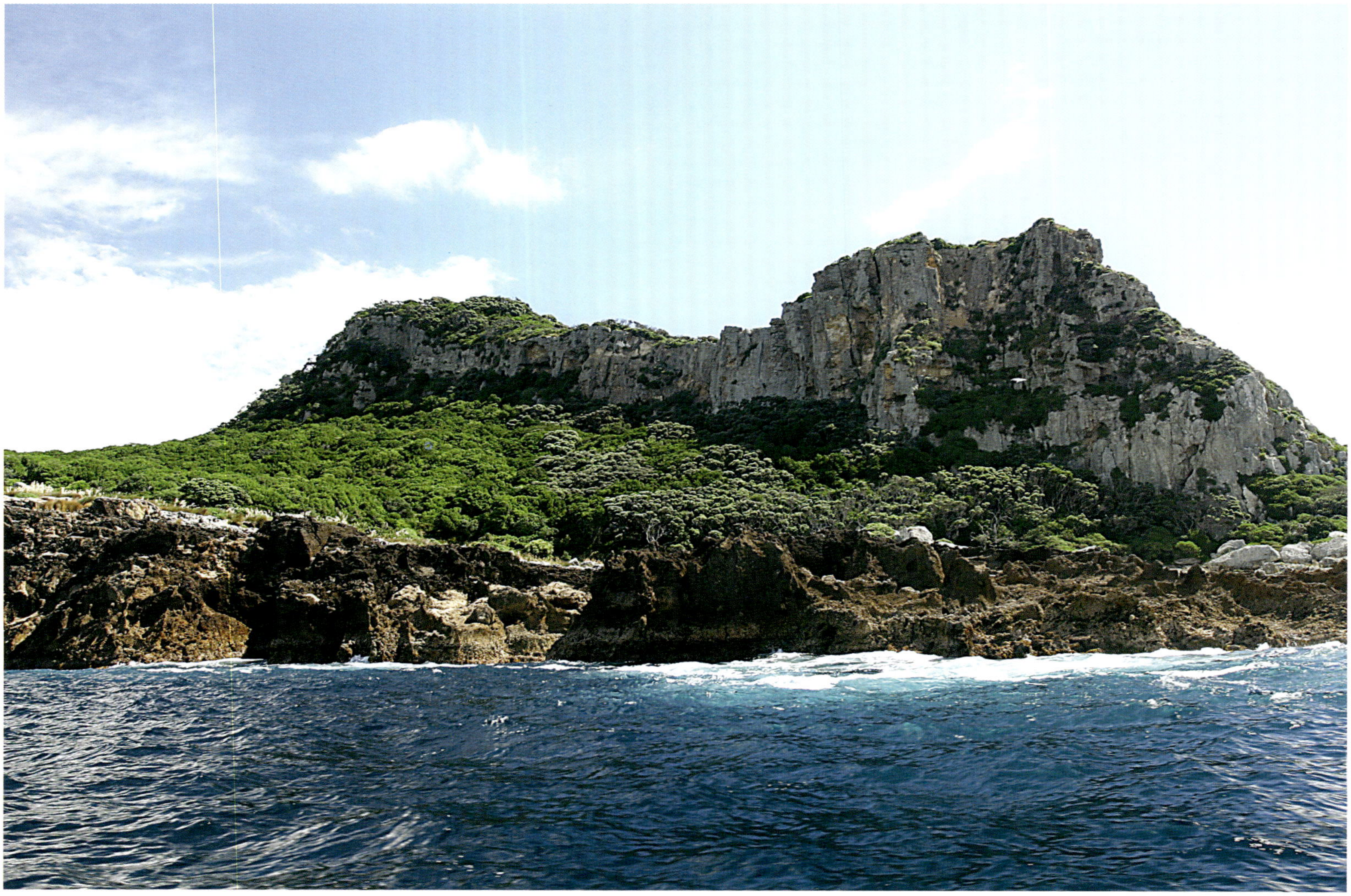

Precipitous cliffs and unusual plants are a feature of the Poor Knights Islands, east of Tutukaka. They also attract divers from around the world who come to explore the marine reserve that surrounds the islands.

They sit off the coast of Tutukaka, north-east of Whangarei, two main islands – northern Tawhiti Rahi and Aorangi to the south – skirted by a collection of isolated rocks and islets. The Poor Squires, also known as the Pinnacles and the Sugar Loaf, are close by. These rugged outcroppings are believed to be the remains of large lava domes, all that is visible above water of an old rhyolite volcano, millions of years old.

There are varying versions of how they got their English name. Some say Cook was eating breakfast as he sailed down the coast and saw these islands. Breakfast consisted of a dish called Poor Knights Pudding, an egg and bread concoction. The raspberry jam was the red of pohutukawa blooms that spread over the islands. Others maintain he saw the islands lying in the sea and likened them to effigies of crusaders, their heads to the north, their feet to the south.

It took us about 40 minutes by fast boat to reach the Poor Knights from Tutukaka. The islands blur in the distance, indistinct dark shapes in the sea as the boat races away from land. When gannets appeared, soaring out from their own private island, we knew we were close to the Poor Knights. But nothing prepared us for the grandeur ahead. Suddenly perpendicular cliffs loomed above the boat. Rough headlands scramble out to sea. Outsize archways form gigantic picture frames. Precious plants thrive, clinging to an impossible rock face, and fissures cut into the cliffs so deeply that sunlight is lost. And then there are the caves. International divers come to explore the depths of Rekoreko (Cave of Sounds), the biggest sea-cave in the world, so enormous that dolphins and whales enter it and divers have seen sharks come in and eat the paper nautilus which live there.

Beautiful the islands may be; memorable without a doubt. But hospitable? Never! Yet people lived on these islands for generations, despite flat land being almost non-existent.

According to stories handed down, about 300 or 400 people lived on Tawhiti Rahi and Aorangi. They were self-sufficient, cultivating arable land and taking abundant seafood from around the coast. Pigs obtained from Captain Cook were bred on Aorangi and were valued as goods to trade with mainlanders for the totara and kauri they used for building canoes and shelters.

But disaster struck in the early years of the 19th century when their chief was away fighting with Hongi Hika. Betrayed by a slave, the islanders left behind were slaughtered by a mainland tribe who held a grudge against them. When the chief returned he was so appalled by what he found, he declared the islands strictly tapu and moved north to Rawhiti in the Bay of Islands. No one has lived on them since.

Bought by the Crown in 1882, they were at first designated a lighthouse reserve, later a scenic reserve and, since 1975, the islands have been fully protected as a flora and fauna reserve, requiring a permit to land.

Most of the original vegetation probably did not survive the Maori occupation and the plants are largely secondary growth. Pohutukawa is prevalent, as are kanuka, kohekohe, karaka and kawakawa. Astelias, flax, toetoe and *Xeronema callistemon*, the beautiful Poor Knights lily, all thrive on exposed headlands and rocky outcrops. The xeronema was not found by Europeans until the early 1920s and in 1924 an expedition went to the Poor Knights from the Dominion Museum expressly to collect the first specimens for cultivation. It is now a popular plant for frost-free gardens.

Millions of seabirds breed on these islands, returning each year in October and re-inhabiting their old nesting burrows. Buller's shearwaters have made the Poor Knights their home, the only known breeding grounds for this bird in the world. Gannets nest on the Poor Squires and several forest birds inhabit the main islands, including bellbirds, fantails and red-crowned parakeets.

Tuatara also live on the main islands, along with several species of skinks and geckos. But perhaps the most formidable inhabitants are the giant tree weta, and the giant cave weta, the latter measuring about 30 centimetres from the top of its antennae to the claws on its hind legs.

Today, the Poor Knights are renowned as a subtropical diving paradise. Where big-game fishing boats used to cruise these waters, now literally thousands of divers come every year from around the world to marvel at the wealth of marine life in the sea surrounding the islands. And when they leave, they take with them images of a rare wonderland.

I'm not a diver, but it's an awe-inspiring experience to enter the great cavern in a boat, to sit there in the dark silence and listen to the echo of the sea; to circle the islands and gaze at the sheer cliffs rising straight out of the ocean; to crane up at the enormous archway that must have been carved out of rock by the gods, and to think that people once inhabited these rugged shreds of land.

How to get there

Motels, homestays and camping are available at Tutukaka, 40 km north of Whangarei and nearby Ngunguru.
Main activities are diving and fishing.
Both chartered and scheduled boat trips available, many daily.
Chartered flights available from Whangarei Aero Club.
Landing on the islands is not permitted. For information go to the Tutukaka marina.

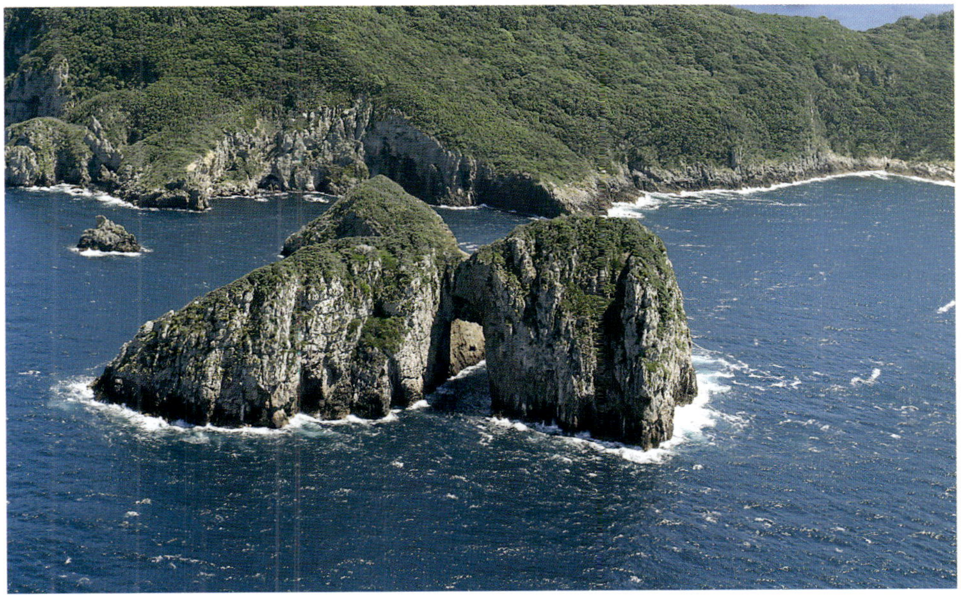

Archway Island, one of the Poor Knights, with the main island of Aorangi in the background.

Marine Reserves

The importance of creating reserves on land to protect endangered or native plants and animals came to be recognised during the earlier decades of the 20th century. In the same way, the idea of creating marine reserves gradually gained support in latter decades of the century as it became apparent that over-fishing and pollution were affecting ocean life.

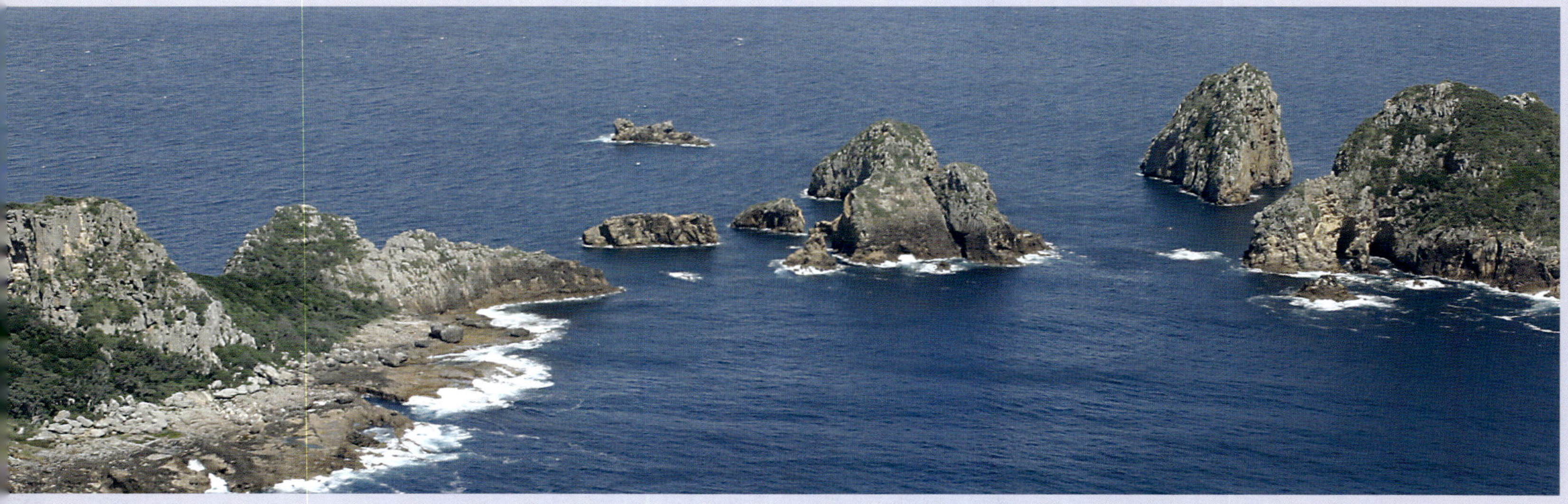

Rocky islands, such as these around the Poor Knights group, attract a huge variety of marine life to their waters. The importance of this kind of environment has been recognised in recent years with the creation of marine reserves.

The idea of 'fencing off' areas of the ocean, seabed and seashore, where species of sea plants and animals are allowed to live natural lives undisturbed by humans and their activities, was first raised in the 1960s – though not without strong opposition from recreational fishers. The Marine Reserves Act was finally passed into law several years later, giving responsibility of management to the Department of Conservation.

New Zealand's territorial waters stretch from the subtropical through to the subantarctic, and it is hoped eventually to form a network of reserves from north to south, where no commercial or recreational fishing is allowed and which enable scientists to study the effects on marine life. Many of the marine reserves are situated in the waters around islands. The largest surrounds the four main islands of the Kermadecs north of New Zealand. This is New Zealand's only subtropical reserve and features a range of fish, corals and other species, some unique to this area.

The first marine reserve was established in 1975 at Goat Island, a tiny speck of land a mere stone's throw from the mainland near Leigh, north of Auckland. The reserve extends along a coastline of approximately five kilometres and the total area is about 500 hectares. The island itself is so close to the mainland that at low tide, when rock shelves extending out from the shore are exposed, it appears almost as a peninsula.

Snorkellers and scuba divers visit Goat Island to explore the underwater environment, and although feeding the fish is not allowed it attracts families who enjoy swimming and fossicking around the shore.

The island itself is not open to visitors. Since 1971, when it was purchased by the Crown from its Maori owners, it has been administered as an island reserve. Formerly farmed and ravaged by goats and pigs, the bush cover is now returning and seabirds nest there, including blue penguins that come home at dusk and clamber ashore to their burrows.

In 1981 the second marine reserve was established in the ocean surrounding the Poor Knights Islands off the East Coast of Northland. Jacques Cousteau, the famous French underwater explorer, claimed the Poor Knights as among the 10 best subtropical dive locations to be found worldwide. Squadrons of stingrays cruise the waters in summer and many subtropical fish living in the reserve are washed to the Poor Knights on the warm East Auckland current. Some of these are found nowhere else in New Zealand. Both novice and expert divers find a visit to the Poor Knights a memorable experience.

Less dramatic but also important is the marine reserve that exists within Auckland's city limits. The Motu Manawa Marine Reserve occupies about 500 hectares of the inner reaches of the Waitemata Harbour, around Pollen Island. It includes mudflats, tidal channels, mangrove swamp, saltmarsh and shellbanks. The intertidal mudflats are an important feeding ground for wading birds. Godwits, knots and sandpipers are international visitors; the South Island pied oystercatcher and the wrybill come visiting in autumn and winter, and white-faced herons, pukeko, spotless crake as well as the endangered banded rail all find the area a rich source of food.

The Coromandel's first marine reserve was officially designated in early 1993 and incorporates offshore areas including Motukorure Island in Mercury Bay and Mahurangi Island off the coast of Hahei.

Further south, Kapiti Island on the west coast north of Wellington boasts two marine reserves in its waters, each protecting a distinctive and contrasting habitat. On the western, exposed side of the island, the shoreline is rocky and there is luxuriant seaweed cover, with many bottom-dwelling creatures such as sea sponges. On the east, the seabed is sandy and gravelly. The waters are much more sheltered and shellfish are common.

In Queen Charlotte Sound in the Marlborough Sounds, Long Island, Kokomohua Islands and a charted rock to the north-east create a chain of islands and submerged reefs that form the nucleus of a marine reserve. This protects all marine life within its boundaries, including seabirds and animals like penguins, dolphins (both Hector's and bottlenose) and seals. The first marine reserve in the South Island, it came into being in 1993 with help from the Department of Conservation after members of Marlborough dive clubs voluntarily stopped taking fish from the area and began promoting the idea of protecting ocean life.

Just offshore from Abel Tasman National Park, and visible from the track at Onetahuti Beach, is Tonga Island Marine Reserve. It covers an area of 1835 hectares, extending around the island to Awaroa Head in the north and Bark Bay in the south. Sandy beaches, bouldery headlands and reefs abound in this area. Snorkellers can see grazing animals in the water, such as sea eggs (kina) which munch on kelp, and turban shells. Crayfish and the occasional conger eel lurk in the caves and crevices of the reef, and sea kayakers enjoy the company of dolphins who are frequent visitors.

New Zealand's second largest and newest marine reserve is approximately 460 kilometres south of the South Island, stretching for 12 nautical miles around the Auckland Islands. It protects not only relatively shallow inshore areas, but also deep ocean environments down to 3000 metres and forms part of the Subantarctic World Heritage Area. The Auckland Islands are the primary breeding ground for the New Zealand sea lion, the southern right whale breeds and calves there and 30 percent of the world's population of the yellow-eyed penguin (hoiho) also breed there, not to mention a great diversity of seabirds.

Other island marine reserves have been established around Mayor Island, north-east of Tauranga, and Te Matuku Bay on Waiheke Island, while applications are in place for more reserves at Paterson Inlet at Stewart Island, Taputeranga Island off the Wellington south coast and at Great Barrier Island.

As the number of marine reserves increases so too will the preservation of threatened species and their establishment around various islands near the New Zealand coast enhances these already special environments.

Goat Island, about an hour's drive north of Auckland, was the first marine reserve to be established in New Zealand waters.

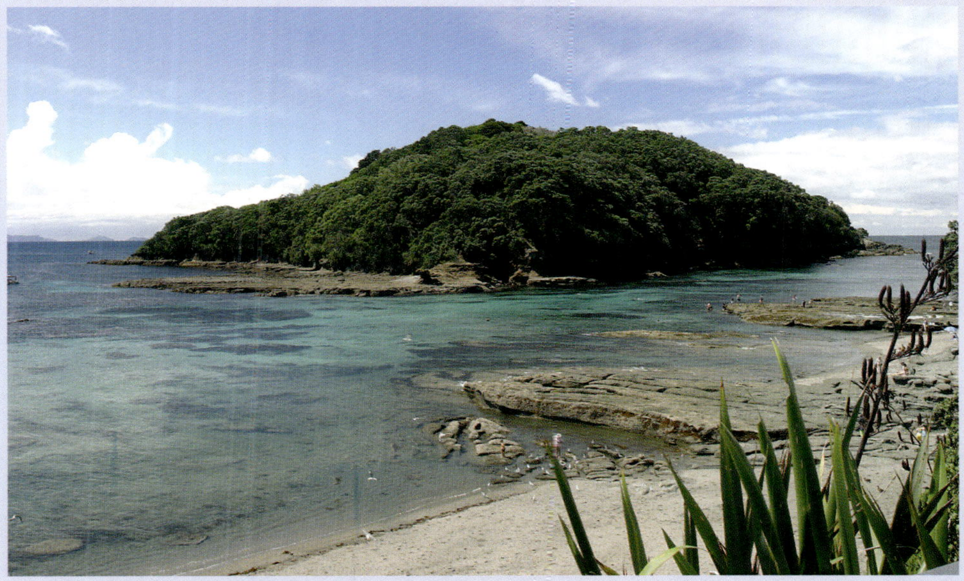

Islands of the Hauraki Gulf

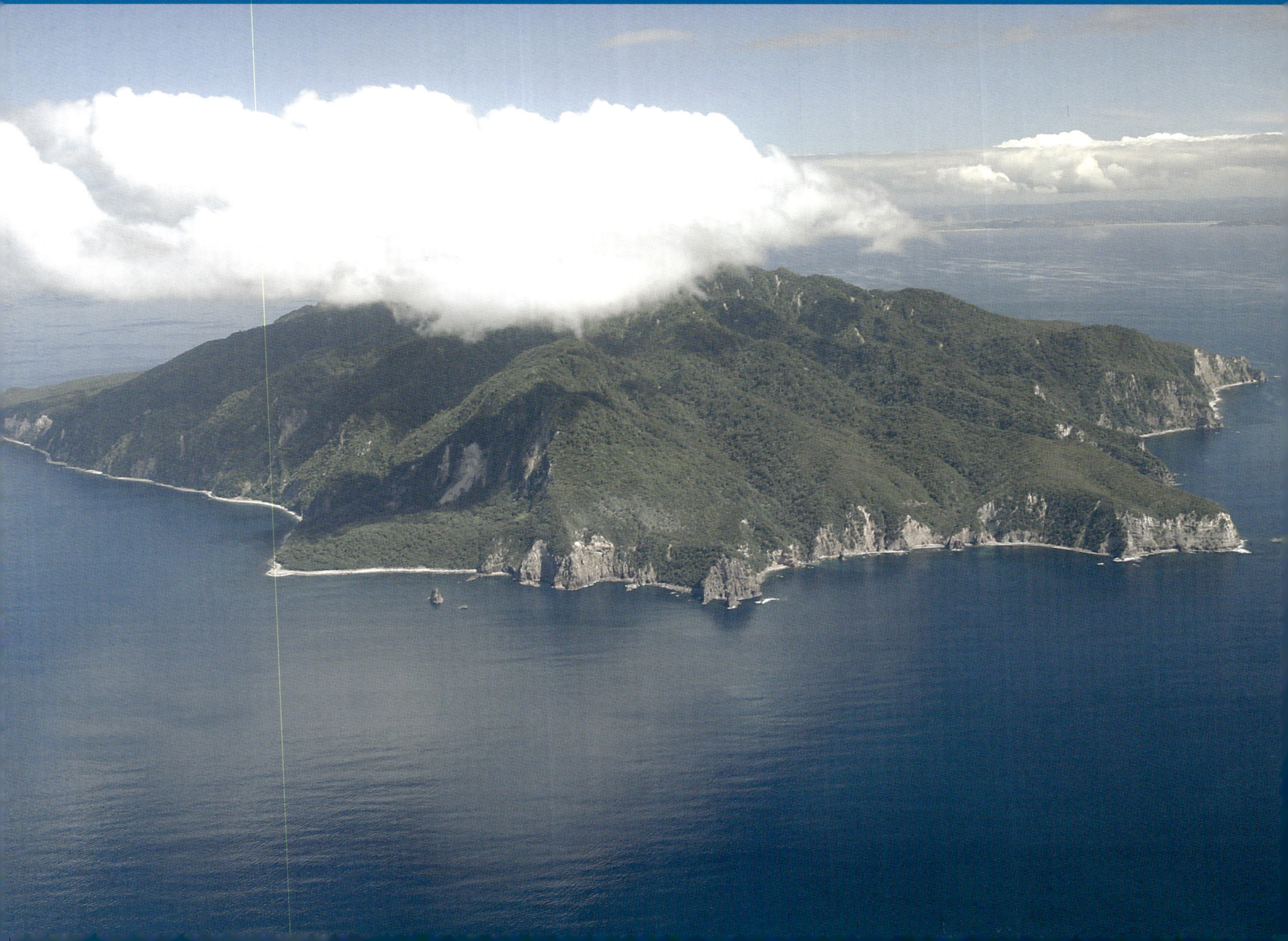

Ask a sailor to list the requirements for his or her boating paradise and it's certain the Hauraki Gulf would contain most of the chosen criteria: sheltered anchorages, wind, long stretches of open sea and a city at its doorstep with marina berths close by. These advantages exist in lots of sailing destinations but it is the profusion of islands in the Hauraki Gulf and everything they represent that make it such a special place for boaties of all persuasions and island lovers in general.

The islands of the Hauraki Gulf lie within the embrace of the Coromandel Peninsula in the east and the long indented isthmus of Auckland along the west. Once the Hauraki Gulf was a low-lying plain, its smoothness punctuated by occasional hills. At the end of the last Ice Age, the glaciers melted, the rivers flooded and the plain was drowned. Only the hilltops remained – to become the islands we know today.

Big, small, smoothly rounded or sharp and craggy, they come in many guises. Since people first arrived on our shores, even the tiniest slivers of these sea-surrounded remnants of land have been inhabited for periods of time. For centuries people have exploited the islands, leaving their imprint for others to discover after they moved on. There were the early moa hunters who preceded the eruption of Rangitoto; the flourishing Maori communities of the 17th and 18th centuries; the European sealers and whalers at the beginning of the 1800s and the traders, ship builders, fishermen and farmers who followed later in the century.

And what of the people who have lived on some of these remote outposts in modern times, because of the jobs they offered?

Until the last two decades of the 20th century when communication systems took an enormous leap forward, life on the really isolated islands such as Little Barrier demanded unusual fortitude. Merely keeping the kitchen cupboards stocked required advance planning. Running a guesthouse at Kawau Island in the 1920s, for example, could be complicated if the steamer did not arrive with enough provisions. Schooling was always an issue. Sickness and accidents posed serious problems. Consider the agony of the curator's wife on Little Barrier in 1910. She and her husband had not long arrived on the island when he became sick and died. All alone, she had to dig his grave and then light signal fires on the beach to attract attention from a passing ship.

In this age of private helicopters, fast ferry services and regular air traffic, such privations are hard to imagine. It was not until the end of the Second World War that the far-flung island-dwellers could depend on regular communications.

Later a Marine Department boat started regular rounds. It brought news, the occasional visitor, maybe the district nurse to check up on the island children and, in its hold, everything from mail and crates of chickens to bottles of wine, batteries and bolts of fabric. If the weather was bad, the boat would be delayed.

When legendary pilot Fred Ladd and his wife Mabel started their Tourist Air Travel Company in 1954 with just one Grumman Widgeon amphibian aircraft, he brought a sense of connection to the people of the Gulf. With the ability to land on almost any beach, he also added assurance of help in times of emergency.

Now when people move to an island in the Hauraki Gulf, they frequently do so to avoid the hectic life of a city, but they know that true isolation is a thing of the past.

Left: More often than not, Little Barrier Island is capped in cloud, creating a microclimate at the summit which supports its own distinct blend of native cloud forest. The island was designated as a wildlife sanctuary at the end of the 19th century.

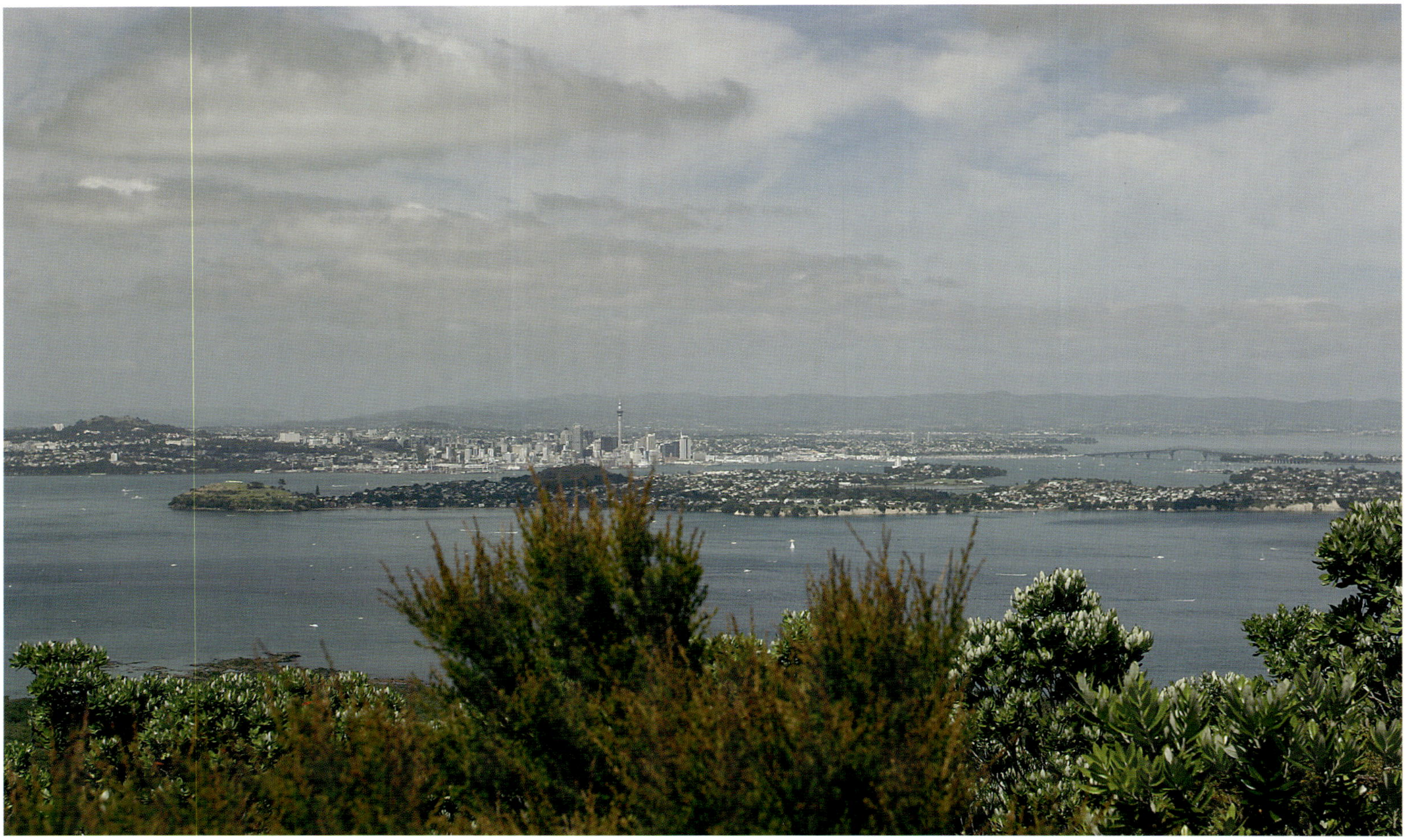

Auckland City, viewed from the summit of Rangitoto Island, only a 30-minute trip by modern ferry.

For decades, Aucklanders have treated the islands in the Gulf as pleasure haunts – venues for picnics, parties, yachting regattas, camping or tramping. More recently, awareness of the need to preserve such special places for the future has increased and the attitude towards the whole Gulf area has become preservation oriented.

In 1967 the Hauraki Gulf Maritime Park was set up with the aim of protecting the existing values of the various islands. In 2000, the Hauraki Gulf Marine Park Act was enacted, replacing the maritime park. It recognises that the islands, beaches and coastal waters contain some of New Zealand's most important conservation icons and that it encompasses our most important marine playground.

The aim of the Act is to achieve integrated management across land and sea and co-ordination between all the decision makers whose actions impact on the area set aside by the Act. It should also 'sustain the life-supporting capacity of the soil, air, water and ecosystems' of the park. Commercial activities like shipping and fishing are allowed, but not at the expense of maintaining conservation and recreational values.

The park extends over a broad area and includes many islands within the normal confines of the Gulf as well as several on the eastern side of the Coromandel Peninsula. It reaches as far north as the Mokohinau Islands and as far south as the Aldermen Islands, but it includes only public land, whether mainland reservation or island territory.

The attitudes that led to the passing of this legislation have also expressed themselves in more local situations. Groups such as the Motutapu Restoration Trust, Rangitoto Island Heritage Conservation Trust and the Motuihe Charitable Trust have been formed by people living in and around the Gulf. Their aims are to preserve the historic and conservation features of individual islands. With the aid of volunteers they have become involved in planting, weed and pest eradication and other restoration tasks on the islands, in association with the Department of Conservation.

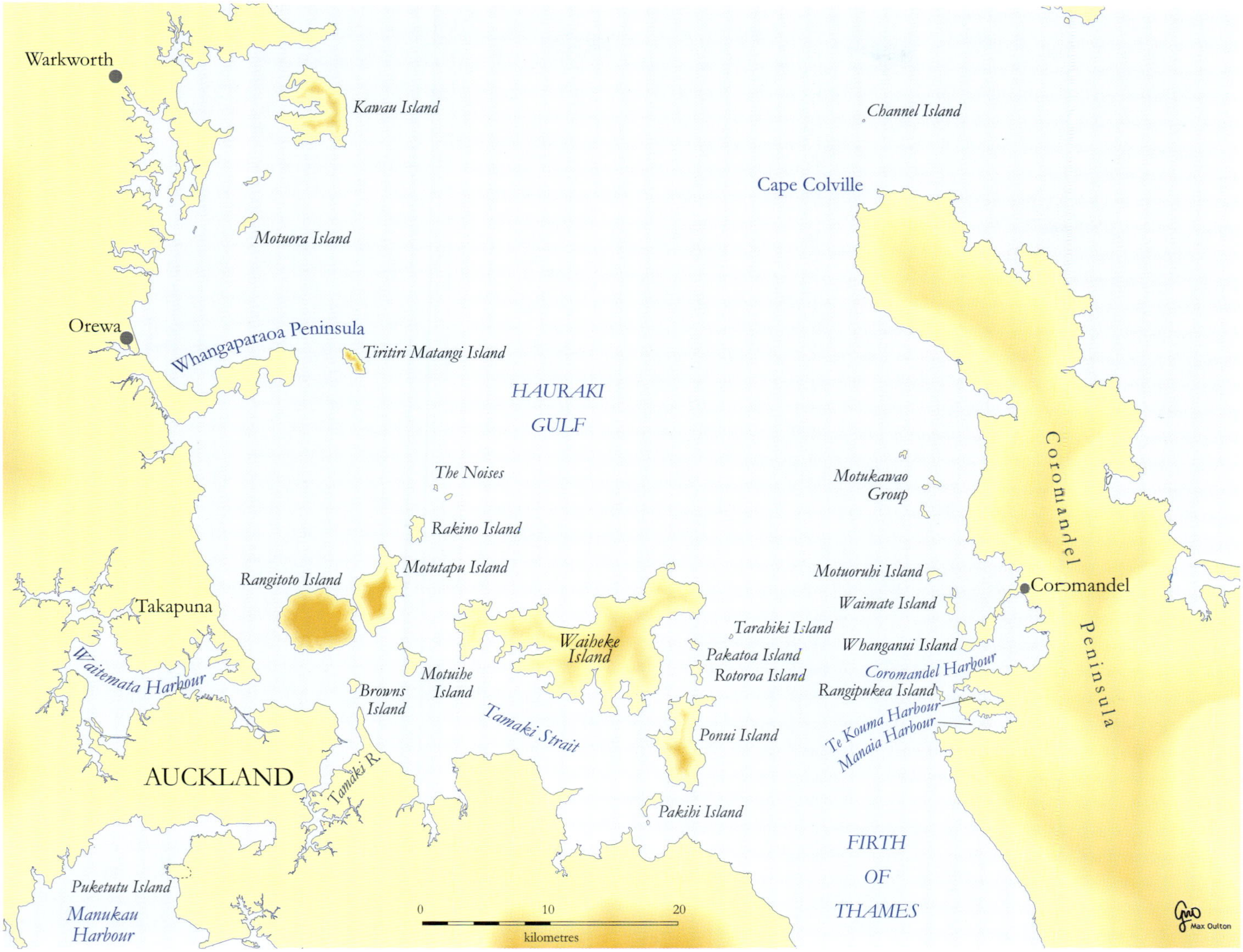

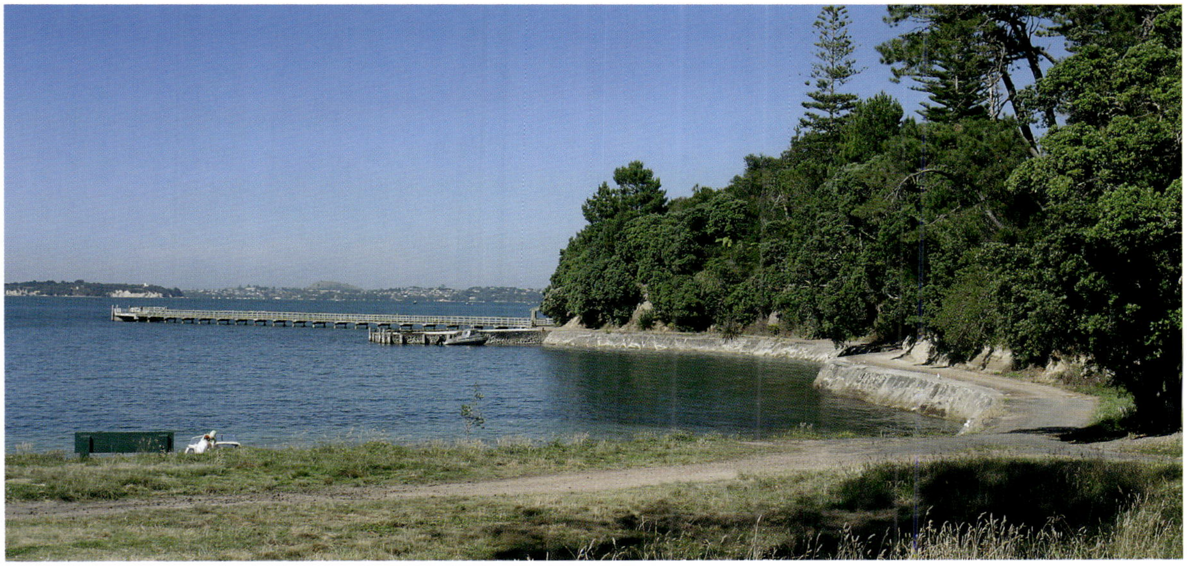

Once designated as a quarantine island, Motuihe is one of the jewels of the inner Hauraki Gulf and has been a popular picnicking spot for Aucklanders for decades.

The inner islands

CLOSE TO AUCKLAND AND LYING IN A CHAIN EAST OF THE MAINLAND IS A GROUP OF ISLANDS ALL WITHIN EASY REACH OF A DAY'S OUTING. EACH HAS ITS OWN PARTICULAR CHARACTER, ITS OWN STAND-OUT FEATURE THAT SETS IT APART FROM ALL THE OTHERS.

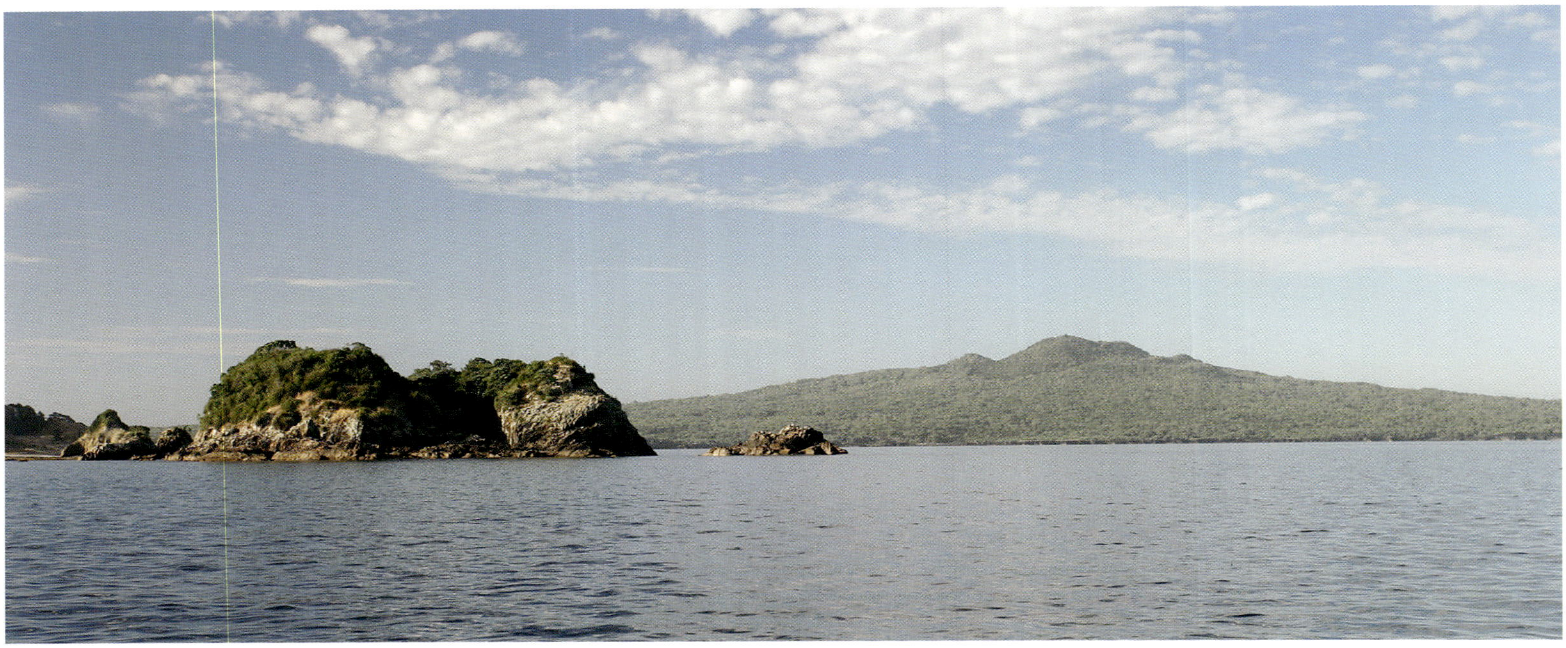

Rangitoto Island is the result of a volcanic eruption somewhere between 600 and 800 years ago. It is remarkable for its symmetry and looks similar no matter where it is viewed from.

A young tree struggles to establish a foothold among bare scoria.

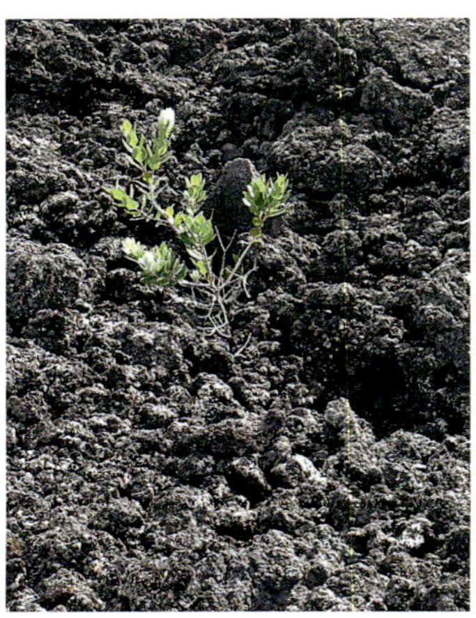

Rangitoto

Rangitoto Island, with its symmetrical cone shape spreading outwards on the sea, has become a symbol of Auckland. Viewed from across the water, the nature of the island is deceptive. A forest of pohutukawa hybrids flourishes – a cross between tough northern rata, *Metrosideros robusta*, and pohutukawa, *Metrosideros excelsa* – and clothes the island in a soft coat of olive-green. Think forest and you probably think of layers of deep peaty soil for the trees to sink their roots into. But the forest on Rangitoto grows in harsh layers of scoria, dry and apparently as barren as coke from a furnace. For Rangitoto is the result of an underwater volcanic eruption and is a mere youngster in the scale of geological time. It has only existed for an estimated 600 to 800 years, a new neighbour among the other islands in the Hauraki Gulf. Underfoot the island is anything but soft. The open patches that remain without plant cover are a mass of black cinders that scrunch as you walk across them.

So where do the trees find their nourishment? Some scientists say there is a vast pool of fresh water beneath the surface of the island that has leached down through the scoria and remained trapped there. Under the canopy of pohutukawa other plants grow on this inhospitable terrain – astelias with slender silver leaves (*Astelia banksii*) push up through the scoria, olearias, griselinias with huge decorative leaves and a variety of hebes all line the track that leads to the summit. It looks in places like a carefully planned rock garden.

As recently as 150 years ago visitors remarked on how barren the island was and how loose ash could still be stirred up. Sketches from the 1860s show only a margin of trees. I climbed to the top recently, the first time in several years, and marvelled anew at the stunning views of Auckland and the Gulf. When I turned to look down at the island it appeared covered in bush. The crater and its tuff ring is still obvious, the vegetation within it mainly manuka and kanuka with the occasional *Knightia excelsa* or rewarewa, standing tall above the rest.

The Crown bought the island in 1854 and 40 years later set it aside as a recreation reserve. For

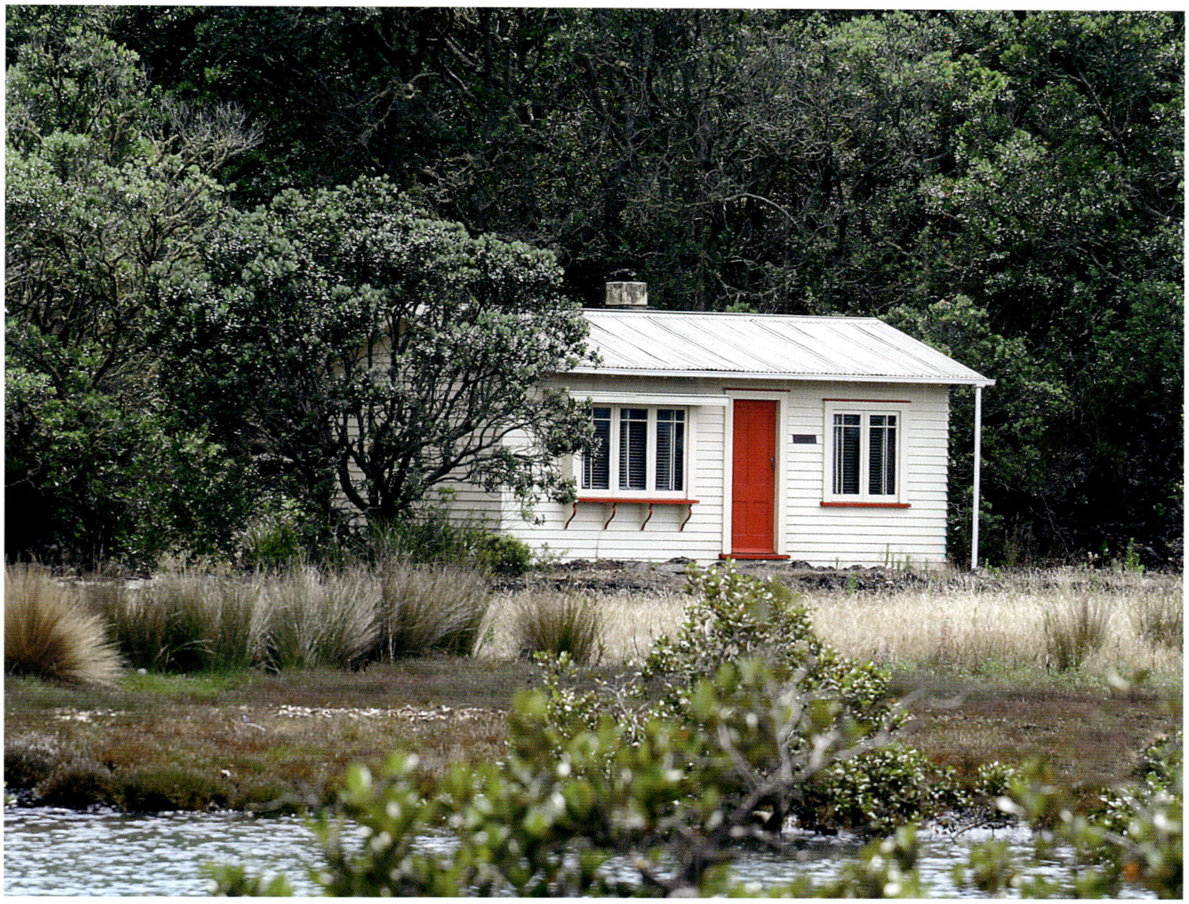

There was a time when city dwellers were allowed to build holiday cottages on the island. Now only a few remain, a reminder of the days before Rangitoto became a reserve under the auspices of the Hauraki Gulf Marine Park.

How to get there

Fullers Gulf Ferries service many of the islands in the Hauraki Gulf, departing from the Auckland Downtown Ferry Terminal, Quay Street. A Harbour Cruise operates daily.
Charter boats are also available by arrangement.
Auckland Visitor Information Centre is open 7 days.
A regular, daily ferry service between downtown Auckland, the Devonport Wharf and Rangitoto allows time for visitors to explore on the island.
A ferry service to Islington Bay, giving access to walking tracks on Motutapu, runs Wednesday and Sunday; more frequently during holiday periods.

over 30 years the volcanic scoria was quarried and shipped to Auckland.

There was a time when city people built holiday baches on Rangitoto and a fringe of cottages sat precariously just above the shoreline. A road of sorts runs from the wharf and continues on around the shore, built by prisoners in the 1920s. They also laid out the track to the summit. There was a swimming pool, walled with scoria, and a shop near the wharf where day visitors now disembark but when the Hauraki Maritime Gulf Park was established privately leased property was phased out. A few cottages remain as a historical record.

There was a time too when vessels leaving Auckland for the high seas would pull into Islington Bay, or Drunken Bay as it was known, on the eastern end of Rangitoto. There they would await crew who had missed departure time while giving those on board time to sober up after a riotous shore leave. It is not far from there to Wreck Bay where several old ships' hulls have lain over the years, rotting out in the shallow waters between Rangitoto and its very close neighbour, Motutapu. At low tide, a causeway links the two and it is possible to walk from one to the other.

Motutapu

Within the Hauraki Gulf, Motutapu is one of the earliest Maori settlement sites and had been home to several generations of moa hunters before the Rangitoto eruption. To what god or taniwha did those long ago inhabitants of Motutapu attribute the disastrous upthrusting of another island alongside them? Everything was buried by the all-enveloping ash – the kauri forests, their cooking sites, even the moa bones from their last meal. Archaeologists have traced evidence of the settlements and gardens of those early settlers. Somewhat eerily there are even human footprints and those of a dog, imprinted forever in the volcanic residue. Modifications to the landscape show where terraced house sites and fortified pa were built.

Farming began in 1840, when the Maori owners sold part of the island to Europeans, and has continued ever since. Its rolling green hills of pastureland present a total contrast to the spreading volcanic cone of Rangitoto. In 1869 the island was bought by two brothers named Reid. Scottish farmers, they established homesteads at Emu Bay – named for a family of emus that was liberated there in the 1880s – and Home Bay, which was to become a favourite anchorage for boat enthusiasts. Only the Reid Homestead at Home Bay exists today. Built around 1901, a conservation plan has been developed for the homestead and initial work on the exterior of the building has been completed.

The Reid family owned and farmed the island for the next 80 years. They introduced other exotic animals such as deer, buffalos, ostriches, wallabies and waterfowl, and they were renowned for their hospitality. Groups of hunters would sail over from Auckland on wallaby-shooting expeditions and the island became famous for its picnic parties. Shirley Maddock reports in her book *Islands of the Gulf* that 'a real Motutapu picnic might see 5000 people in the bay and ferries tied up on both sides and the full length of the wharf'. She adds that a paddle steamer would come all the way from Thames with a group of gold miners and their families on board, bound for Motutapu and their annual picnic.

During the Second World War, the army bought Motutapu and it became an important link in the defence network. Roads and barracks were built, a gun battery and observation post set in place and underground stores were created for ammunition. Fortunately none of these facilities were needed and what remains of the military installation is now home to an outdoor education camp that hosts more than 8000 school children a year.

At Home Bay there is a DOC campsite with basic facilities that is adjacent to the beach and accessible by private boat.

The Motutapu Restoration Trust has produced a 50-year working plan that involves the ecological restoration by community groups of those parts of the island that are unsuitable for farming and which have already been assessed by archaeologists. Panels interpreting the history and ecology of the island have been set up at the gun emplacements and carvings commissioned to mark the entry point to Motutapu.

Motuihe

Nearby, Motuihe is easily recognised by the long stand of Norfolk pines that march along the spine of the land. Tall and stately, they distinguish the island long before visitors arrive at one of the curving white-sand beaches that make it a popular picnic spot for Aucklanders in summer. In days

Motuihe is easily recognisable by the line of stately Norfolk pines along its spine.

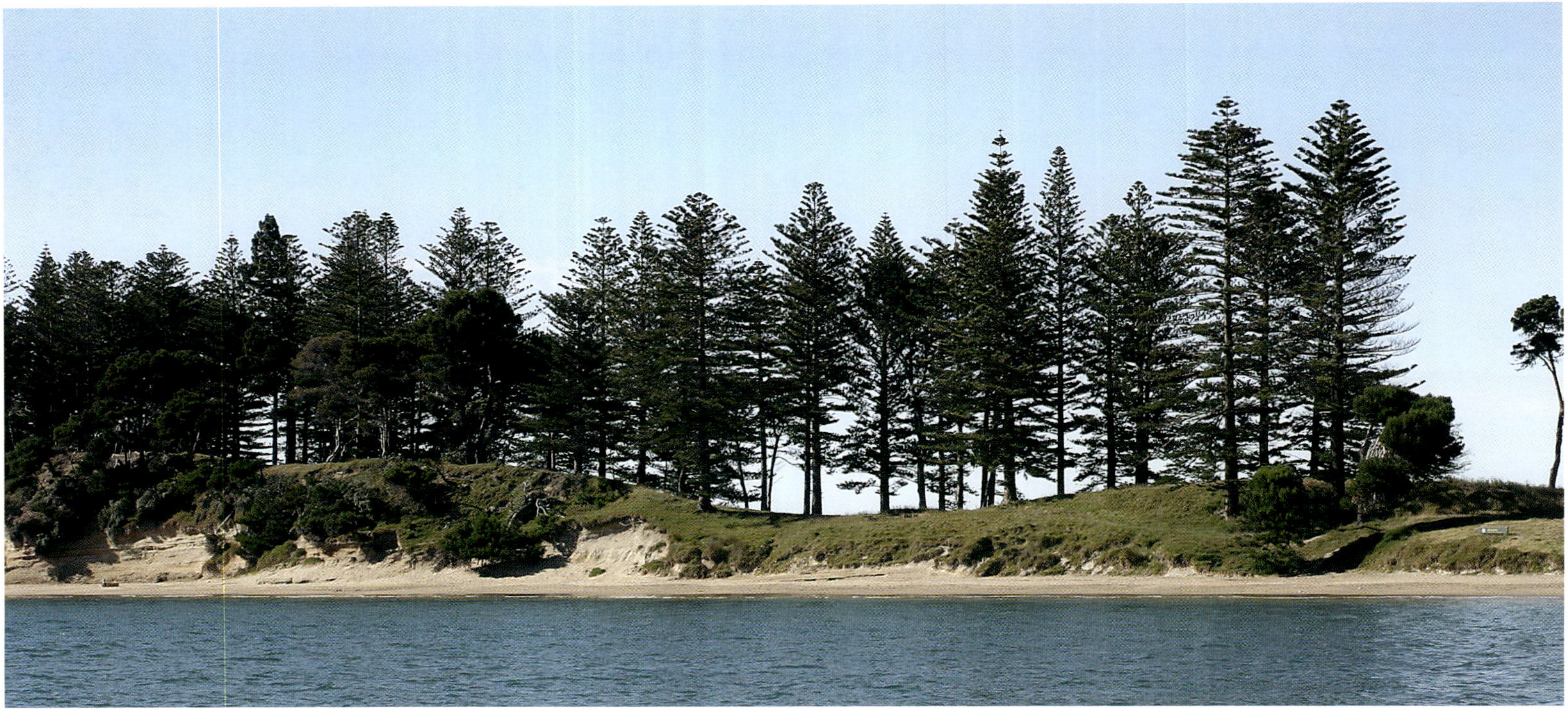

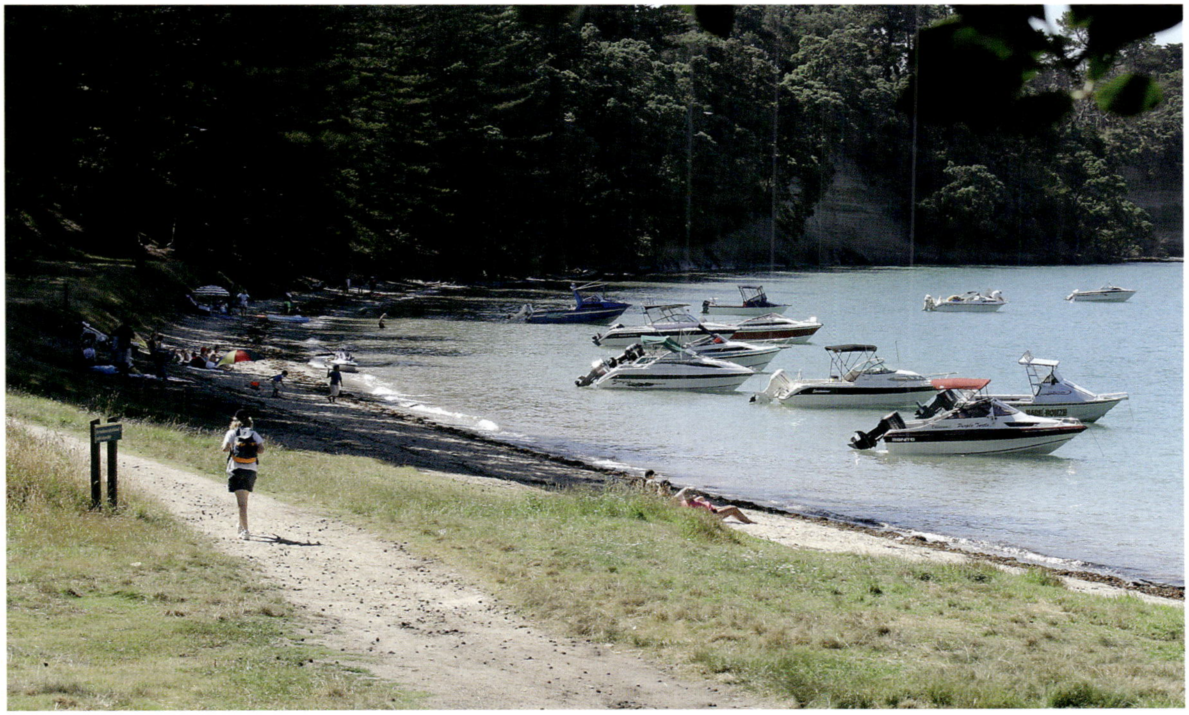

In summer, numerous picnickers find their way to Motuihe by kayak, yacht, launch and power boat and the beaches on both sides of the island are favourite sun-bathing locations.

gone by a regular ferry service tied up at the jetty and landed large groups on the island for a day's outing. Ferries still operate as part of a 'mail run' service to Rakino Island, and private boats, dinghies or kayaks pull up on the shore.

The island is a recreation reserve, administered by the Department of Conservation and was originally purchased by Europeans for farming. William Brown and John Logan Campbell, two of Auckland's best-known early businessmen, bought the island in the 1840s. It was farmed, with the help of Maori workers, by a manager whose diary records the building of stockyards, fences and a fowl house, the tending of pigs and goats and the planting of crops. Later it was reported to be well stocked with horses, cattle, sheep, pigs, turkeys and geese 'without number'.

In 1873 it was bought by the Crown for use as a quarantine station but its period of greatest fame came during the First World War. Along with Germans living in New Zealand at the time, the charismatic German Lieutenant-Commander Felix von Luckner and several of his crew from the *Seeadler* were detained there. Imprisonment seems too strong a word for they managed to charm their guards into a very relaxed mode of surveillance. They conveniently ordered the stores for the island, cached a supply to sustain an escape, fashioned a sextant out of bits and pieces, and one dark night made off with the island's launch. Heading east they hi-jacked a coastal scow, kidnapped the crew and set sail for Samoa. Unfortunately for them, but fortunately for our national reputation, they were apprehended when they had reached the Kermadec Islands. When von Luckner returned to New Zealand just prior to the Second World War on his private yacht, he was feted and fawned over.

Motuihe had a role to play in the Second World War also. The complex of buildings on the island was converted into a naval training base, remaining in use until 1963 when the island reverted to public use.

Part of the island is still farmed and in 2003 the Motuihe Project was launched which, in concert with the Motuihe Trust and DOC, aims to restore indigenous forest to the island while protecting the significant historical sites and ensuring that visitor opportunities are enhanced. Project Crimson is growing thousands of pohutukawa from seed collected on Motuihe and the first of these trees were planted in 2003. A nursery, manned by volunteers, has been established to produce plants and with the island predator-free, the next step in the restoration process will be the introduction of selected threatened bird species. Already the New Zealand dotterel and the variable oystercatcher make their home on the beaches.

How to get there

Return ferry service from downtown Auckland to Motuihe Saturdays, Sundays and Public Holidays; daily December 26 to January 5. Limited camping available on the island.

How to get there

There is no regular ferry service to Browns Island. Charter boats and kayaks available. A mail boat service to Rakino Island runs Friday, Saturday and some public holidays.

Browns Island

Browns Island seems well-named in late summer when its normally green hills are burnt golden. But in fact its European name came from William Brown who bought the island and took up residence there with John Logan Campbell in 1840. They started a pig farm while biding their time until the site for Auckland was firmly established and they could move there and set up in business.

Motukorea, its Maori name, meaning island of the pied oystercatcher, seems much more evocative. Motukorea was once extensively settled and historical sites – remains of Maori pa sites, gardens and settlements – are evident to the trained eye on much of the island. It is estimated that the earliest of these dates to at least 500 years ago. Remnants of stone-walled gardens can be seen in a valley on the south-east side of the hill and these were still in use when the first Europeans arrived in 1820.

The Devonport Steam Ferry Company owned the island in the early 1900s and, as with several other islands in the Gulf, it was a popular destination for picnickers. Later it was bought by Sir Ernest Davis, a local businessman, who gifted it to the people of Auckland in 1955.

One of the best preserved volcanoes in the Auckland area, the island is now managed by the Department of Conservation as open pasture, allowing the contours of human and volcanic modification to remain evident. It lies a short distance offshore and in good weather is easy to reach by kayak or small boat. When you climb to the top it is easy to distinguish the main scoria cone that has survived in good condition with a distinctive, well-shaped crater.

Rakino

Neither conservation estate nor part of the Hauraki Gulf Marine Park, Rakino is mainly

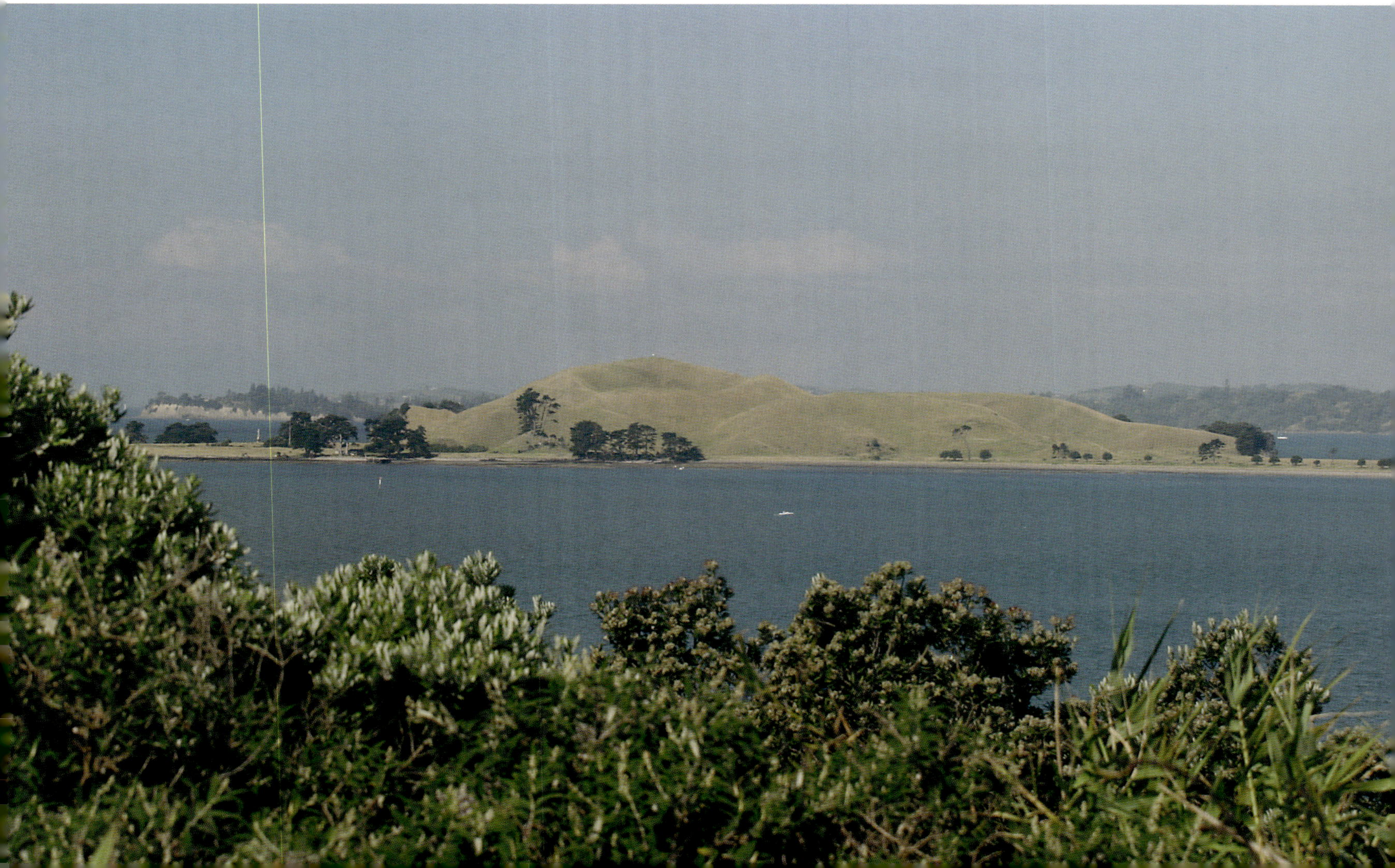

Browns Island in the Hauraki Gulf is visible from many points around Auckland's eastern suburbs.

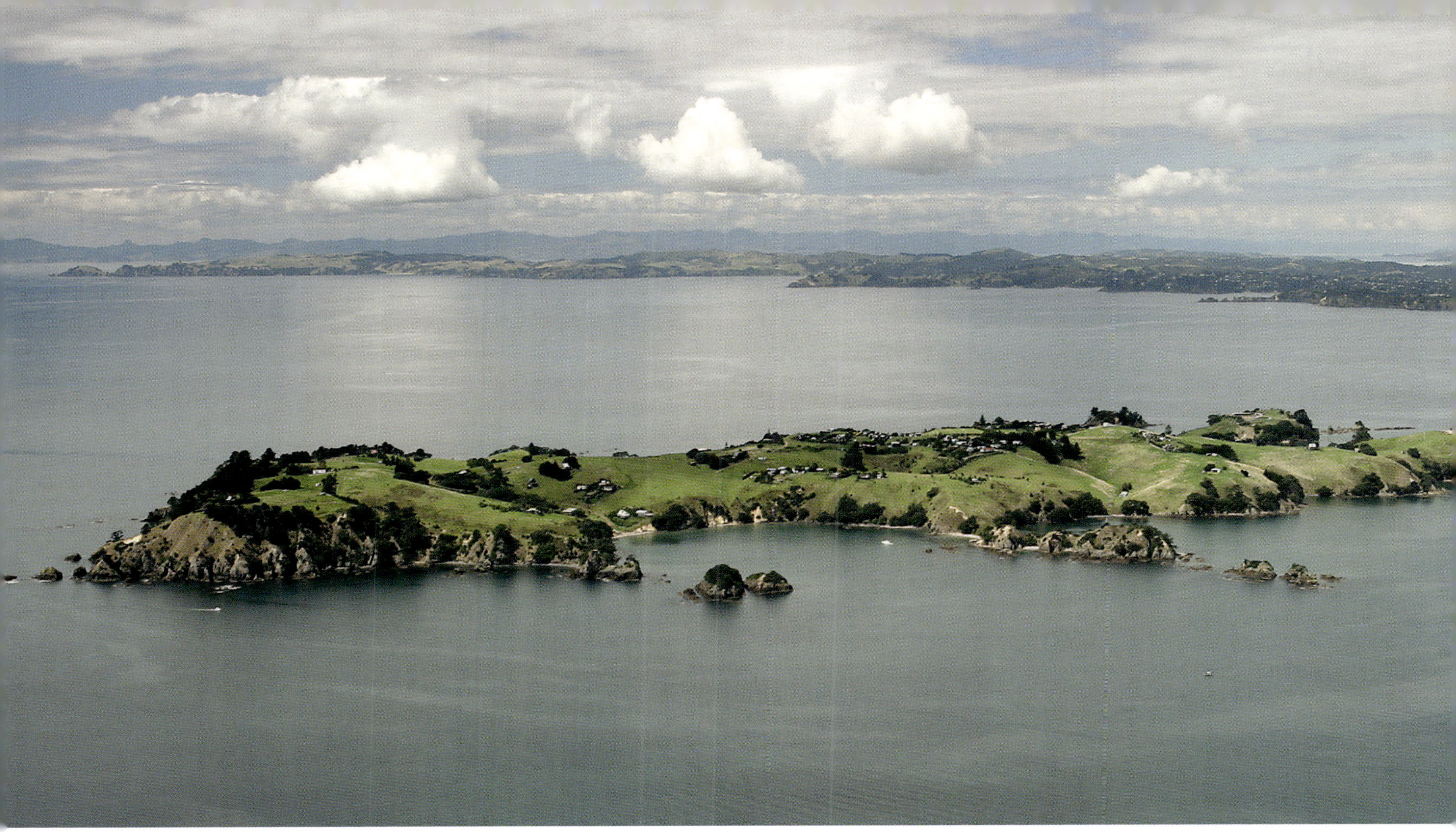

privately owned and in recent years has become a holiday retreat for Aucklanders, especially the well-heeled. Interesting architecture stands out against the skyline of this rugged little island girt with rocky cliffs and jutting, jumbled headlands. Pohutukawa grow on the edges of the land, their roots clinging and exposed, holding on where no other roots would dare to try. As suburban subdivision comes to the island there is a move to preserve the coastal vegetation and make provision for coastal walkways with public access.

Sir George Grey was one of the early European owners of Rakino but he lost interest in it when Kawau became available. In 1874 Albert Sanford leased the island, later buying it from Sir George Grey. His family lived there for nearly 80 years, first working as fishermen, later involved in farming. Island life must have been anything but idyllic when their livelihood was wrested from the sea and the daily catch rowed to Auckland for selling. They built their homestead out of kauri, rafted from the Coromandel: today's helicopter transport would no doubt seem an easy option.

In the 1960s a scheme was hatched for establishing a philanthropic community on Rakino. It sputtered and died and the island has remained a peaceful haven for those who own their own patch of paradise there.

Rakino Island is almost entirely privately owned and many of the holiday homes, such as those at Woody Bay below, command fine views across the Hauraki Gulf to Auckland and the nearby islands.

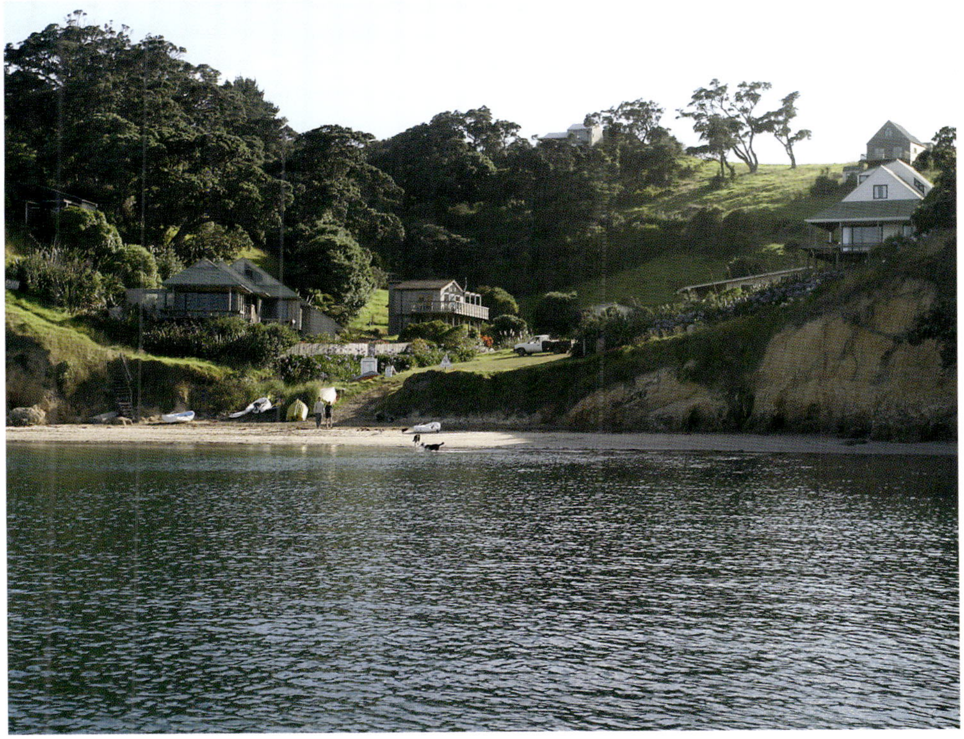

Waiheke

Right: Boats moored in Rocky Bay, Waiheke Island.

The wine industry is well established on Waiheke Island and many of its rolling green hills are patterned with vineyards, such as this one with a fine view to Rakino Island.

Once known as Te Motu Arai Roa (The Long Sheltering Island), in European times it became known as Waiheke, which translates as Cascading Waters, both evocative names for a beautiful island. Situated 19 kilometres from downtown Auckland at its closest point, Waiheke Island is the second largest island in the Hauraki Gulf, after Great Barrier. Sweeping resort-style beaches, secluded anchorages, rolling green hills, rocky headlands and magical views around the Hauraki Gulf – these are the attributes that make Waiheke Island a sought-after location for Aucklanders and others from further afield.

It has always been this way. Long before European ships first nosed their way into these waters, the island was prime real estate. It is a natural stopping place for travellers from the north and it gives easy sea access to the Firth of Thames. For early Maori it was a convenient staging post on the way from the Coromandel Peninsula to the west coast via Tamaki Strait and the Tamaki River. With abundant fish and fresh water, plentiful timber for making canoes, dwellings or palisades and excellent defensive positions, the island was home for generations of Maori. It was also the scene of numerous battles and centuries of tribal jostling for supremacy and territory. It seems that tribe ousted tribe in regular succession from perhaps as long ago as 700 years.

And then the Europeans arrived with new ideas, new tools and an age-old greed. It was their muskets that armed Ngapuhi in 1821 when they swept down from the north and from there the story of Waiheke is the story of New Zealand in microcosm. The northern tribes attacked and killed many residents on the island. Those who

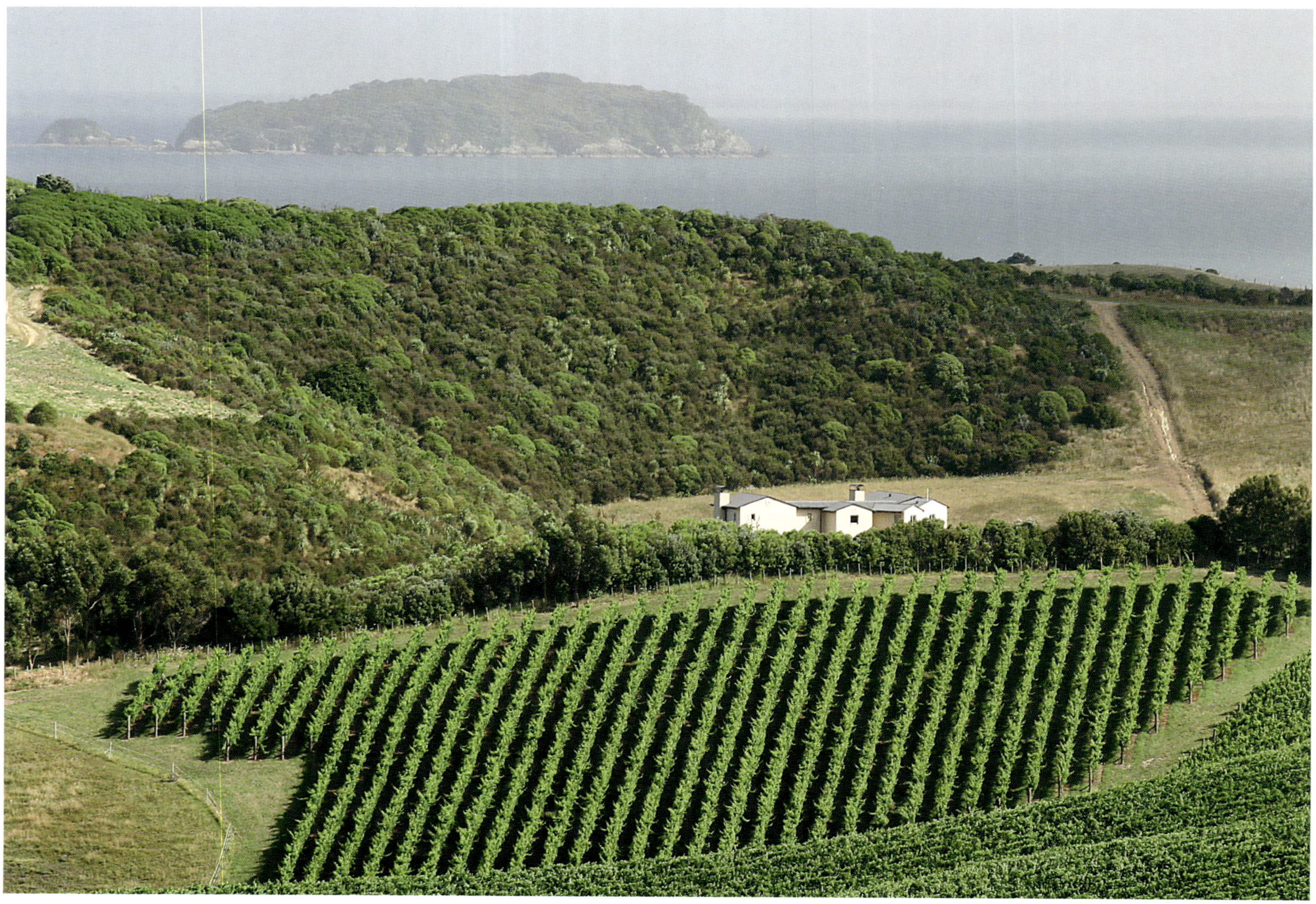

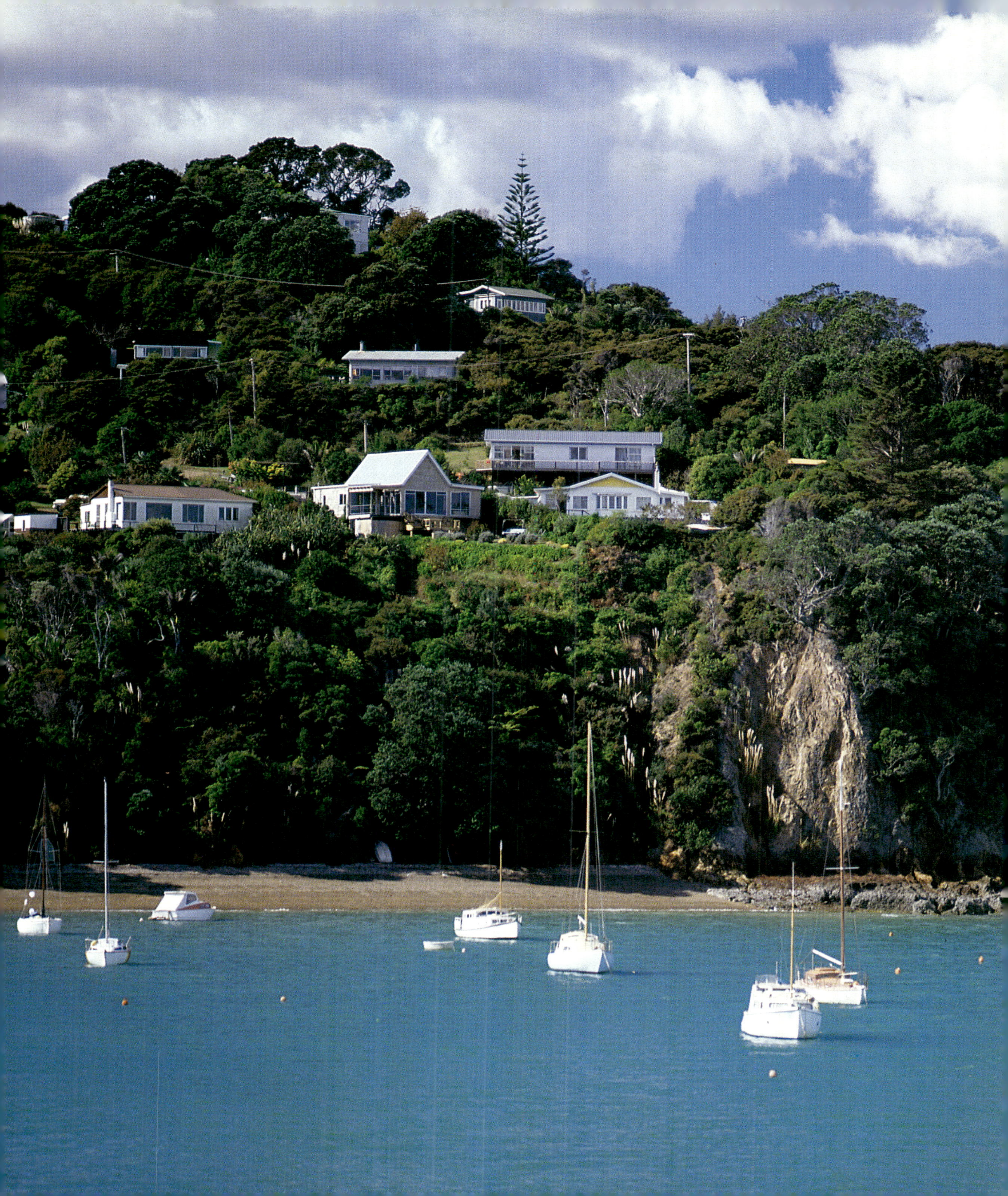

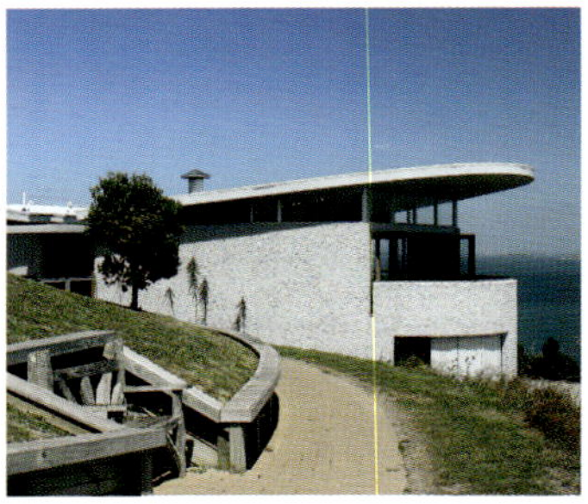

The café at Te Whau Vineyard looks out across Kennedy's Bay.

were captured were taken north as slaves; those who survived and escaped, fled southwards.

French explorere Dumont D'Urville saw the island in 1827 and commented on its magnificent forest and 'sites that would be admirably fitted for settlement'. Few people today would argue with the latter comment. When a small number of Europeans – traders, timber workers, and settlers – arrived at Waiheke in the 1830s, drawn to this crossroads territory for much the same reasons as voyagers in earlier centuries, they found a largely depopulated land.

Maori who still lived there were quick to see the advantages of trade with the newcomers. They helped the settlers bring kauri logs out of the forest and stories were told of the ritual and the chanting that the leader used to urge on teams of workers who dragged the logs out of the bush with primitive ropes. They were quick also to cultivate the crops the Europeans brought with them and sell them for profit. Land seemed like a simple swap for a few English pounds, some axes and blankets and maybe a musket or two. Such an easy transaction laid the ground for countless appeals and counter-appeals as the decades rolled by.

Settlers came, intrepid would-be entrepreneurs, turning their hands and their minds to whatever would make them a living. Those were the days of a few isolated communities mainly concentrated at the eastern end where boats could find ready shelter on their voyages to and from Auckland or Sydney via the Waiheke Channel and the Tamaki Strait, protected by the Long Sheltering Island. Gradually holidaymakers gravitated to this end also, from the mining town of Thames and, later, Auckland. The western end of the island in the early Auckland days was inhabited mainly by Maori and a few European farmers.

Kauri while it lasted was cut and exported for spars. When the kauri was gone, the settlers took kanuka for firewood and gravel to make concrete for the booming town of Auckland. The island was serviced by brig, barque and brigantine, which took cargoes offshore. Whatever needed to be taken from mainland to island and back again – wood, wheat, stock, gravel, sand – went on scows, those workhorse boats that plied the coasts of New Zealand until the Second World War and occasionally later. For a period of more than 25 years manganese and copper were intermittently mined and exported, but never became the profitable industry their promoters imagined.

With increasing prosperity in Auckland, so the emphasis of life on Waiheke began to change. In the 1870s a steamer service started, signalling the beginnings of an excursion business that burgeoned as the years went by. From 1881, the ships of the Northern Steamship Company, driven by propeller or paddle wheel, plied the Gulf, bringing supplies to settlers and city workers to the islands for excursions. The era of annual company and club picnics began. From the evidence of old photographs, the ladies' long skirts and wide-brimmed hats of the day were no deterrent to picnicking on the sand.

Yachting became popular. Eyewitness reports told of yachts arriving from the city on Friday nights. It was evidently not uncommon for as many as 30 or 40 yachts to be moored at Matiatia. Yachting, it seems, has long been a favourite pastime of Aucklanders.

With Waiheke's attraction for day-trippers, the next step in the recreational development was the establishment of boarding houses. Visitors could spend a weekend away from the city rather than just a long day trip. Today most resort accommodation is at the western end of the island, but at the end of the 19th and beginning of the 20th centuries, the sheltered eastern bays where European settlers had first set up their homesteads were more popular for holidays.

In 1896 an advertisement appeared in the *New Zealand Herald* luring prospective visitors to the first guesthouse at Cowes, a boaties' paradise at the eastern end of the island, owned by Innes Parris, who also ran the post office there. 'The Best Summer Resort in the Province' it read, offering 'boating in a safe bay, fishing, bathing and airy bedrooms'.

Photos taken at Cowes show steamers lined up at the wharf, the beach crowded with families demurely 'promenading' on the sand and numerous small boats active in the bay.

With a wharf already erected at Cowes Bay, a second wharf was built soon after at Awaroa by the Northern Steamship Company. It wasn't long before passenger ships were a common sight in the summer holidays carrying excursionists to Cowes and Awaroa. Between 1916 and 1924 a further seven wharves were erected at Orapiu, Ostend, Surfdale, Palm Beach, Onetangi, Omiha/Rocky Bay and Matiatia, the last to be built in 1924.

Over the years the passenger trade gradually increased but after the Second World War, the

How to get there

There is a frequent daily ferry service between downtown Auckland and Matiatia, Waiheke Island. Fullers ferries are serviced by bus on the island, to and from all sailings. Weekends and public holidays ferries make the return trip several times daily from Halfmoon Bay, Auckland to Waiheke. There are all grades of accommodation available on the island.

For all Waiheke accommodation bookings and information on scenic bus tours, vineyard tours and other activities, contact the Waiheke Island Visitor Information Centre, Oneroa.

era of steamship excursions waned and the way was open for a different kind of tourism, that of the holiday bach. Until the late 1980s, however, ferries still took nearly one and a half hours to reach Matiatia – travelling at 10 to 12 knots – a long commute time for anyone who wanted to live the island life but work in the city.

In 1987 the 600-passenger, 26-knot catamaran *Quickcat* entered service, covering the distance to Matiatia in just 35 minutes. At the commissioning ceremony, the then Minister of Transport, Richard Prebble, remarked that Waiheke commuters could now travel to their city jobs 'in less time than it takes many mainland dwellers who face frustrating delays in Auckland's peak hour traffic jam'.

The arrival of a fast, comfortable, modern ferry between Waiheke and the mainland had an enormous effect on island life, stimulating commuting and development. Allied with the development of electronic communications in the 1990s it enabled many more people to live on Waiheke and continue city jobs.

Transport, however, was only one side of the equation. Available land in small lots was also necessary. By the beginning of the 20th century there were several landowners with relatively extensive runholdings on the island and only a few pockets of population. At Te Huruhi, near Kennedy Point, there was still an area of Maori-owned land. Much of this was gradually sold and Fred Alison, whose father founded the North Shore Ferry Company, bought Oneroa and Matiatia in 1901. He cleared the scrub and put animals on the land. When he sold Oneroa in 1922, the process of suburban subdivision started, resulting in the western end of the island becoming the most populated area.

A road linking the two ends of the island was only completed in the 1960s and the eastern end of the island is still dominated by open grazed land, punctuated by vineyards and an occasional olive grove. Now and again secluded tourist accommodation is tucked into a hillside with a fabulous view. Rare are the days when no boats are visible in the numerous bays along the coast.

Stony Batter, the impressive remains of a Second World War Counter Bombardment Defence Battery, sits high atop a hill above Man o' War Bay at the eastern end. When war broke out in 1939, with it came the fear of invasion and the Waiheke battery was planned as part of a larger coastal defence system protecting Auckland Harbour and its approaches from enemy ships.

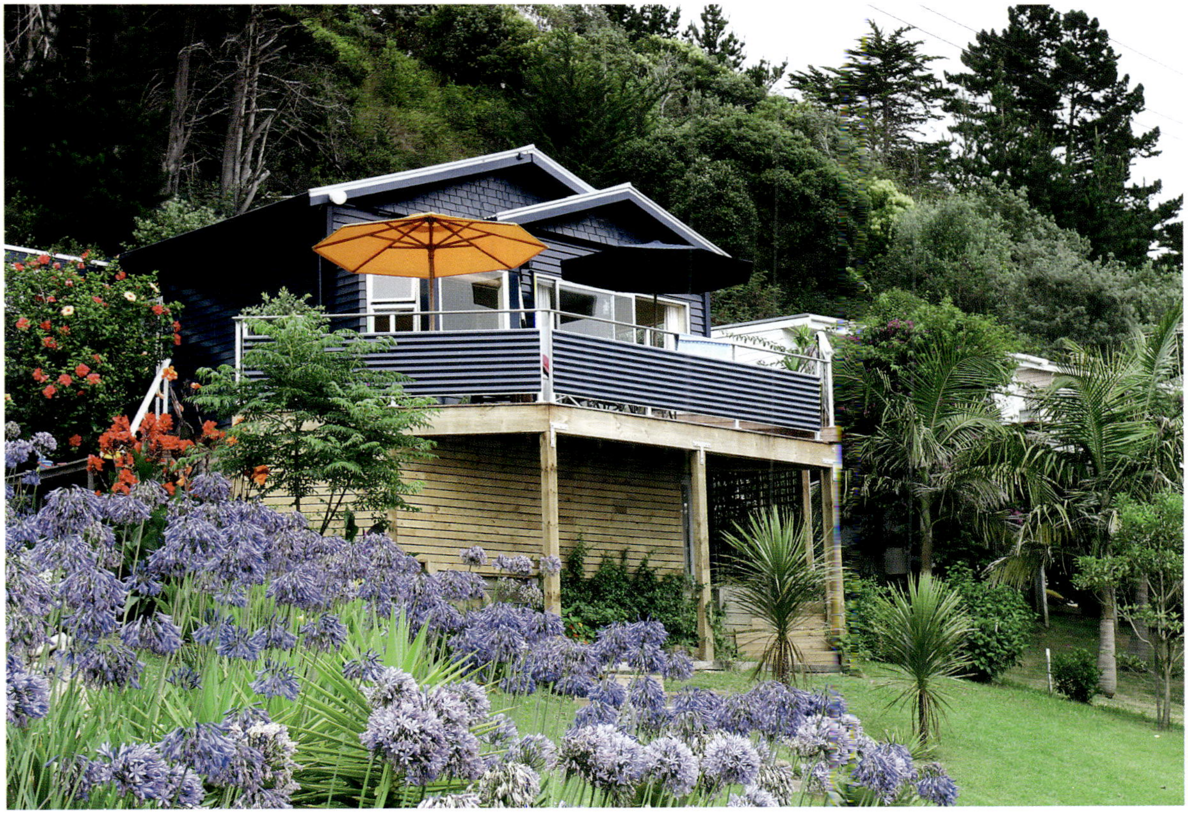

This cheerful blue holiday house fits right in with the relaxed atmosphere of Waiheke Island.

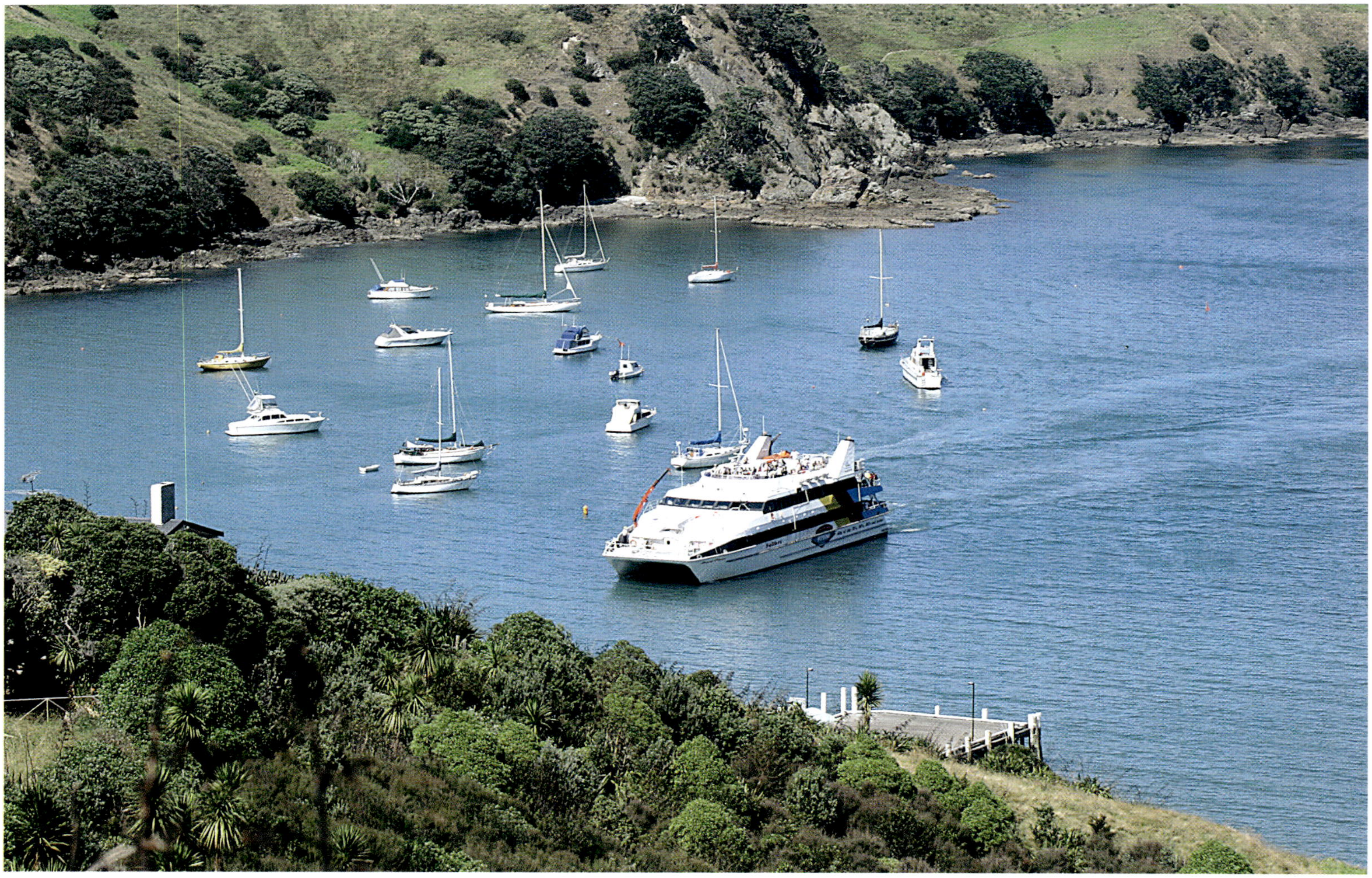

The fast ferry prepares to berth at Matiatia on Waiheke Island.

Two other counter bombardment batteries were located at Whangaparaoa and Motutapu Island. A network of observation posts was established at Rangitoto Island, Tiritiri Matangi Island and other locations. Three concrete gun emplacements and an extensive system of underground chambers connected by stairs and tunnels still exist on Waiheke and are open to visitors.

Construction started in the early 1940s but only two guns were ever installed, the third being cancelled as the threat of invasion became less imminent. The tunnels and underground chambers were dug by hand and aggregate used in the concrete was obtained from the rocky outcrops that give Stony Batter its name. The guns were only fired once, in 1951. They had the capacity to fire a shell over a distance of more than 30 kilometres and were finally cut up for scrap in 1960–61.

Never used to repel invaders, the gun sites did, however, cause a different kind of battle. For 19 years from the early 1980s the public waged a right royal fight with the landowning Spencer family over the issue of access on a road that leads to what most people regard as a national museum piece. With the outcome finally decided in the Privy Council, the Spencer family had to concede and the public now has unimpeded access to an engineering site considered to be of international significance.

The Stony Batter Protection and Restoration Society was formed with the aim of restoring the Stony Batter defence complex.

The Department of Conservation and the Auckland Regional Council maintain park lands and reserves on the island while the Forest and Bird Society looks after a reserve at Onetangi.

Now more than 8000 people live on Waiheke permanently. Some people see it as a suburb of Auckland, but I found it retains an atmosphere all its own. Narrow country roads still traverse the land, rambling up and around the rolling hills. In summer bright agapanthus, cannas, bougainvillea and the occasional banana palm add a tropical ambience. Phoenix palms and tall Norfolk pines add stature among acres of manuka scrub and flax. Traditional kiwi baches and modern condominiums sit comfortably side by side in the western bays and on the beachfronts. Modest suburban houses drift along the valleys and millionaires' mansions compete with vineyards on the hilltops.

Waiheke of the new millennium is a mobile community of artists, farmers, tourism providers and city business people. The goat farming and beekeeping of the hippie years have given way to olive growing and winemaking. There are burgeoning communities at Oneroa, Palm Beach and Onetangi on the northern side with more development at Rocky Bay, Ostend, Surfdale, Blackpool and Kennedy's Bay (where the vehicle ferry arrives) on the southern coast.

No longer the sleepy resort of 20 or 50 or 80 years ago, the island has become a mecca for tourists. Both New Zealanders and overseas visitors go to the island to play golf, fish, dive, visit the wineries, enjoy the restaurants and cafés, mountain-bike, sea-kayak or lie on the wonderful sandy beaches and swim. Accommodation ranges from top-class resorts to beach lodge, backpackers or youth hostel. Restaurants and wineries provide seasonal work.

I stopped at a café where I sipped city-strength coffee and admired never-ending rows of grapes. I talked to a student, only too happy to find summer work at the Mudbrick Café, and chatted to a former Waiheke Islander who has returned there to live. 'Has the island been spoiled?' I asked.

'No. I don't think so,' he replied. 'The biggest change is probably the amount of traffic. A car used to be a luxury. Now every family has a couple. But we get our beaches back after the flush of summer visitors,' he added. 'And there's work here now.'

Waiheke has its own chain of sheltering islands at 'the bottom end', the whole area being a favourite weekend haunt of Auckland yachties. Ponui is the largest of these, an island of gentle hills and green pastureland, owned and farmed since 1854 by the Chamberlin family. The original homestead, shipped out from England in sections, was built at Poroaki or Home Bay and still exists. Other houses were built on the island as the family expanded and it is still farmed by descendants of the two brothers who first settled there. North Harbour, also known as Chamberlin's Bay, is frequented by boats and the Chamberlin family has always been tolerant of visitors walking on their island. School parties and Christian youth groups hold regular camps at Motunau. With the co-operation of the

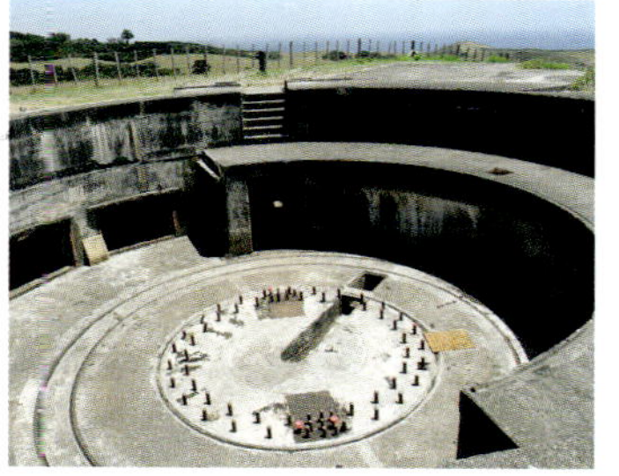

The remains of the gun emplacements of the World War II Counter Bombardment Battery sit on the hill high above Man O'War Bay at the eastern end of Waiheke.

It would be hard to imagine a more attractive setting for an olive grove than this site looking out across the water to Matiatia.

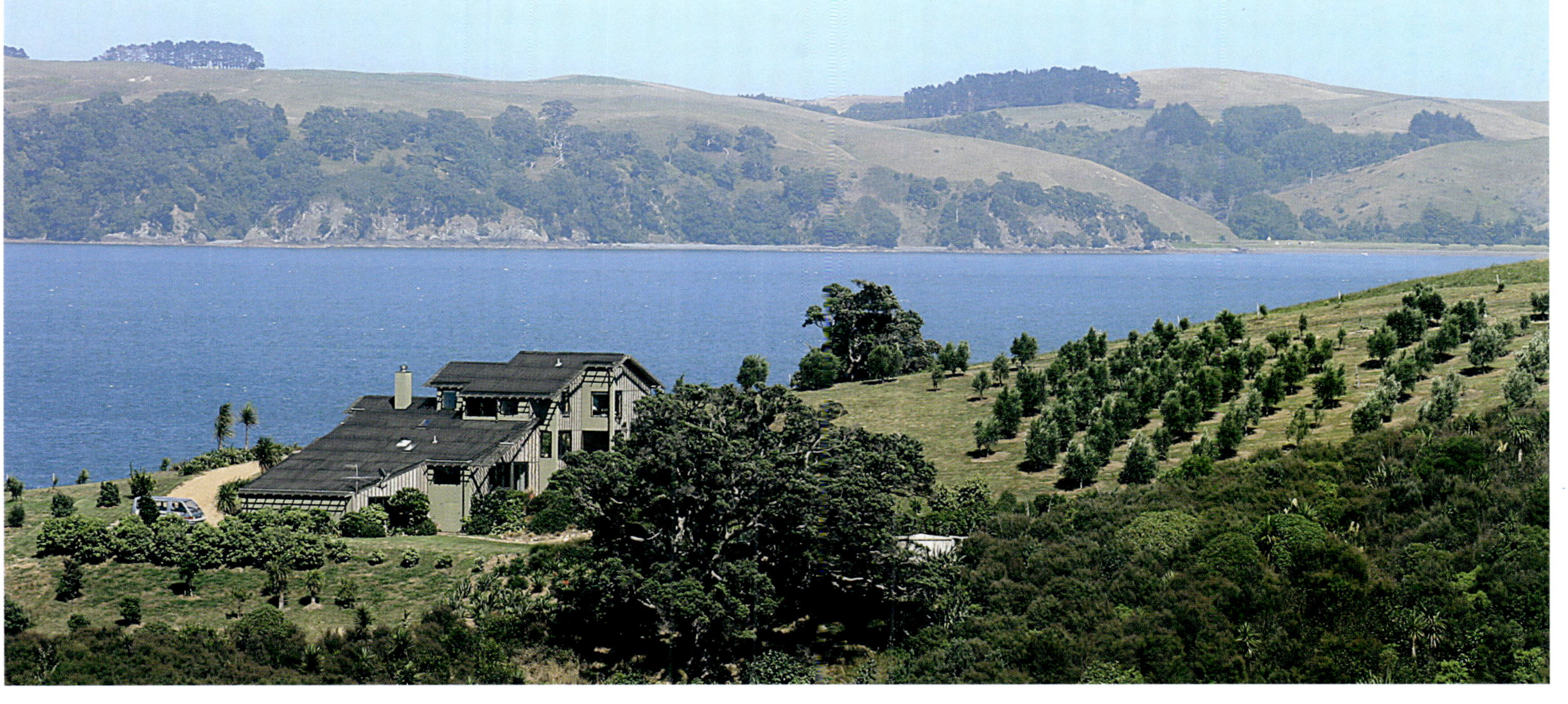

landowners, weka have been released on the island and are doing well.

Above Te Kawau Bay the contours of Maori terraced pa sites are visible and there's a cliff where gold quartz was discovered in 1869.

A herd of feral donkeys exists on the island, descended from two or three originally taken there in the early days of European settlement. Like the feral sheep of Arapawa Island in the Marlborough Sounds they are genetically pure (if inbred!) and rare-breed enthusiasts are trying to build up a viable population of these sturdy, docile animals.

Just north and nearby is Rotoroa Island, owned by the Salvation Army since 1907 and used as a treatment centre for recovering alcoholics. In the cemetery there is the grave of one lonely First World War soldier.

North of Rotoroa, and even smaller, is Pakatoa Island, which in the 1960s was developed as a private resort island. Only 24 hectares in size, half of it is golf course but weka released there in 1996 are thriving and visitors are charmed by them, even if occasionally they are mistaken for kiwi.

We used to go to these islands often, anchor in their bays, swim at their beaches. We would sail into Sandy Bay at Waiheke, walk up to a friend's bach and sit on their deck watching the sun set. Or join them for an evening's revelry. Or take them on board and sail down to 'the bottom end'. I remember rambles over the hills, children in tow, and the perfume of pennyroyal mixed with grass. One memorable night we dropped anchor in lonely Hook's Bay after 12 stormy hours sailing from Great Barrier. The Long Sheltering Island indeed.

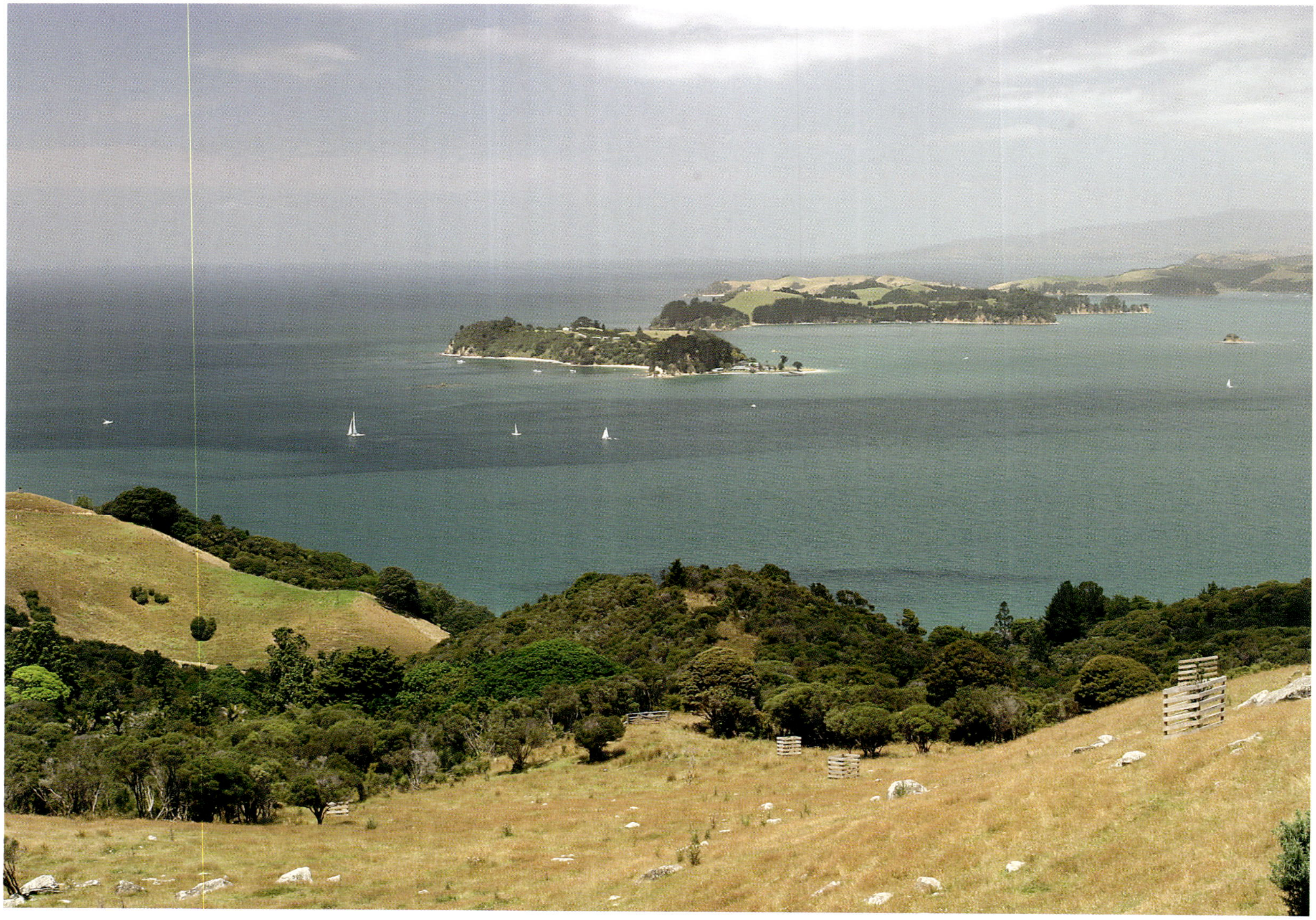

Pakatoa, Rotoroa and Ponui Islands seen from the 'bottom end' of Waiheke.

Kawau and its neighbours

Of all the islands in the Gulf, Kawau Island has probably seen more boats in its bays than any other. Scows plied back and forth with myriad building and mining materials in the 1840s, and lots more ferrying of building supplies took place in the 1860s when Governor Grey established his residence there.

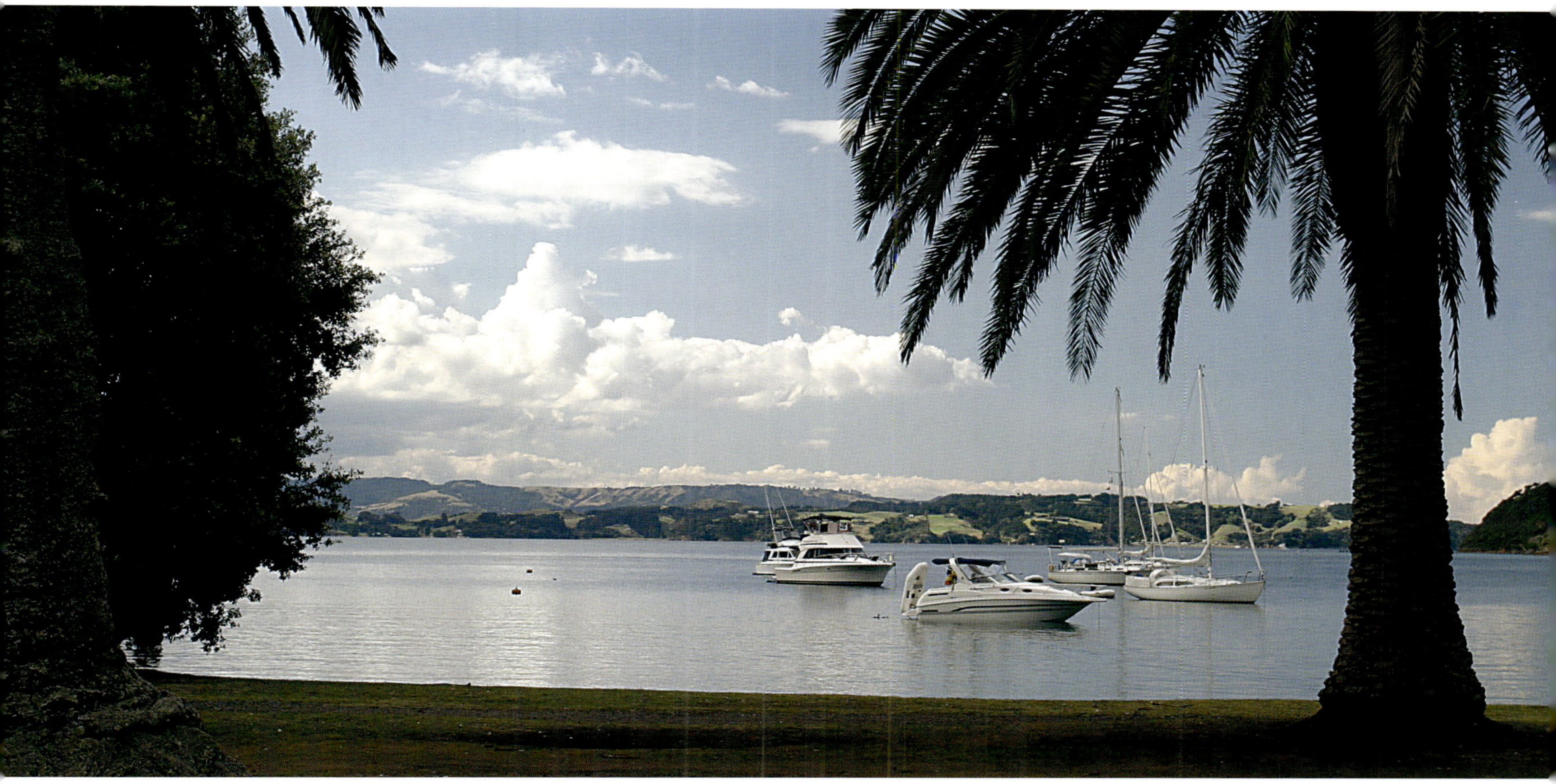

During the 1920s when yachting became a popular pastime for Aucklanders, Bon Accord Harbour was host to dozens of boats and their crews and in the years leading up to the Second World War it was not unusual for the navy to go visiting there. Today, the most popular image of Kawau is of Mansion House Bay with boats riding quietly at anchor. Add an elegant house from yesteryear surrounded by old trees and an air of grace and beauty that persists from an earlier time. Blissful, in the words of one writer.

Blissful it may be on a calm day and as a haven in a storm but the island's reputation was less than blissful in the days before Europeans came to these shores. Maori 'pirates' are said to have lived there once. Fast-moving canoe people they were, who kept a sharp-eyed watch over the waterways between their island home and the mainland, and raced out from their stronghold to prey on travellers from the north.

Vengeance came in the early 19th century when Hongi Hika and his allies came south with their muskets. They set out to avenge their dead, attacked the island tribes and put an end to Ngati Tai predations once and for all. When Europeans arrived on Kawau all that was left of the former

Phoenix palms planted by Sir George Grey at Mansion House Bay on Kawau add an exotic appeal to this island off the coast of Sandspit, an hour's drive north of Auckland.

Right: A popular spot for Auckland boaties at Mansion House Bay.

inhabitants were heaps of whitened bones and the remains of terraced pa sites. Nowadays the boats that race in the vicinity of Kawau are sleek yachts. It is a rare weekend when no sails can be seen on the water between Auckland and the island.

Since the arrival of Europeans, the history of Kawau divides neatly into three eras. For a little more that 20 years, from the early 1840s, the main focus on the island was the mining of copper and manganese. Governor Sir George Grey purchased Kawau in 1862 and for the next 25 years spent a fortune developing the island as his home and private fiefdom.

When he sold out in 1888, Kawau had already earned a reputation as an idyllic holiday retreat and since then it has been a playground for Aucklanders, in particular Auckland boaties and for increasing numbers of overseas tourists.

In the height of the copper-mining days, there were up to 300 Europeans living on the island, many of them Cornishmen with their wives and families. Some Maori had returned to live on the island and they also worked in the mine. Copper traces still stain the rocky cliffs in the vicinity of the mine and copper ore litters the beach in the vicinity of the Kawau mine. The mineral apparently existed in worthwhile quantities. But the seam ran under the sea, the ore was difficult to extract and the workings had to be constantly pumped free of water. The best year was 1846 when 1200 tons of ore were exported. Difficulties then arose shipping the ore to Europe. To alleviate the problem, a smelting works was built on the north side of Bon Accord Harbour at Smelting House Cove.

Then the mine flooded, fierce rivalry flared between the two separate companies working the same seam and the initial promise of copper extraction was never realised. When the mine closed in 1851, many of the miners left, tempted by better prospects of copper mining at Great

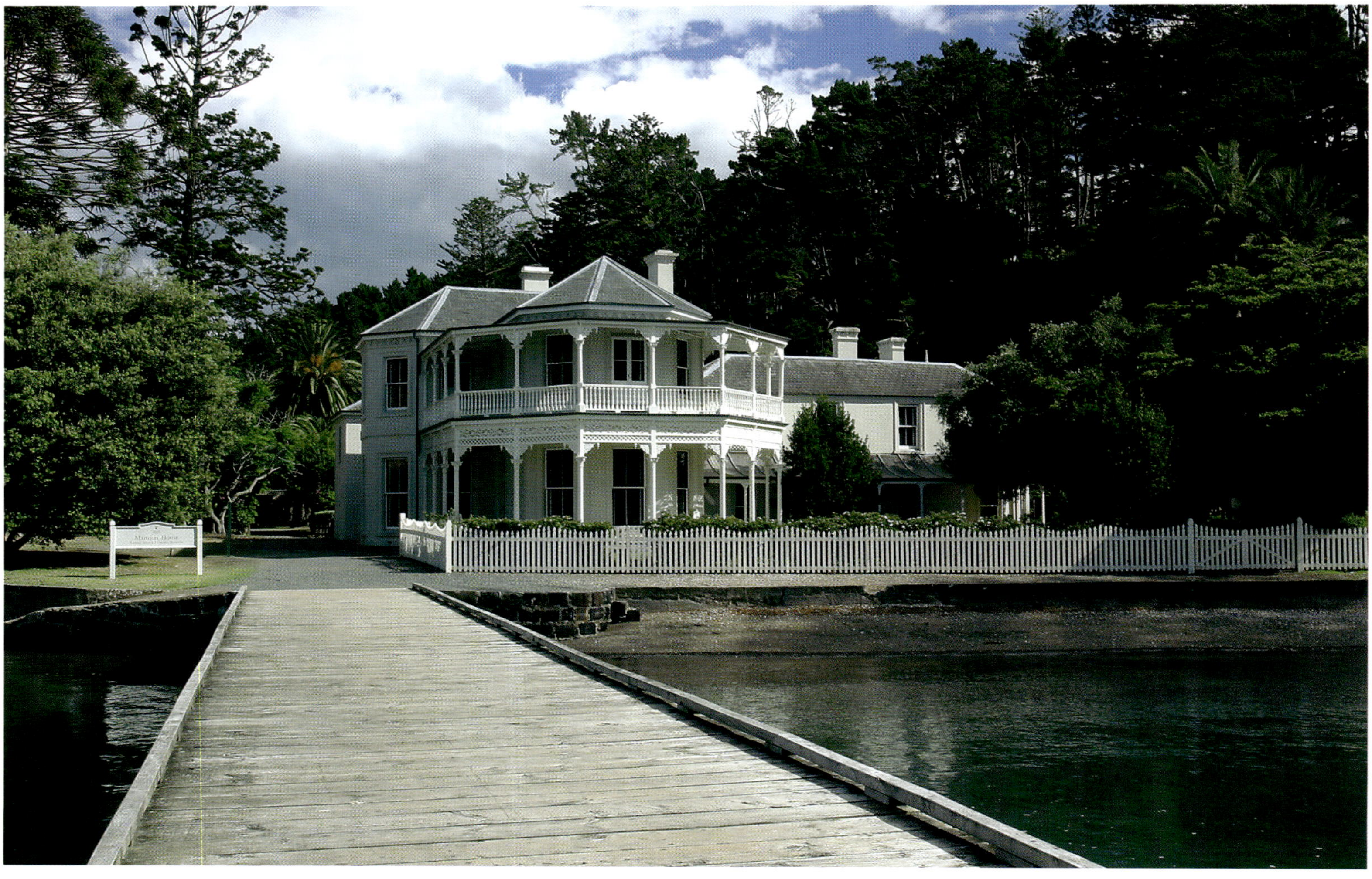

Stately old lady of the Hauraki Gulf, Mansion House was built by Sir George Grey as his private demesne. After he returned to England it was managed as a boarding house for many years. Now it lives on as a museum, administered by DOC.

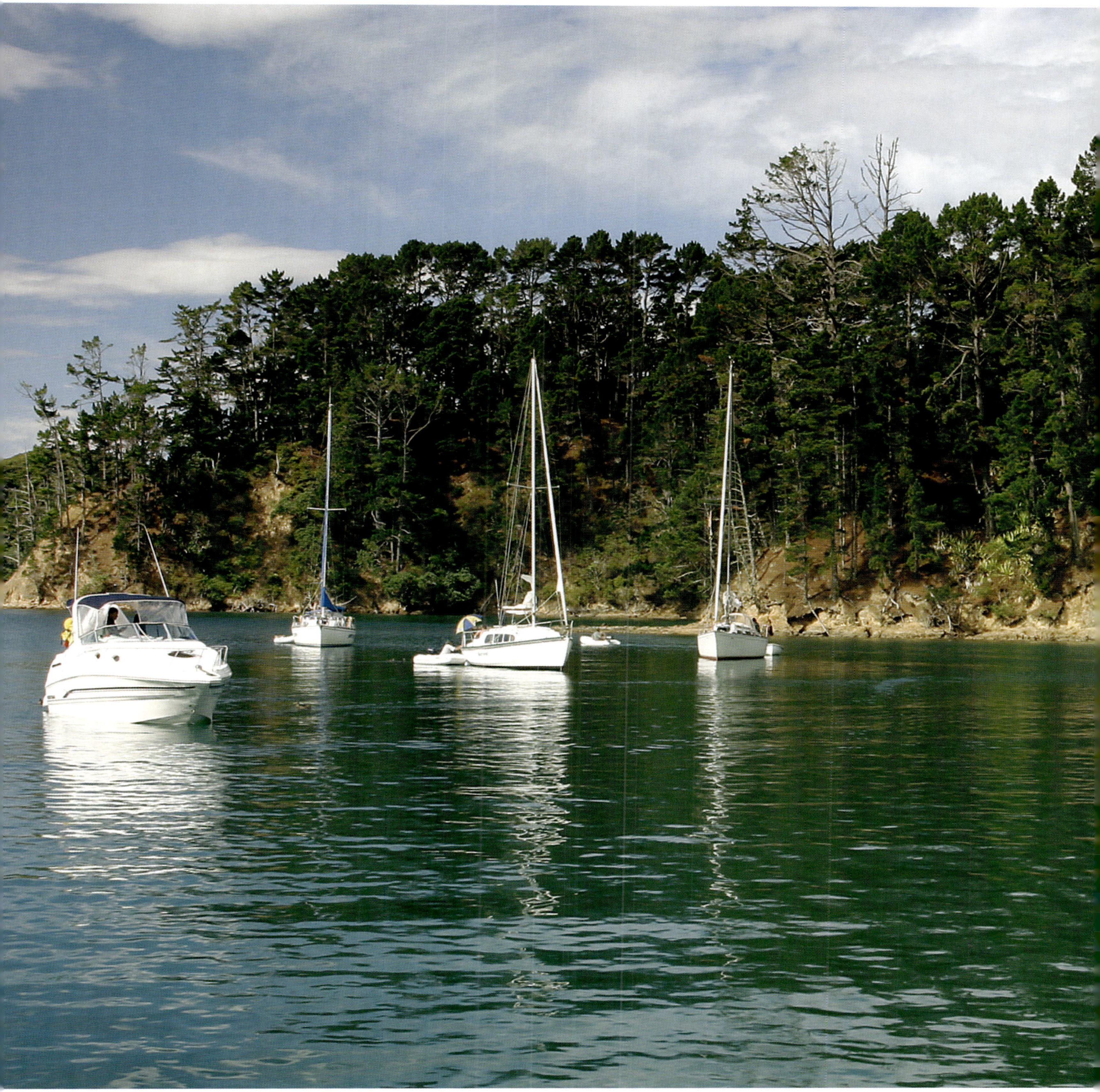

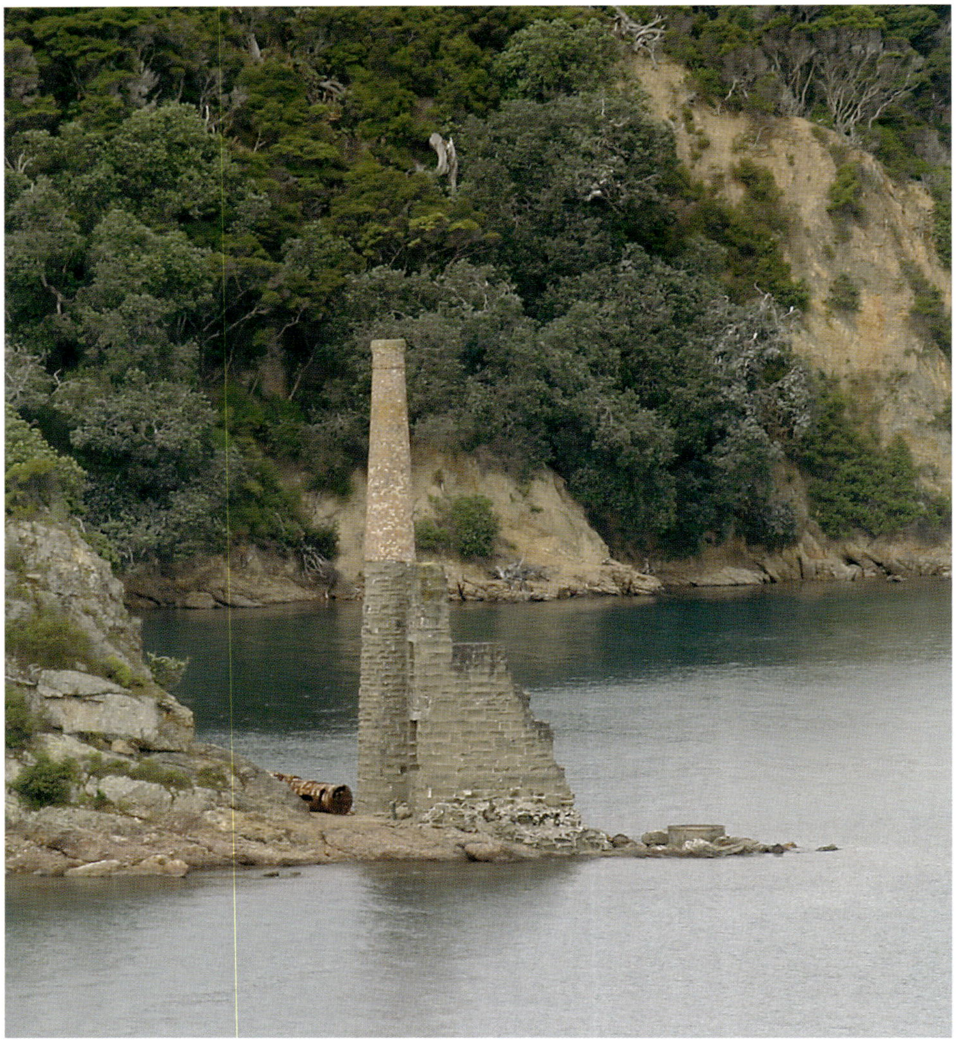

Copper was mined on Kawau in the 19th century and the remains of the pumping station are a reminder of its brief success.

A cheeky weka scrabbles for insects near Mansion House.

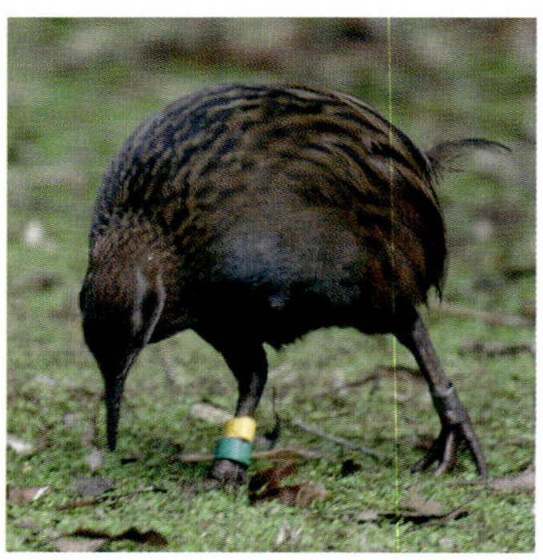

Barrier or the lure of gold in Otago. Further attempts to reopen the operation in 1854 and 1900 were unsuccessful. Now a tall chimney and the ruins of a sandstone building which housed the steam engine and pump are all that remains of the short-lived copper-mining industry on the island, its greatest legacy the name Bon Accord. One of the mining companies was Scottish, reputedly established in Aberdeen and the city's motto is 'Bon Accord' – 'happy to meet, sorry to part, happy to meet again'.

When Sir George Grey bought Kawau, a somewhat austere but nevertheless imposing house already existed on the island. It had been the home of the copper mine manager. Sir George Grey remodelled it, adding 20 more rooms to the original ten. Kauri logs that had been pit-sawn on the island were sent to England for turning and they came back to their island home as elegant impressive pillars for the interior. Kauri panelling lined his library. He developed the house into a gracious mansion where he later retreated from a public life that had turned sour but where he still entertained lavishly. One of his guests was Prince Alfred, a son of Queen Victoria. Learned men came from Europe to discuss art and philosophy and returned home amazed to have found such a cultured retreat in a raw colony. Over summer excursion steamers made the trip from Auckland packed with picnickers, bound for a day's pleasure – and no doubt bubbling with curiosity to see the former Governor's island sanctuary. On Boxing Day 1883, three boats arrived bringing 1000 visitors who were free to wander through the gardens and the house, admiring the accumulated treasures.

Keenly interested in natural history, Grey planted hundreds of different species and the gracious, exotic flavour of Mansion House Bay today is due in part to some of the trees he planted. These are now wonderfully mature – the enormous Moreton Bay fig that spreads its shade near the shore and the elegant Chilean wine palms (*Jubaea spectabilis*) behind the house are only two among many.

Not only plants are a legacy of his time on the island. He also introduced a number of exotic animal species, among them a pair of zebras (which did not thrive), kookaburras and peacocks whose raucous cries still make visitors jump, and the ubiquitous wallaby with which the island is infested. They may look cute as they hop around the lawns of Mansion House, nibbling as they go, but like their Australian friends the possums, they are a terrible pest and have laid waste much of the vegetation on the island. Nora Creina Wilson, in her book *Memories of Mansion House*, talks about wallaby drives in the 1880s when as many as 180 would be shot in one day. The aim now is to rid the island of the pests by 2005.

The rest of Grey's island was managed as a farm with stock being loaded and unloaded from the copper mine wharf. There were enough families living and working on the island to warrant the building of a school for their children in what became Schoolhouse Bay.

After Grey's tenure, the fortunes of Mansion House fluctuated for the next couple of decades as several owners bought the property and stayed for short periods. In 1904 much of the island was sold, with about 100 acres retained around Mansion House. The next 70 years saw it run as a hotel or guesthouse under a variety of owners.

A few resident families farmed on the rest of the island but from the copper-mining days until 1945, manuka provided the biggest source of income on the island. Initially, huge quantities were cut to keep the smelting house furnaces fed. Mansion House itself needed firewood for heating and cooking and in the Depression years of the 1930s it provided a livelihood for out-of-work men from the city. They came and camped on the island, working all the daylight hours to slash the wood and load it onto sledges, which horses dragged down to the nearest beach. From there it was loaded onto scows and taken to Auckland to be sold. Hard labour for man and beast.

Between the two world wars Mansion House Bay increasingly became a mecca for yachtsmen. The Royal New Zealand Yacht Squadron instituted its first squadron weekend at Kawau in February 1931, a tradition which continues today. Mansion House, owned by the Reeves family at the time and operated as a guesthouse, became renowned for the warm welcome extended to visitors. Many returned season after season. Yachties from the boats anchored in the bay mingled with the land-based visitors and parties were a regular feature of summer life.

By 1967, the days of Mansion House as a private establishment were over. It was sold to the Crown for inclusion in the then Hauraki Gulf Maritime Park and was leased for several years by Dominion Breweries – and the boating parties continued. Those were the days when cynics spoke of the depths of Mansion House Bay being carpeted with broken glass a metre thick.

The pendulum has swung and Mansion House is now restored to much of its former glory. Administered by the Department of Conservation it is operated as a museum, with a café open during major holidays. Publicly owned reserve land around Mansion House is included in the Hauraki Gulf Marine Park, and DOC maintains it as a historic reserve. Lots of the threatened North Island weka live on Kawau and are often seen in the gardens around Mansion House. Walking tracks traverse the island, many of them following paths established in Grey's time.

The island will always be a boatie's paradise. With a fair breeze in the right direction, it's a short day's sailing from the city and boats regularly pour out of Auckland Harbour on a summer Friday afternoon, headed for Kawau. Squadron Weekend attracts hundreds of racers and revellers. Bon Accord Harbour on New Year's Eve can be so crowded it's difficult to find space to drop anchor and when a storm blows up in the Gulf, boats scuttle in for shelter. We spent a week in Bon Accord once, waiting for a storm to abate. Even the waters of the harbour were rough and rowing to a friend's bach on shore was a risky manoeuvre, especially with small children on board. But desperate times demand desperate measures and six people cooped up on 10 metres of boat get more than cabin fever! Our two year old at the time was enchanted with the wallabies

How to get there

Ferries and water taxi depart daily for Kawau from Sandspit, an hour's drive north from Auckland, and less frequently from Gulf Harbour, Whangaparaoa. Homestays are available on the island.

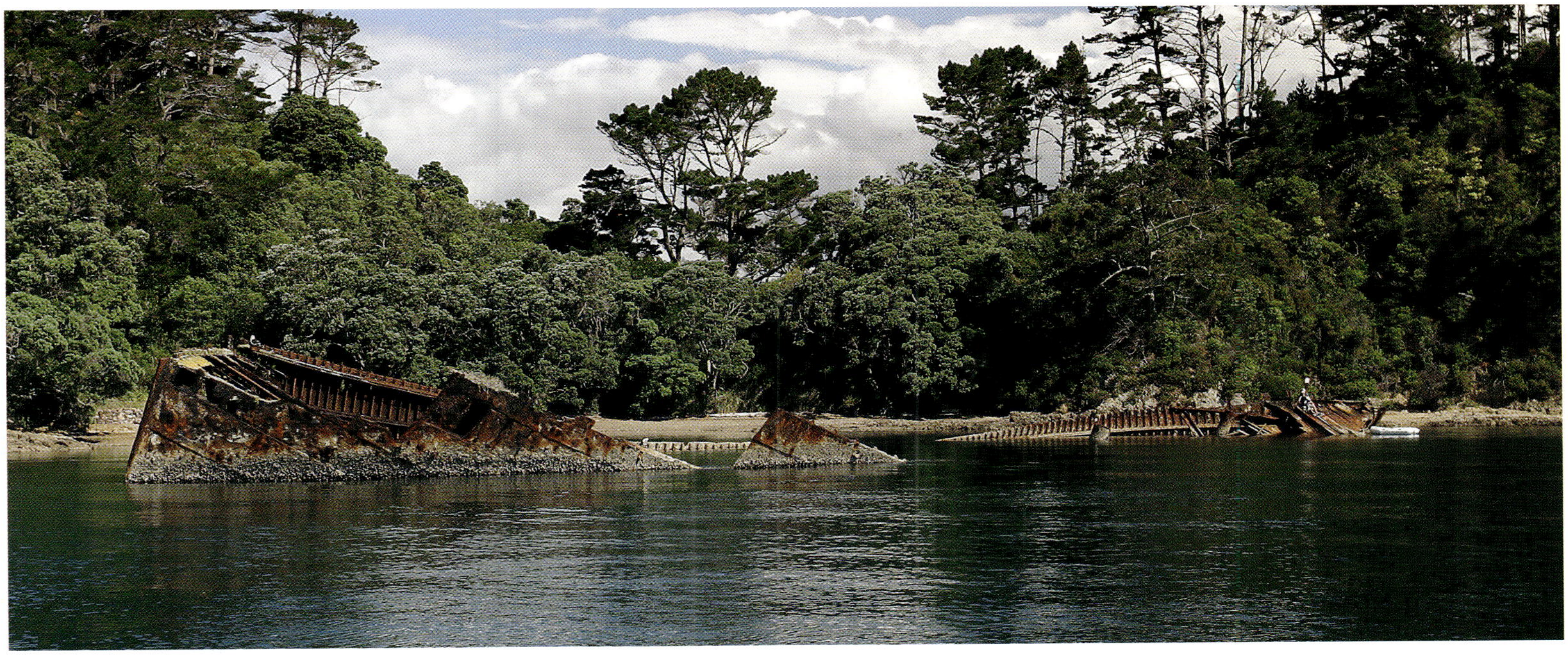

An exotic south sea island – not far from mainland North Island. Moturekareka, complete with its own wreck, lies close to Kawau Island.

Beehive Island

Close to Kawau is 'The Beehive', a rocky outcrop full of character with its skirt of white sand beaches and 'hair-do' of craggy pines. A great place to picnic at low tide, the island is home to New Zealand dotterels, which nest on the beach. Visitors need to tread carefully as the birds' nests blend into the surrounding beachscape.

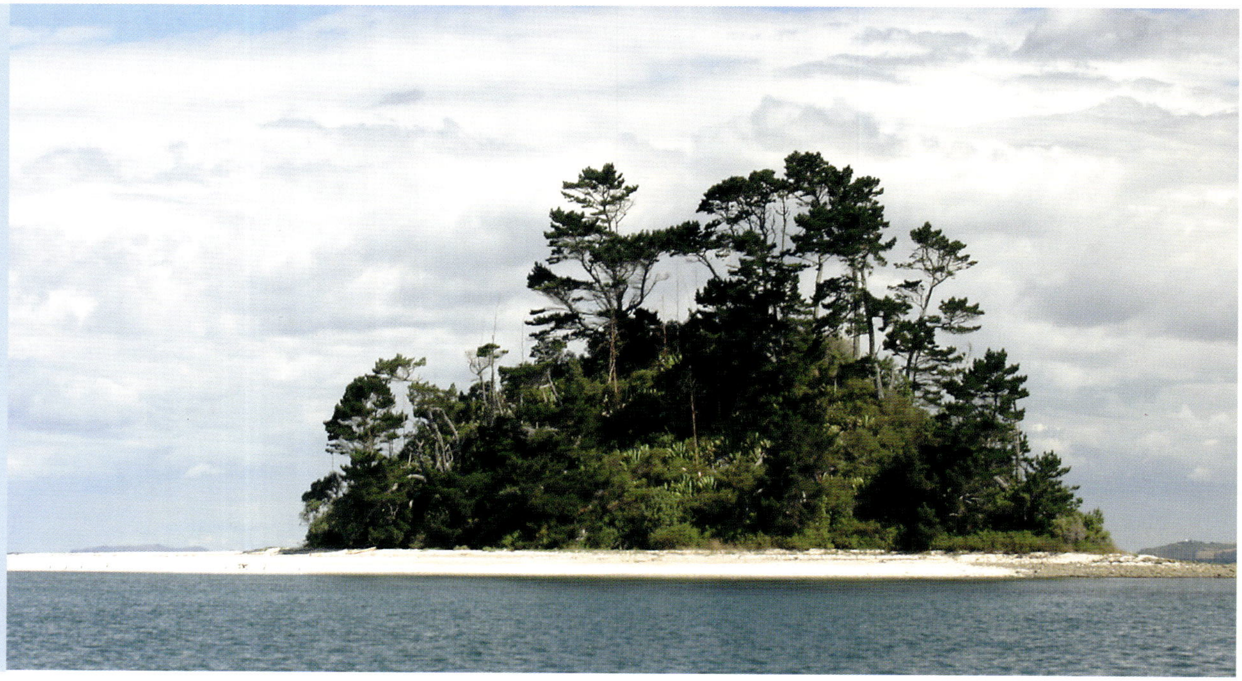

that bounced out of the bush to inspect us.

Although few people live permanently on the island, houses dot the bays around the shore of Bon Accord Harbour and several other bays. Scattered baches peer out from clearings in the bush. Nowadays the mailboat calls once a day and letters are delivered to each private jetty. When changes to local government legislation put the administration of Kawau under Rodney District jurisdiction, the mainlanders had trouble understanding that the islanders had no need of footpaths. They go visiting by dinghy, or wander along the shore at low tide. For the residents of the island, the sea is both their motorway and their main road.

Kawau Island is itself the main act with a supporting cast of tiny islands that introduce seafarers to this idyllic part of the northern coast. It's about half an hour by water-taxi to Kawau from Sandspit on the mainland. On the way you pass Goat Island and Rabbit Island, both privately owned and cherished as get-away retreats.

Not far away are Motuketekete ('Food Basket Island'), and Moturekareka or Pleasant Island – so pleasant in fact that in 1930 it attracted the renowned hermit Charlie Hanson there to live on this merest sliver of land. He bought the hulk of a steel sailing ship, *Rewa*, had her towed to the island and beached there, with the idea of making a breakwater that would create a secure anchorage for small boats. But the hull settled in the wrong place, leaving little space for a harbour. Now snorkellers visit to explore around the wreck.

Motutara ('Gull Island') is connected to Moturekareka by reefs. Once there was a quarry on it and ruins of the jetty, where scows pulled in to take away their loads of gravel, still exist near the shore.

Further south is Motuora. These islands now are all devoted to conservation and replanting of native species is making good progress. Shore plovers have been brought to Motuora and zoo-raised kiwi have been released there as the first step in acclimatisation before being returned to the wild in Northland.

Whitehead on the open sanctuary of Tiritiri Matangi.

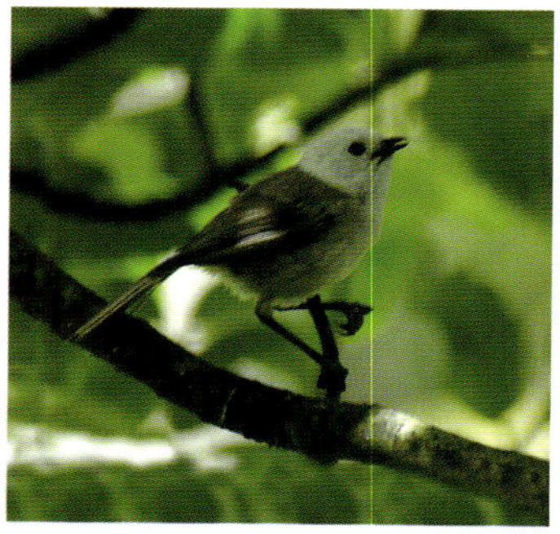

Tiritiri Matangi

A low, flat island, sitting barely four kilometres off the coast of the Whangaparaoa Peninsula, Tiritiri Matangi Island is one of the big success stories of island conservation. Twenty years ago this island was bare, abandoned farmland. Now it resounds with the call of birds that, not so long ago, were on the point of disappearing from our bush forever. Unusually for a scientific reserve, it is, like Ulva Island at Stewart Island, open to visitors, the only restriction being imposed by the days when the ferry visits the island.

For more than 100 years, from 1850, Tiritiri was farmed. Only a few pockets of native forest remained in 1972 when the island was included in the Hauraki Gulf Maritime Park. The farm animals were removed, the grass and bracken were left to their own devices and only the lighthouse keeper and his wife remained. With no other pests on the island apart from kiore or Pacific rats, a few endangered species of birds were released on a trial basis. Red-crowned parakeets (kakariki), spotless crakes and bellbirds survived.

Gradually plans were developed to create a wildlife sanctuary on the island and when the light was finally automated in 1984, Ray Walter, the lighthouse keeper, stayed on to manage the plant nursery established to facilitate the reforestation project. His wife Barbara worked initially as a volunteer, co-ordinating work parties and managing the small gift shop which catered to the needs of visitors. Barbara now works as a ranger for the Department of Conservation, who has overall responsibility for the administration of the sanctuary. An important feature of its success are the Friends of Tiritiri Matangi, and widespread community involvement through the North Shore Forest and Bird Society, service clubs, outdoor recreation clubs and hundreds of school children who have all worked on the island.

For anyone interested in bird watching, Tiritiri Matangi is a dream come true. Day visitors are welcome and can wander freely on the tracks and boardwalks that lead up to the lighthouse, through the developing forest and along the spine of the island. Birds are everywhere, totally oblivious of human interest in their daily lives. Stitchbirds sip nectar at special feeding stations built for them; saddlebacks forage in the undergrowth. They have flourished and birds bred on the island have been transferred to Mokoia Island in Lake Rotorua and into the Karori Bird Sanctuary in Wellington.

The fluid notes of tui and bellbird ring out, and lucky visitors may hear the deep, organ tones of the kokako. The North Island robin is another bird that has adapted readily to life on Tiritiri and is breeding successfully.

Seabirds also abound on and around the island. The Kiwi Conservation Club has built small perspex-topped tunnels where little blue penguins can hatch their chicks in safety.

How to get there

Ferry service once per day from downtown Auckland and Gulf Harbour to Tiritiri Matangi, Thursday-Sunday (incl.). As visitor numbers are limited on the island, reservations are essential. No accommodation available on the island.

Tiritiri Matangi, once bare farmland, is now a bird sanctuary where visitors are welcome to wander and see rare native birds, such as saddlebacks, stitchbirds, kokako and the magnificent takahe.

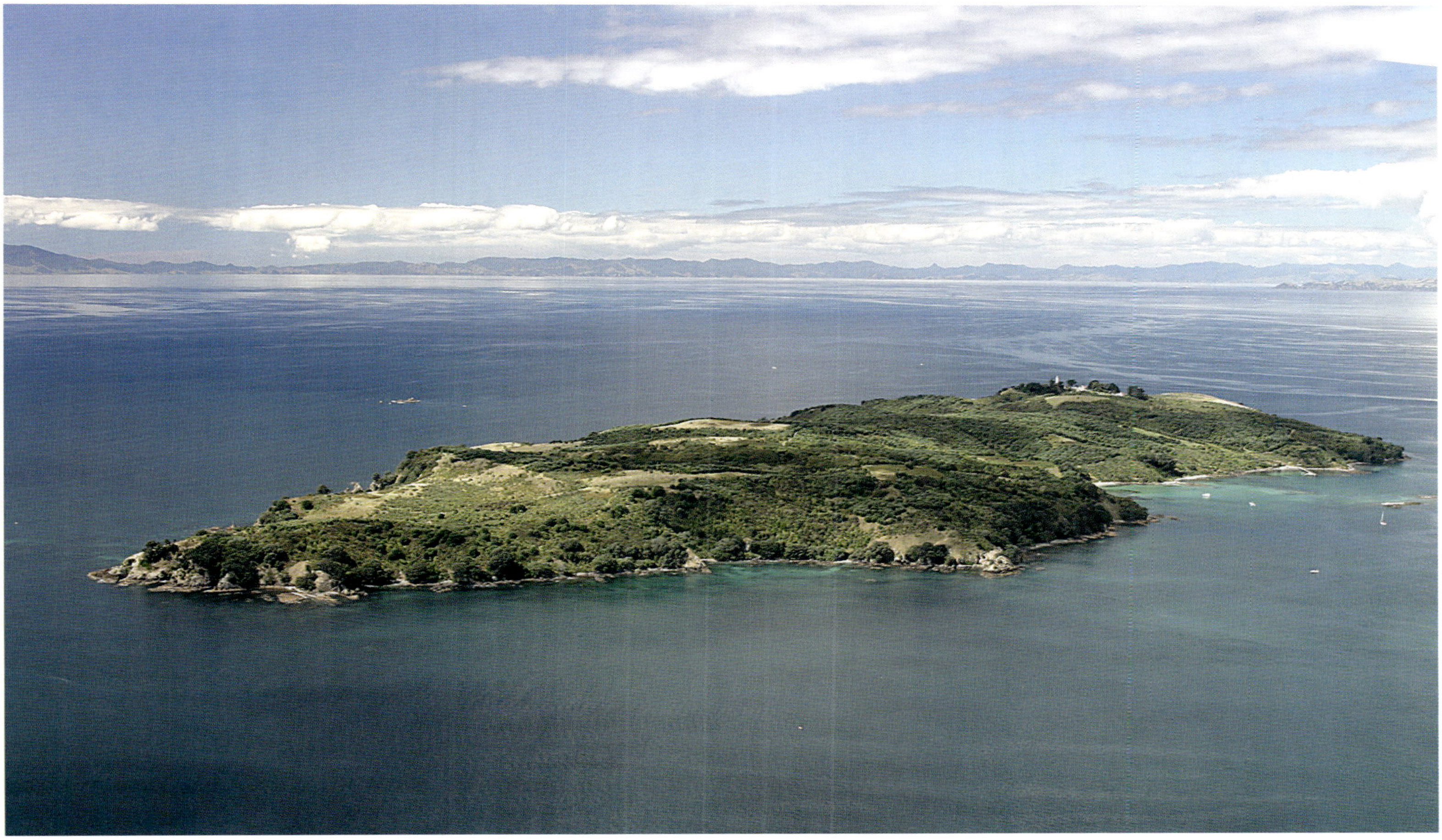

The outer islands

Right: Port Abercrombie, Great Barrier, looking out to Kaikoura Island.

WHEN CAPTAIN COOK FIRST SAILED TOWARDS THE HAURAKI GULF AND SAW A CHAIN OF ISLANDS STANDING GUARD AT ITS ENTRANCE, HE NAMED THE LARGEST ONE GREAT BARRIER AND ITS SMALLER SIBLING LITTLE BARRIER. THEN AS NOW, WITH THE HEN AND CHICKENS AND THE MOKOHINAU ISLANDS, THEY MARK THE OUTER EDGE OF THE GULF.

Great Barrier Island

I remember clearly the first time I arrived at Great Barrier Island. We had been living overseas for 11 years, most recently in Europe. Our children spoke French better than they spoke English. Now we had brought them back to become New Zealanders. Shortly after Christmas we sailed out of Auckland Harbour, bound for the Barrier. We arrived at Port FitzRoy on one of those idyllic New Zealand days when you can see forever. Bush-covered hills dropped to the shore. Water lapped at the thin line of rocky beach. A tin shed was just visible in the bush. Solitude reigned.

I stood on deck and breathed a sigh of pure pleasure. 'This is why we came back to New Zealand. To enjoy beaches like this,' I said.

'You call this a beach?' our 12-year-old daughter asked incredulously. 'There're no boutiques, no apartments, no people. That is not a beach,' she declared emphatically and went below.

Imagine then how Emilie Malcolm must have felt when she arrived at Great Barrier with her husband and several children in 1859 to start a new life on the island. She had been brought up in a genteel home on the banks of the Thames in England, surrounded by flower gardens and art.

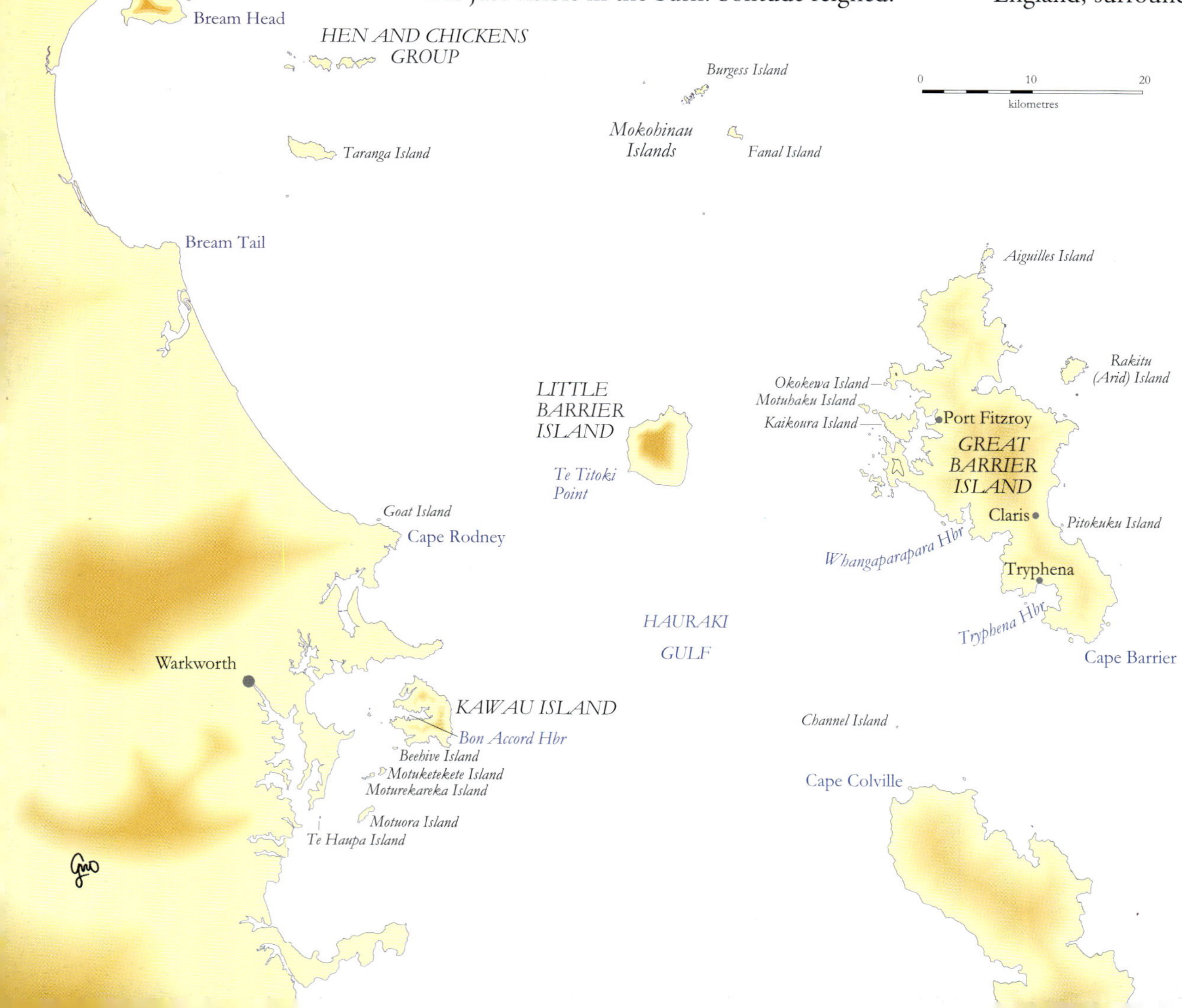

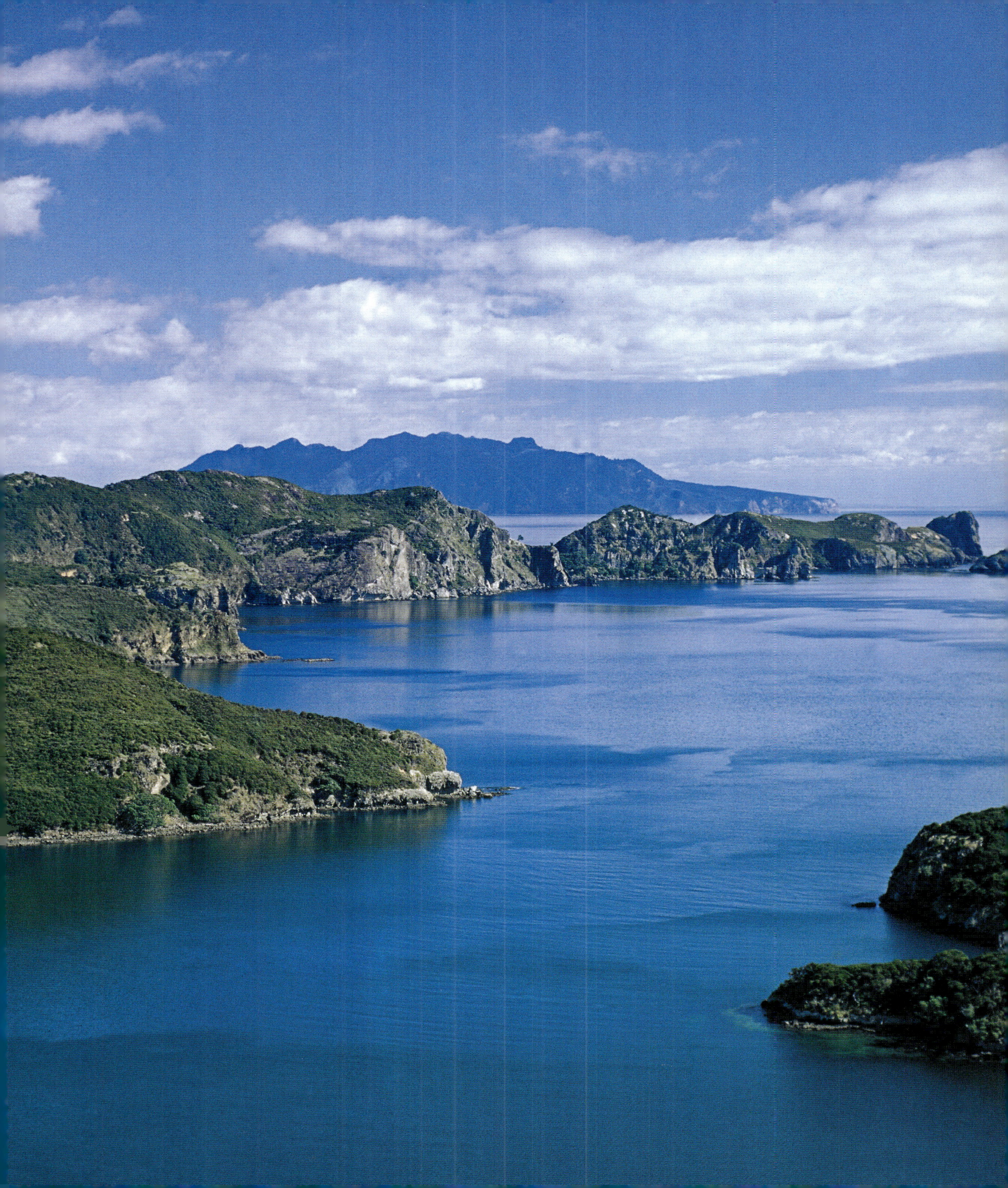

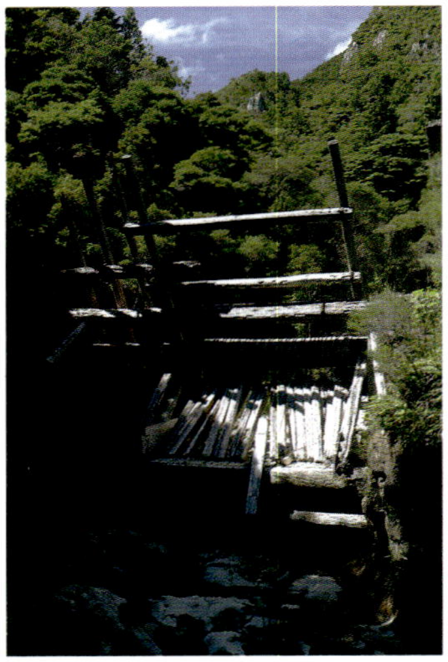

The remains of an old kauri dam on the track from Kaiarara Bay to Mount Hobson are a reminder of the vast amounts of timber that were taken from Great Barrier over a period of nearly one hundred years.

Married to a lawyer at 18, she had expected her life of comfort to continue in the old country. Their circumstances changed, they decided to start a new life in the colonies and after several false starts on the mainland they leapt at the opportunity to start farming on Great Barrier. She wrote many years later about their arrival. 'Arriving at night, we were awestruck with the sombre magnificence of the forest, its stillness, its seeming desolation, its fine harbour of considerable extent, with little bays and sandy beaches wooded to the water's edge.'

The Malcolms settled at Rosalie Bay, for a long time the only Europeans in that area. Their first dwelling was a raupo hut. Later they built a cottage, adding more rooms as more children arrived. They cleared bush, established a thriving garden and enjoyed the bounty that the island provided. Emilie went on to have 13 children, eight of them born on Barrier. All the children were well-educated, a task she and her husband – who spoke five languages – shared. One of their daughters, Fanny Osborne, lived a remarkably similar life on the island – to the extent of also bearing 13 children – and eventually finding time to become a well-known botanical painter, using the plants of Barrier as her subjects.

Maori had lived on the island they named Aotea for centuries before the arrival of Europeans, but had never tamed the rugged landscape. The highest peak of Mount Hobson (Hirakimata) rises to 621 metres and narrow ravines, steep cliffs and jagged pinnacles are a feature of the inland area. A few inhospitable islands scatter themselves around the shoreline and rocky stretches of coastline have spelled doom to ships in the past. In 1894 SS *Wairarapa* was wrecked at Miner's Head with the loss of around 140 passengers and crew.

Remains of a farm shed on the eastern side of Great Barrier.

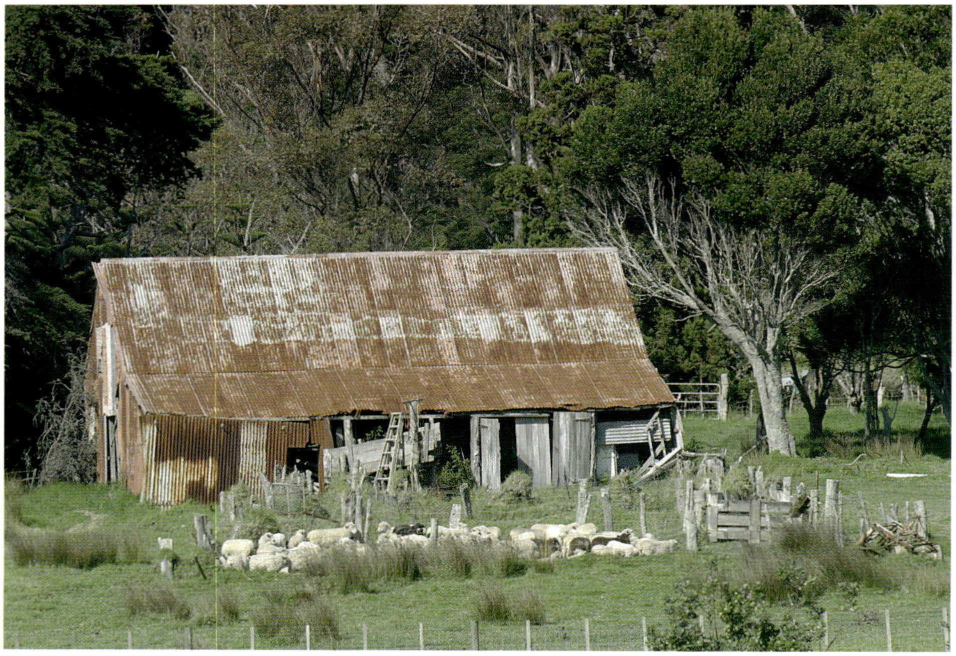

In the decade preceding, and during the years that the Malcom and Osborne families flourished on the Barrier, great changes came to this barely touched wilderness. Its natural resources attracted several 'boom and bust' industries. Copper was discovered in the remote northern part of the island in 1841, leading to the establishment of the earliest mine workings in New Zealand in 1842. Later it was silver and gold that lured men. But none of the mining ventures produced much revenue, although the 'gold rush' resulted in the establishment of the town of Oreville. At the end of the 19th century it boasted 500 inhabitants. Now only the remains of the stamping battery, built to extract the gold and silver from the rock, are visible in the regenerated bush and on the whole island there are only 800 permanent residents.

While the mining enterprises produced no long-lasting results, the timber industry changed the face of Great Barrier. It is thought that Europeans may have taken kauri from Great Barrier for ships' spars as early as 1794. But the real industry of felling the age-old giants and extracting them from the rugged landscape did not get going until about the middle of the 19th century, continuing for nearly 100 years. When you have tramped the narrow, steep-sided Kaiarara Valley, the mind boggles at the enormity of the task of extracting the timber – even with the aid of the dams which harnessed the power of water to drive large booms of kauri logs downstream.

The dams were impressive engineering feats. They were constructed by loggers, of kauri, pit-sawn on the spot and set in foundations of rock. As trees were felled they toppled into the stream and built up behind the dam, which had to be strong enough to withstand the pressure of tons of water and timber. When the gate was tripped, the logs were swept through and downstream.

Around 3000 of these dams were constructed in New Zealand in the 19th and 20th centuries. Kaiarara Dam, built in 1926 by George Murray of the Kauri Timber Company, is one of the largest at 40 metres wide by 14 metres high. It is known as a 'rafter flume dam'. Rather than having a solid gate it was built with loose gate planks that hung vertically – a concept thought to be unique to New Zealand. While this method of moving large quantities of gigantic timber was a practical solution, it was also enormously wasteful. Many logs were smashed or lost. Once at the coast, the logs were either barged to Auckland or sometimes floated across the Gulf in great rafts.

In the early days of the 20th century there was a large sawmill at Whangaparapara that processed logs rafted by sea from the Coromandel and Northland as well as from the island. The remains include a steam tractor, cast iron chimney stack, concrete pads and stone walls. Some of the walking tracks in this area follow the early tramlines used by the logging industry.

Much of the island is now covered with second-generation bush, dominated by kanuka. Pohutukawa fringe the shoreline and a few areas of the original kauri forest survived and are regenerating. New plantings of kauri were made by the Forest Service in the 1970s and, in another 200 years, maybe Great Barrier will resemble the island its previous inhabitants knew.

We've spent idyllic boating holidays at Great Barrier – once the long sail to get there is over. We've gathered cockles in Kaiarara Bay, searched for sea-urchin shells along the beaches of the little islands on the western side of Barrier and sailed into Port FitzRoy through the narrow, fiord-like Man o'War Passage. When I think of Barrier I see steep hills wooded to the waterline; deepwater anchorages and curving bays with the occasional house tucked into the bush; Tryphena Harbour and its small village; Smokehouse Bay with its tin shed for smoking fish; the rope swing on the beach there where our son fell and concussed himself!

But visitors flying into Okiwi or Claris, on the eastern side of the island, gain a different impression. The small population is clustered in a few settlements around the coast and the plane sweeps over the spine of the island to follow a line of gently sloping hills that lead to endless swathes of curving white sand. These are the beaches, open to the vagaries of the Pacific and all its moods, that give the Barrier its reputation as an unsurpassed surfing location.

A bad day's fishing is better than a good day at the office, especially on the beach at Awana Bay on the east coast of Great Barrier.

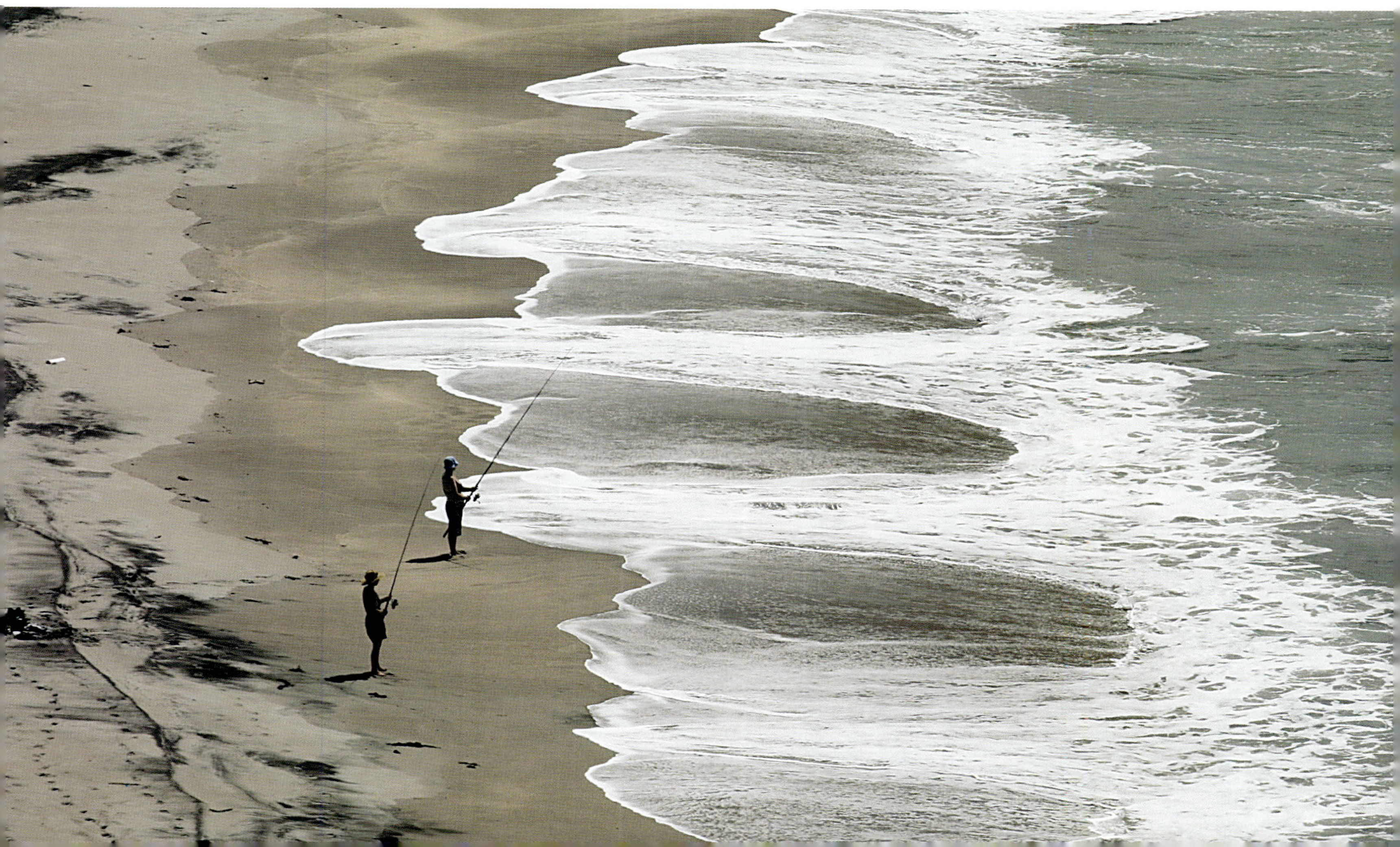

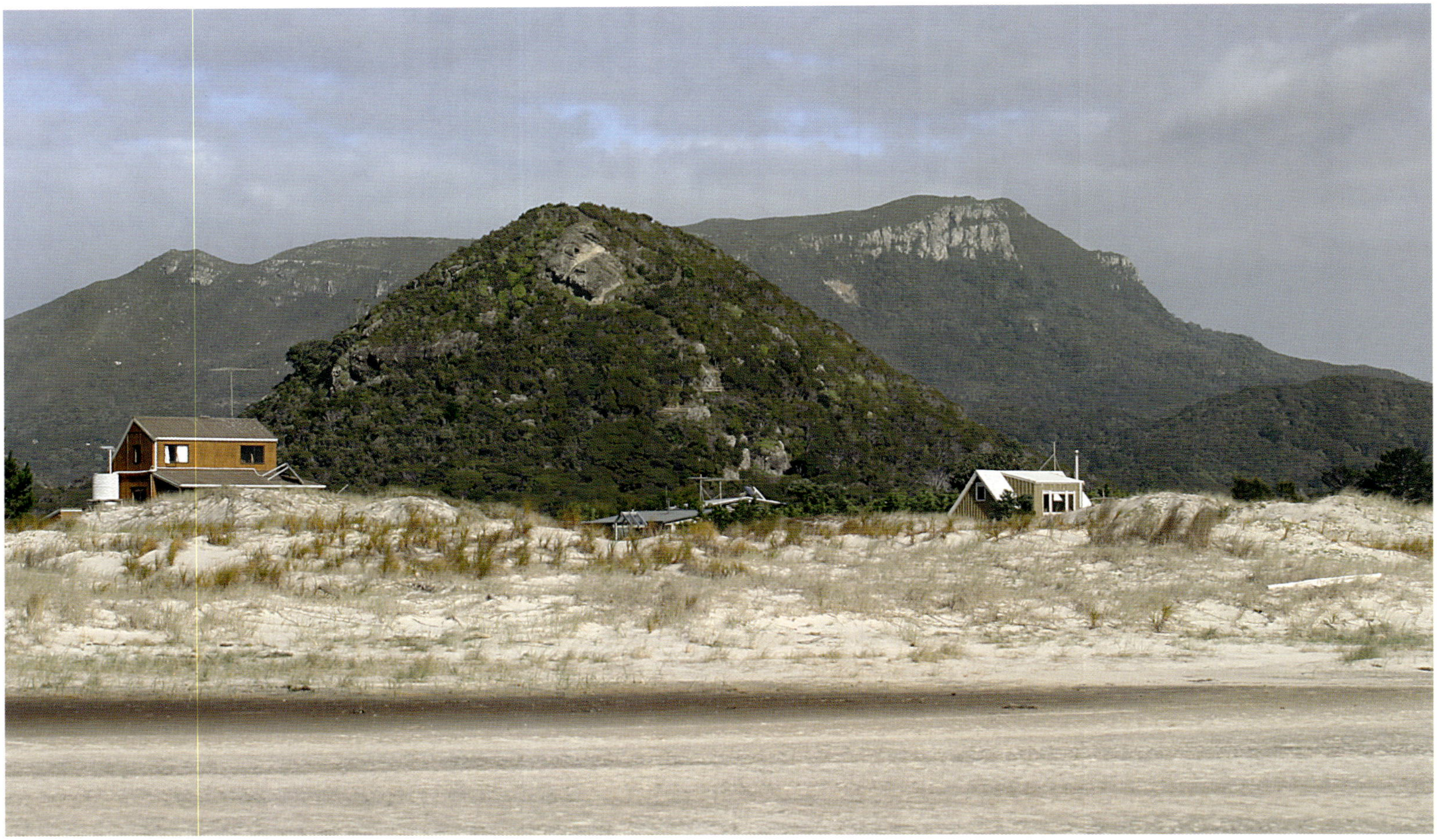

Lonely it may be, but these houses at Kaitoke on Great Barrier have a view of the sea that goes on forever.

Driving around the island is different again. There's little space for vehicles to pass and the locals have God on their side. Narrow, twisting gravel roads follow the steep contours of the land and if the driver has time to take his eyes off the road there are wonderful views of luxuriant bush and sharply upthrusting rock pinnacles.

Why do people today come to live on an island that is undoubtedly far-flung and, unlike Waiheke for example, has few of the comforts of mainland living? Debbie has lived on the Barrier for eight years. 'It's a great place to live,' she said. 'It's stress-free, we have clean air and clean water. But, if you're a woman you have to learn how to do things you've never done before – like changing the oil and water in the car and bleeding water lines. Because if your partner's in town, you can't wait till he gets home to have water in the house. The first time I turned on the generator I screamed, just because it made so much noise.'

Debbie also had to throw away her electrical appliances when she moved to the island – there is no mains electricity. People choose their own power system. Some opt for windmills, most decide on solar panels to heat water and LPG for cooking and refrigeration, with a generator as back-up. 'I found it hard at the beginning,' Debbie continued. 'But now I'm used to it. When I go into town and stay with a friend, I go around her house turning off lights if no one is in the room. We get very good at conserving power here. And water. We're all on tanks.'

Keeping occupied out of office hours is no problem. 'I go walking with girlfriends. We go to the hot pools and spend time on the beach. My partner goes diving, plays golf or goes fishing. The fishing's great. We'll miss it when we leave.' Debbie and her partner have bought land on the island for the future. Right now they plan to move to the mainland for a period, to earn money. Later they hope to retire to the Barrier.

There are three primary schools but lack of a secondary school means that families tend to move off the island when their first child leaves primary school. The permanent population has decreased from 1200 to 800 in two short years.

More than 60 percent of the island is owned by the Department of Conservation. Additional public land is in the hands of the Auckland Regional Council, which recently voted to contribute funds towards the purchase of Kaikoura Island in Port FitzRoy. This island has enormous possibilities for ecological restoration and as a habitat for the endemic brown teal. Some years ago DOC bought Arid Island off the eastern coast of Barrier and it is now maintained as a conservation area with a permit needed to land on it.

Views about the role of DOC vary among islanders. Debbie is enthusiastic about the emphasis on conservation and believes that tourism is the way forward for Great Barrier. Mary has lived on the island for 40 years and feels more equivocal. 'We moved here because I met an Irishman and he had to have land,' she said. 'We could afford it here. He bought the land and I bought the cattle. Every day is Christmas,' Mary continued. 'We have scenery that is comparable to any on earth. It's paradise. But the internal political system is lethal. The minute I step out of the gate I'm in the thick of it.'

I get the feeling Mary is a battler and that she will go in to bat on any issue that threatens Great Barrier. Since the Auckland Regional Council took over the responsibility for the administration of the island Mary feels that islanders have lost their say. While not totally opposed to DOC she feels that their priorities and the restrictions imposed by the Resource Management Act are eroding the way of life that people went to the Barrier to enjoy. 'You can't chop down native trees without permission and you can't go and dig out gravel. The price of land has shot through the roof and only rich people, who come in by helicopter, can buy it now. Everyone still goes to

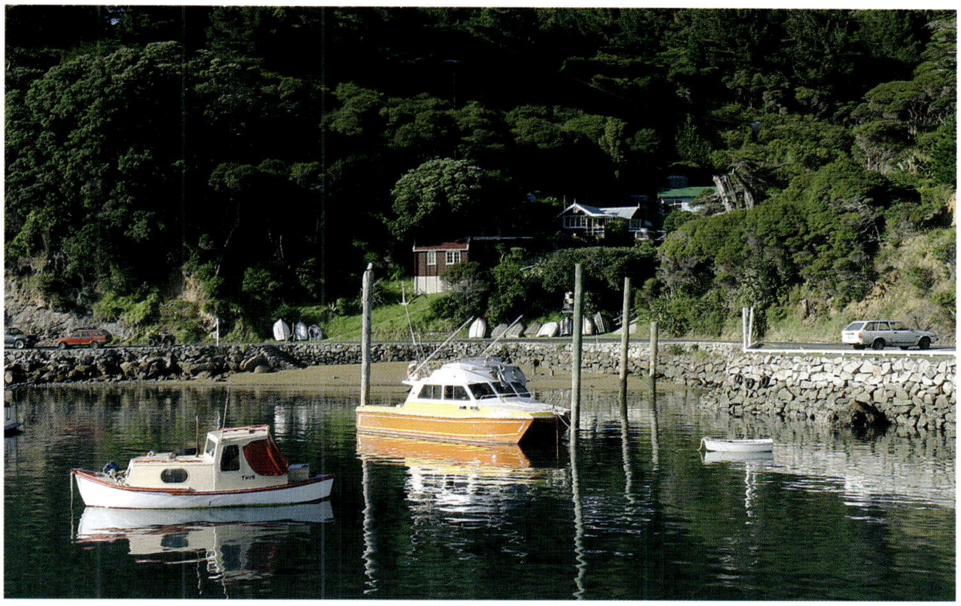

A 'Barrier Morning' at Tryphena Harbour, Great Barrier.

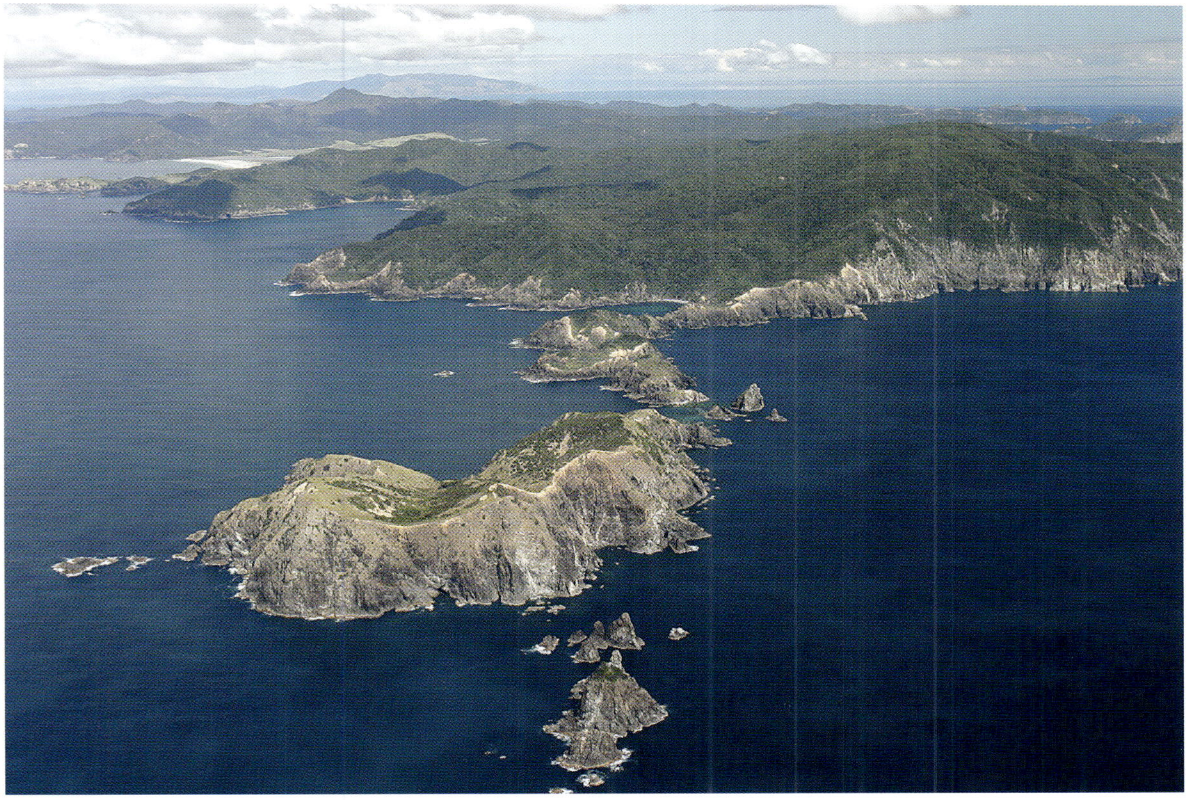

Aiguilles Island and Needles Point wander out into the sea beyond the northern tip of Great Barrier Island.

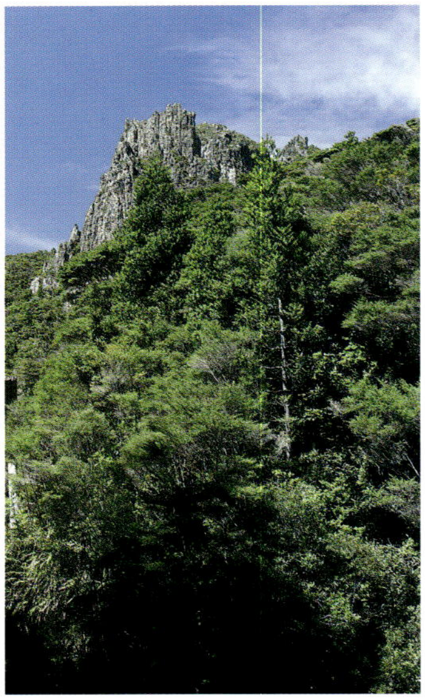

Tracks inland provide access to Great Barrier's dramatic volcanic landscape and forests.

the weddings and the funerals but the sense of community is eroding.'

What about the Barrier's reputation for alternative lifestylers? 'The subterranean folk are still here,' she chuckles. 'They're part of the scene. They're not causing any trouble. Every now and again the police go looking for plants.' The police have got tougher on the registration of cars too, though it doesn't stop dead cars being abandoned in the bush.

Mary talked about their life when they arrived on the island. 'We had the first solar shower. It was a piece of black alkathene pipe draped over a tea tree and we washed under that. We baked bread in a camp oven.' Mary taught her two children by correspondence during their primary years, but as parents do now, she spent a few years living in Auckland while her sons went to high school there.

Getting supplies is much easier than it once was. Mary and her husband order from a supermarket near Mangere Airport in Auckland and their supplies are flown out to the island. Boats make the trip six days out of seven.

Times have changed from the days when the Osbornes lived on the island at the beginning of the 20th century and the fastest communication was by carrier pigeon on a regular service set up in 1898 between Great Barrier and Auckland. (The stamps used on the flimsies are now sought after by philatelists.)

'But it's still a man's world,' said Mary. 'You have to be strong if you're a woman here.'

As always for visitors, the issues of local government are of little concern, and they enjoy the relaxed lifestyle offered by the island. Accommodation varies from exclusive lodge through bed and breakfasts to backpackers or DOC camping ground. Young overseas tourists come to tramp the tracks that traverse the island and enjoy the beaches on the eastern coast. Mountain biking is a popular pastime with visitors, as is diving.

'The diving in the local waters is as varied as the island itself. There are wrecks, drift dives, deep dives, huge crayfish, and scallop beds. There are caves and walls, kelp forests and schools of fish like you will only see at the Poor Knights,' comments one diver. 'On my very first dive at the Barrier I dived a wreck and found a complete teacup. I caught the biggest crayfish I had ever seen, let alone touched. I dived with a pod of dolphins numbering about a hundred and I saw a whale.' Paradise indeed!

How to get there

There is a regular ferry service from Auckland. A one-way trip takes approximately two hours. Charters by arrangement. Daily flights to Great Barrier leave from Auckland International, Dairy Flat, Whangarei, Whitianga and Matarangi airports. Homestay, camping, luxury lodge and private accommodation available on the island. For all Great Barrier accommodation bookings and information on tours and activities, contact the Great Barrier Island Visitor Information Centre, Claris.

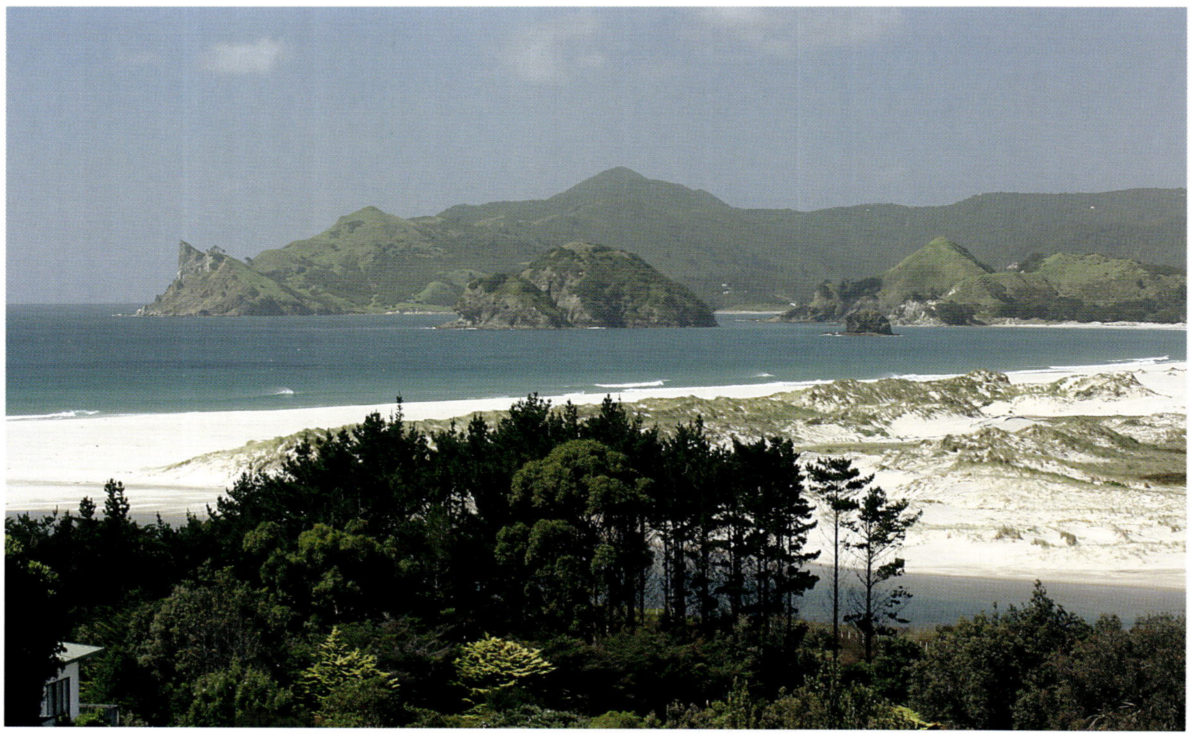

Boaties rarely sail around to the eastern side of Great Barrier and its superb beaches are the preserve of island residents and surfers.

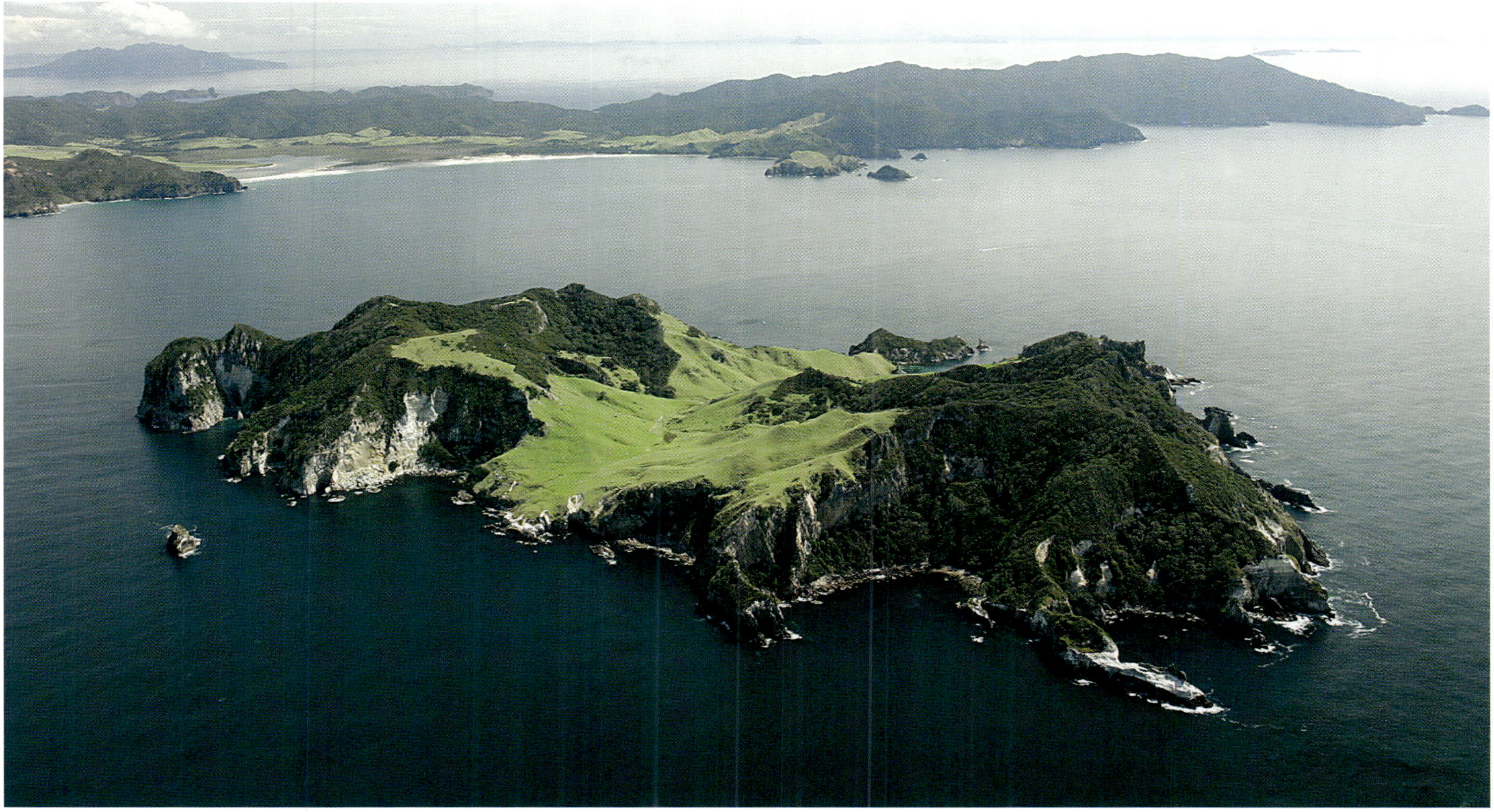

Little Barrier Island

If Great Barrier Island is paradise for visitors and residents alike, then Little Barrier is paradise for birds and insects. They live a protected life on this island that sits aloof, sheltered under a duvet of cloud most days – dwelling place of the fairies in the folklore of early explorers. Mountainous and densely forested, Little Barrier is one of the lonelier islands in the Gulf. The Maori call it Hauturu, 'resting place of the wind'.

Roughly circular in shape, the island covers nearly 3000 hectares with a succession of crumpled ridges and valleys leading up to the centre where Mount Hauturu sits more than 700 metres above the sea. Titoki Point, a rare area of flat land, extends from the island like the pointed tongue of a tuatua extracted from the sand and this is where most human activity has taken place. The rest of the island is ringed with forbidding cliffs and beaches of graded boulders. If the wind is in the wrong direction visitors cannot land, or once on the island may be stranded there for several days before a boat can return to take them off again. Landing on the island is forbidden, however, without a permit, and only carefully scrutinised visitors go to the island, for this has long been a sanctuary for rare native birds and, until recently, for the prehistoric tuatara. This is one of the few places where a magnificent dawn chorus of birdsong can be heard – more muted maybe than the chorus Cook and his crew heard, but glorious nevertheless.

Unlike so many other islands around the country, Hauturu has never been extensively farmed. No goats, deer or cattle have grazed the land and munched the native plants, though cats, which came with the Europeans, decimated birdlife, lizards and finally tuatara. When the last wild feline was exterminated in 1980 the birdlife started to increase dramatically.

The government bought the island more than 100 years ago with the intention of making it a wildlife sanctuary and in January 1897 the first curator went to live there – not, however, before much strife and litigation over the question of rightful ownership. Maori occupied the island in times gone by. Terraces, storage pits, cultivation enclosures and a burial ground all bear testimony to its early inhabitants. By the time the Colonial

Arid Island, bold and cliff-bound, sits off the east coast of Great Barrier.

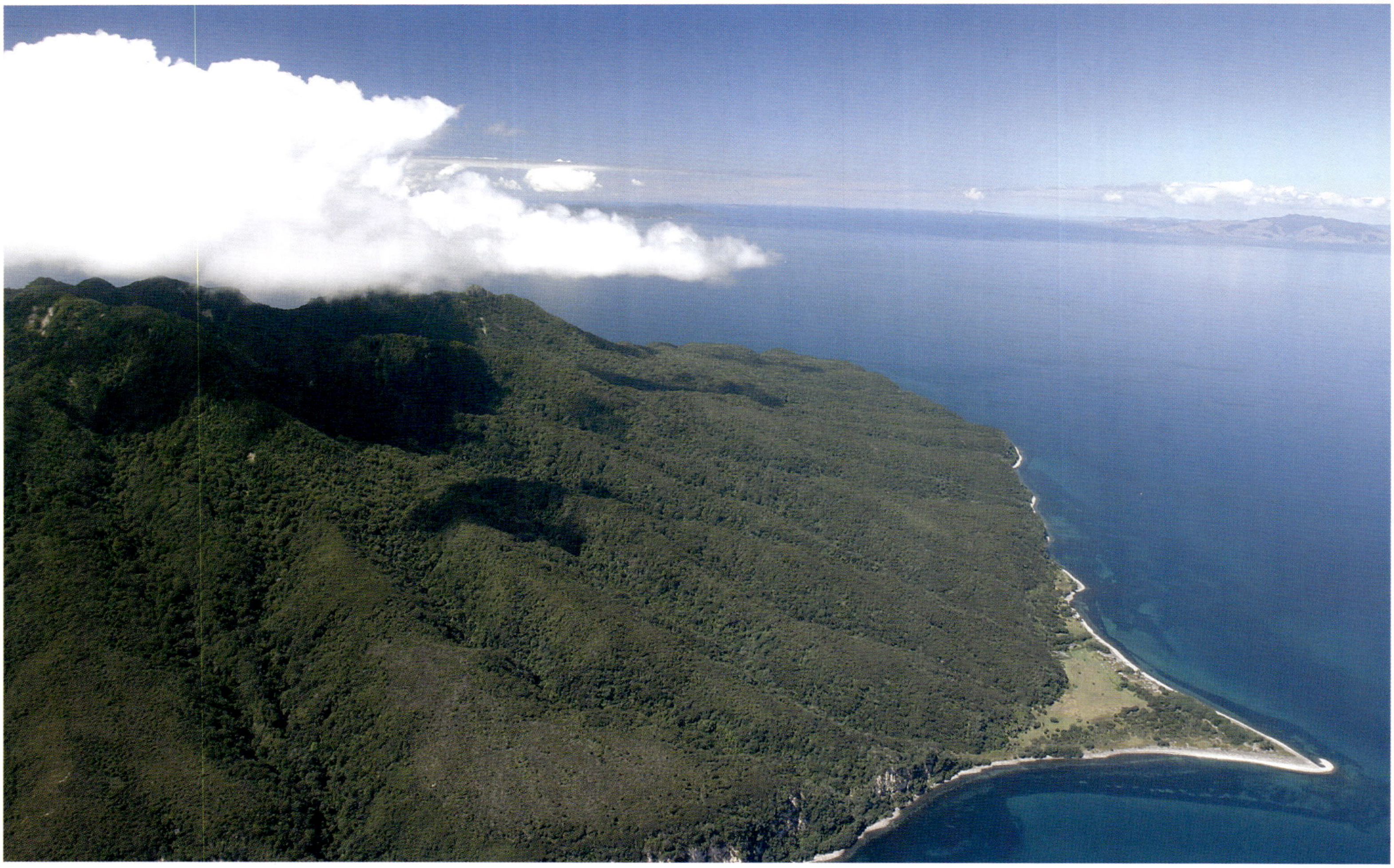

Looking like the beak of a giant bird, Titoki Point at the south-west of Little Barrier is a rare area of flat land on the island. In the past Maori lived there; now the caretaker's home is situated on the point.

Rare native stitchbird, today found only on New Zealand's offshore islands.

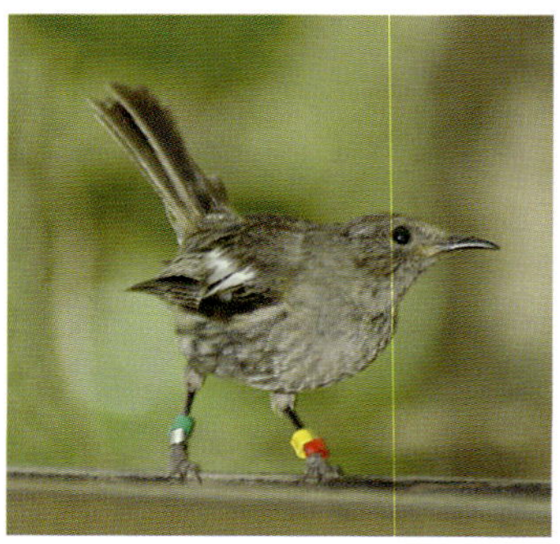

Government expressed interest in buying the land, two tribes were in dispute over ancestral rights. And while it was going through the courts those in possession at the time started stripping out the kauri for a quick profit.

Not only the kauri suffered. Andreas Reischek, an Austrian who spent 12 years in New Zealand towards the end of the 19th century, visited the island four times in the 1880s in search of stitchbirds, by then extinct everywhere else but reputed to live on the island. A so-called naturalist, he succeeded in eventually tracking down the birds and with a greed that is difficult to believe today, shot and killed a total of 150 stitchbirds over a period of three years.

Almost unbelievably, enough birds survived to continue breeding and in the last few years stitchbirds from Little Barrier have been translocated to Kapiti Island, Tiritiri Matangi and to Mokoia Island in Lake Rotorua. Although stitchbirds are raised in captivity at Mount Bruce in the Wairarapa, for the moment the fate of the stitchbird is inextricably linked to the fate of this island, one of the reasons for the restricted entry to Little Barrier.

About the size of a starling, stitchbirds are honeyeaters that nest and roost in dark, confined cavities, hence their susceptibility to predators. Males and females are dressed in different colours. The male sports a flashy plumage of black head with silvery white 'ear' feathers, bright yellow shoulder and breast bands and a white wing bar with a mottled tan/grey body. The female is more subdued with an olive/brown body cover, white flashes on the wings and small white 'ear' tufts. Sir Walter Buller, a 19th-century ornithologist, thought their call resembled the word 'stitch', hence their English name. Maori interpret their call and have named them as hihi.

Once cats had been eradicated, kakapo were transferred to Little Barrier in the hope of building up a breeding colony. But the rugged terrain of the island, combined with the need for intensive management of females before they bred, defeated the best efforts of conservationists to achieve their goal. Finally, in 1998 the remaining kakapo that could be found were removed to a more suitable island.

At least 30 native birds, including brown kiwi, New Zealand pigeons, kaka, both red and yellow-crowned parakeets, fantails, bellbirds and tui are known to breed on Little Barrier while blue penguins and Cook's petrels also nest there.

Less visible inhabitants of the island ecosystem are geckos, several varieties of skink and the formidable giant weta.

Mokohinau Islands

The Mokohinaus are a remote group of rugged islands lying about 50 kilometres east of Whangarei Heads and not far from Great Barrier Island. Burgess Island, the only island in the group open to the public, was a lonely outpost in the past for any lighthouse keeper who lived there. When the light was automated in 1980, however, nearly 100 years after the light was lit for the first time, the last keeper departed and the island has been uninhabited since.

It was suggested at the time that the light contributed to the sinking of the gold-carrying ship *Niagara* off the Mokohinaus in 1940. The German raider *Orion* used the position of the light to work out the route that ships took into Auckland and laid mines throughout the Hauraki Gulf, one of which sank the *Niagara*. Consequently, the lighthouse was closed down until after the end of the war. It's a good walk up to the lighthouse and visitors are likely to see parakeets, lizards and other small creatures that thrive on the predator-free island.

The remaining islands, including Fanal, Flax and Trig Islands, are nature reserves and protected wildlife sanctuaries. They provide a safe refuge for some of New Zealand's smallest endangered species, including the Mokohinau skink, the robust skink and the Mokohinau stag beetle. Several species of burrowing and ground-nesting seabirds make their home on the islands. Maori traditionally visited in the muttonbird season, to take the nearly mature chicks and preserve them for future food supplies and the Ngati Wai tribe retains muttonbirding privileges on the islands.

The Mokohinaus are a favourite destination for boaties – in good sea conditions small

How to get there

Little Barrier Island is a wildlife sanctuary. No accommodation is available. Access only allowed with a permit. To visit the island, contact the DOC office at Warkworth.

All islands in the Mokohinau group except Burgess Island are DOC conservation islands. Land access is by permit only. Burgess Island is open to the public. There are no tracks or facilities on the island. Chartered trips are available out of Auckland.

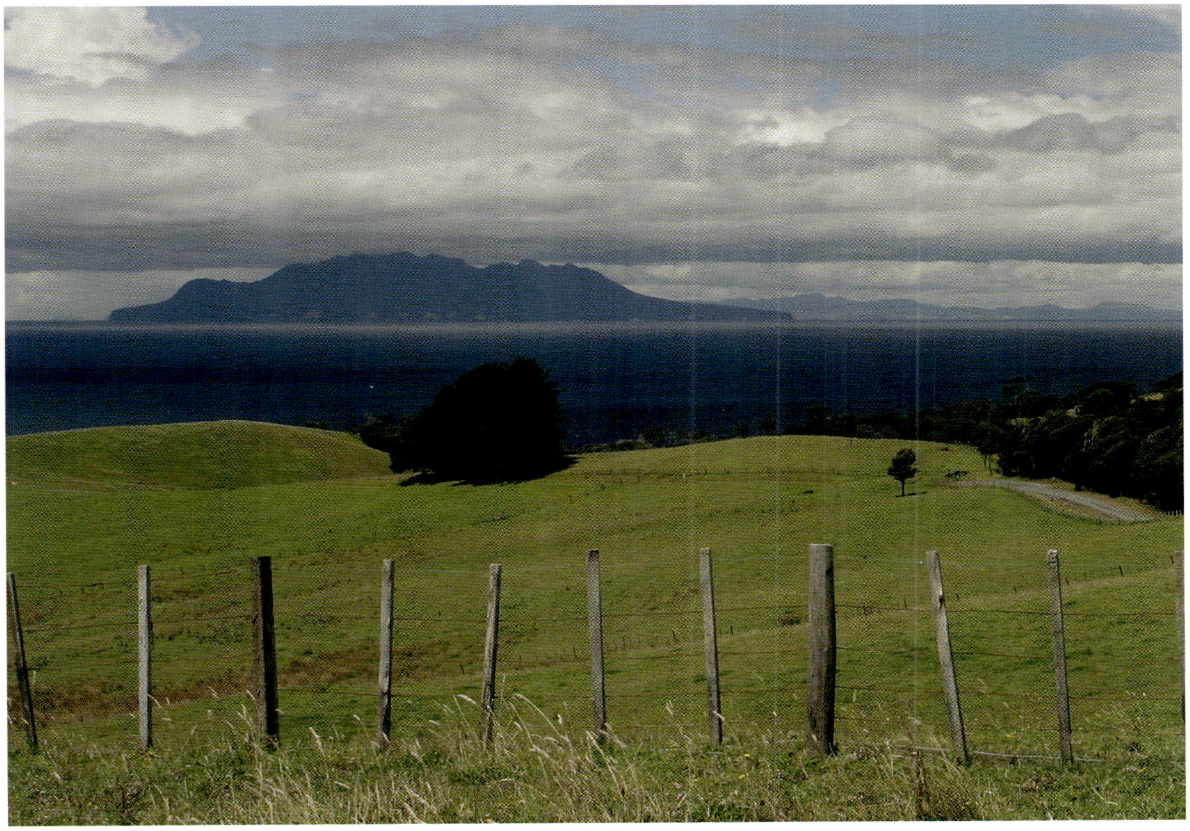

Looking out to Little Barrier Island from the mainland at Cape Rodney, an hour's drive north of Auckland.

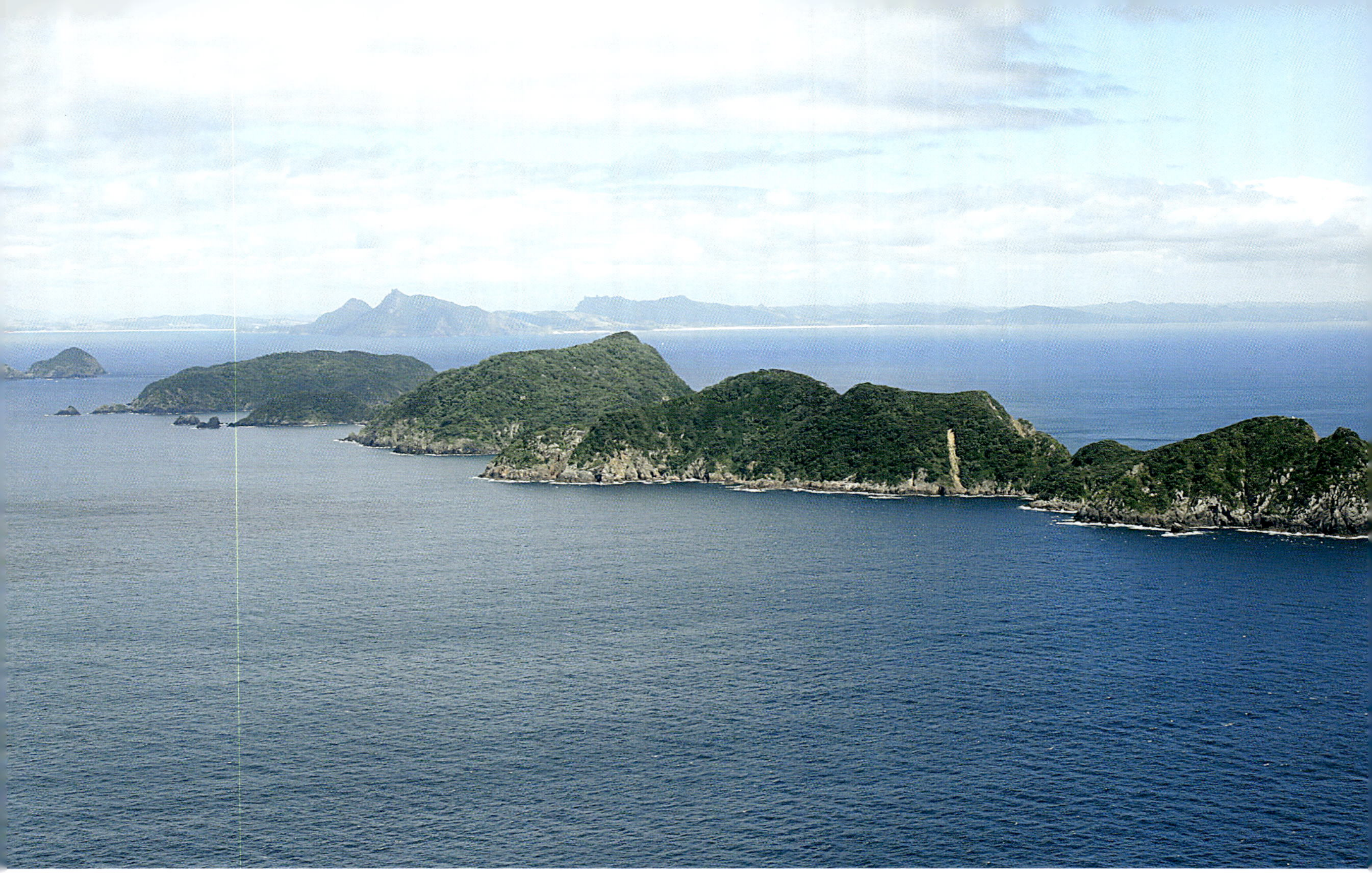

Each with its own character and home to a variety of native plants, this string of little islands are 'the chickens' or Marotere Islands.

powerboats make the journey without difficulty, though normal access is by charter or private boat. Snapper, kingfish and crayfish are plentiful and dolphins often swim in the area. Occasionally whales may be seen. Spectacular coastal formations include high cliffs, stacks, guts and sea-caves. The waters are clear and excellent for diving with exciting wall dives, pinnacles with deep drop-offs and a wide variety of marine life.

Such a wealth and variety of islands is rare in such a small area. No wonder then that visitors for decades have come from beyond New Zealand to experience the boating, fishing and more recently the diving around these and many other of our northern islands.

Hen and Chickens Islands

Much closer to the mainland and clearly visible from State Highway 1 on the route north to Whangarei is the group of islands named by Cook as the Hen and Chickens, known by Maori as Taranga and Marotiri.

Taranga (the 'hen') bulks large on the horizon with two sharp peaks making a distinctive outline. Gordon Ell in his book *Wild Islands* describes Taranga: 'The island seemed to rise vertically above, three giant monoliths of rock standing proud of the bush, towering 427 metres above us. There was no clear land at all and precious little room to come ashore.'

The bush cover is thick, tangled and uninterrupted. It is difficult to imagine how the Maori people who lived there in the past were ever able to find a foothold on the island, let alone living space. But live there they must have for there are numerous remains of stone walls, constructed to hold soil on the steep slopes so food plants could be cultivated. Empty shells are piled up in middens and right near the top of the island occasional caves have been scooped out of the rocks, large enough to hold the bones of revered ancestors.

Long after the resident Maori had departed, Taranga remained as the last refuge of saddlebacks or tieke when they disappeared from the mainland. Gordon Ell was involved in an expedition in the early 1980s to Taranga, camping on the island for several days and helping to capture saddlebacks in order to disperse them to other islands and thus continue the process of encouraging the population to increase. Time has proved the exercise a success. Since 1964 tieke have been moved from here to Little Barrier, Red Mercury, Middle Chicken, Cuvier, Mokoia, Tiritiri Matangi, Lady Alice and Kapiti Islands. In a reverse process, little spotted kiwi from Kapiti and stitchbirds from Little Barrier have been relocated to Taranga. Another survivor on Taranga was the rare Poor Knights lily, *Xeronema callistemon*, which now grows in many mainland gardens.

The Marotere Group north of Taranga consists of several rocky outcrops plus four small islets: Mauitahi or West Chicken, Lady Alice, Whatupuke and Coppermine Islands. The individual islands each have several Maori names and in 1928 when The Chickens became a scenic reserve, Big Chicken was renamed in honour of Lady Alice Fergusson, who had the unique distinction of being daughter, wife and mother of three Governors-General of New Zealand. She also learned Maori and together with her husband visited all the offshore islands of New Zealand.

Unbelievably, cattle were grazed on Lady Alice Island but once they were removed the bush regenerated thickly. Now a mixture of species including kawakawa, kanuka, puriri and pohutukawa clothes the island and in protected valleys puka or *Meryta sinclairii* can be found. It also was a rare plant in the wild but now is commonly found in gardens throughout the country where frost is not a problem.

Each of these little islands has its own character. One may be primarily a home for lizards, another for tuatara and shearwaters, uncomfortably sharing the same burrows.

Coppermine Island is even smaller than Lady Alice. It is the furthest east of the Chickens and remnants of the 19th-century mining enterprise can still be seen there.

All these islands are protected, landing is only allowed with a permit, but the sea surrounding them is renowned for its fine fishing and diving.

How to get there

Hen and Chickens Islands are DOC conservation islands and land access is by permit only. Chartered fishing and diving trips are available out of Auckland and Whangarei.

Taranga Island, otherwise known as the Hen, gives its name to the Hen and Chickens Islands. Surprisingly, given its steep terrain, Maori lived here once and burial caves exist near the top of the island.

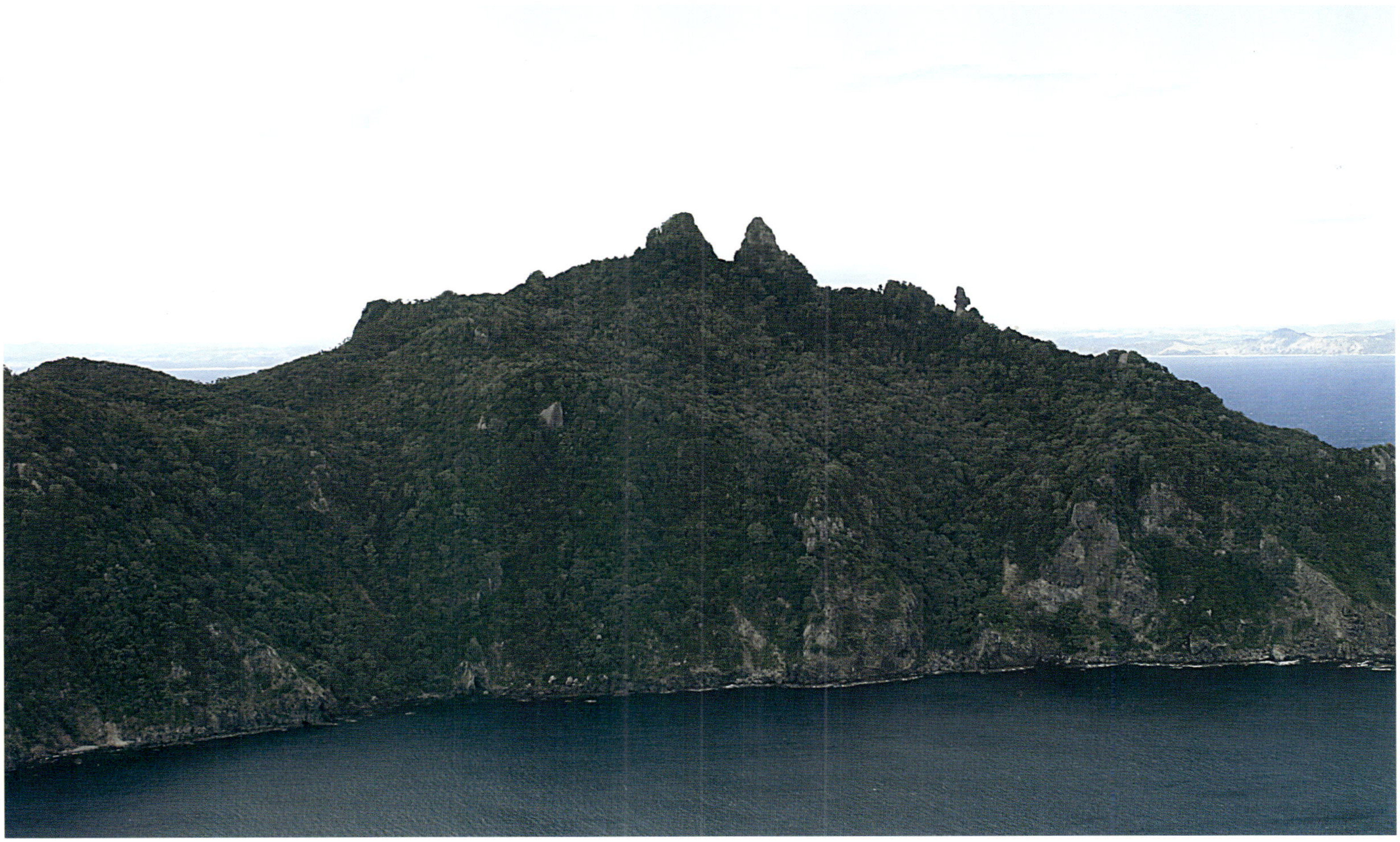

Lighthouse Islands

'My family is reduced to a state of starvation. All such stores as flour, oatmeal, tea, coffee, cocoa, potatoes, preserved meat, butter, rice, etc. are entirely exhausted. ... Distress signals have been flying for several days but no vessel has come within signalling distance ... hence my appeal to you in the hope that a shorter and more regular service will be established.'

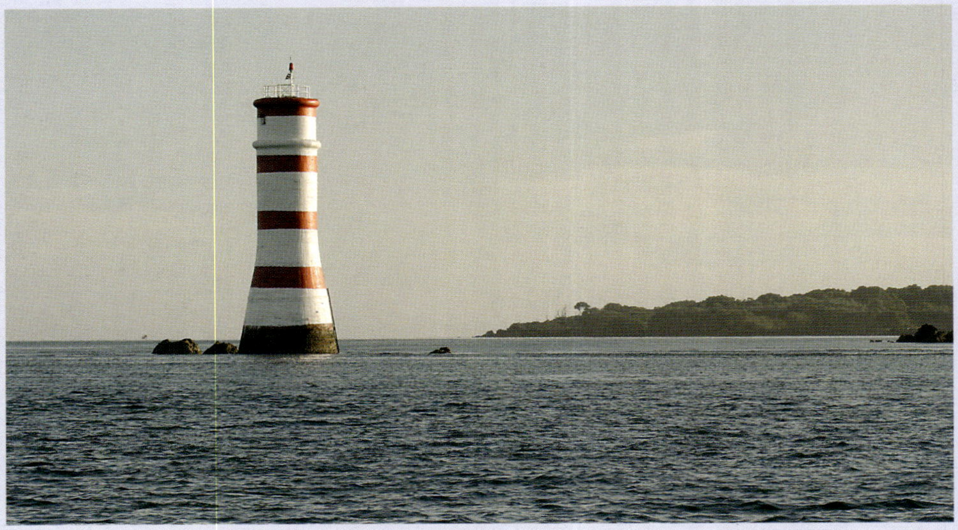

Known by boaties as The Barber's Pole but more correctly as the Rangitoto Light, this lighthouse sits on its own 'island' and serves as a marker for all the shipping that enters Auckland Harbour.

This was the plea from a distraught lighthouse keeper at the Mokohinaus to a Cabinet Minister in the early days of the lighthouse service. In spite of such obvious distress, matters did not improve immediately for the keepers on this most lonely of lighthouse islands where supplies came only three times a year. The next plea was more successful. One of the keepers fashioned a small tin boat with tin sails, complete with deck and a hatch. In the cavity he placed three letters – one to a friend, one to the Marine Department and one to the nearest general store. The boat was launched in a favourable wind, picked up on a beach and within nine days a boat arrived at the Mokohinau lighthouse on Burgess Island, the middle of the three islands in the group, with stores for the keepers and their families.

It is hard to imagine a life more lonely than that led by lighthouse keepers on some of the islands around the coast of New Zealand in the days before radios became common. Nowadays all lighthouses in New Zealand are automatically operated. Standby units are automatically activated in the event of failure and any faults are relayed to the Marine Safety Authority or the Regional Authority responsible for the light.

The isolation posed a real problem not only because of the difficulties of providing supplies but also because it sent some men 'rock-happy'. The lighthouse on the larger of the two Brothers Islands, a mere rock in the ocean off the coast of Tory Channel in Cook Strait, was notorious for this condition. In 1973 the principal keeper reported that he had to lock a 'joker' in his room for 24 hours because of the way he was acting. This station was deemed unsuitable for women and children because of the hazardous landing by crane and the confined living conditions.

Physical risks were also high. Many of the islands were steep and bounded by precipitous cliffs. The buildings themselves posed dangers and medical help was a long way away, even if a message could be passed on.

In 1883 the principal keeper at Dog Island died after falling down a 23-metre shaft that ran down the middle of the tower. On Centre Island, off the coast of Colac Bay at the bottom of the South Island and one of the last to be automated (1987), the keepers were largely self-sufficient, providing much of their own food as well as servicing the light. In 1943 two keepers drowned on a fishing expedition from the lighthouse.

Tiritiri Matangi has the oldest lighthouse still in operation in New Zealand. Situated in the Hauraki Gulf, it marks the approaches to Auckland Harbour. Construction of the lighthouse was difficult, taking workmen four months to dig through 3 metres of boggy, clay-like mud to reach ground that was hard enough for foundations. Once building materials were shipped to the island they had to be dragged by bullock sled to reach the construction site.

In spite of its strategic location at the western entrance to Cook Strait, just north of D'Urville Island, the Stephens Island light was not built until 1894. So exposed is it and so steep its terrain that, before construction of the light could begin, a concrete landing platform had to be built in the south-eastern corner of the island. Next a tramway was constructed to haul all the materials from the platform up to where the light and the keepers' houses would stand, 600 metres above the sea.

The concrete pier still exists and it is here that visitors to the island land. A former keeper's wife remembers returning home to the island with her 10-day-old baby. Strong winds were blowing at the time and when the transfer 'box' was lowered by crane, it kept blowing horizontal. Her husband jumped in to stabilise it, the baby was handed over and mother followed. It takes strong nerves to accept such living conditions.

The lighthouse on Cuvier Island (Repanga) was the first cast-iron lighthouse to be built in New Zealand. The keepers lived with their families in a small sheltered valley at sea level, near the landing place. To tend the light on the highest point of the island, they climbed a nearly vertical zigzag track, a treacherous trip on a stormy night. On Somes Island – the first harbour light in the country – as at Stephens Island, the accommodation was much closer to the lighthouse.

East Island, the most easterly point of the North Island, was briefly home to the East Cape lighthouse, first lit in 1900. East Island was so steep and the clay soil so inhospitable that the keepers there could not grow their own vegetables or keep livestock. The island was subject to earthquakes and the cliffs kept slipping progressively into the sea. By 1920 the lighthouse itself was in danger and East Cape on the mainland became the site for the lighthouse.

The first lights were fuelled by colza oil, derived from rapeseed. Later this was replaced by cheaper paraffin oil. Initially duties included trimming the wick of the oil lamp, polishing the lenses and rewinding the clockwork mechanism that caused the light to revolve. Signalling ships was an important task for the early keepers, sometimes they were involved in sea rescues and once radios were installed weather reporting became another of their duties.

On some islands, the keepers became interested in the resident wildlife. Keepers on Stephens Island helped scientists with their research on the tuatara and the last keeper at Tiritiri Matangi, withdrawn from lighthouse duties in 1984, became the first conservation officer when the island was designated an open wildlife sanctuary within the Hauraki Gulf Marine Park.

Automation came gradually to the various lighthouses around the country. The light on Somes Island was automated in 1924; that on The Brothers, in 1990, was the last. For more than 100 years, lighthouse keepers on islands around the coast and in other lonely outposts helped keep the sea safe for mariners. When the lighthouse keepers' service finally closed, an era in the saga of New Zealand's islands ended.

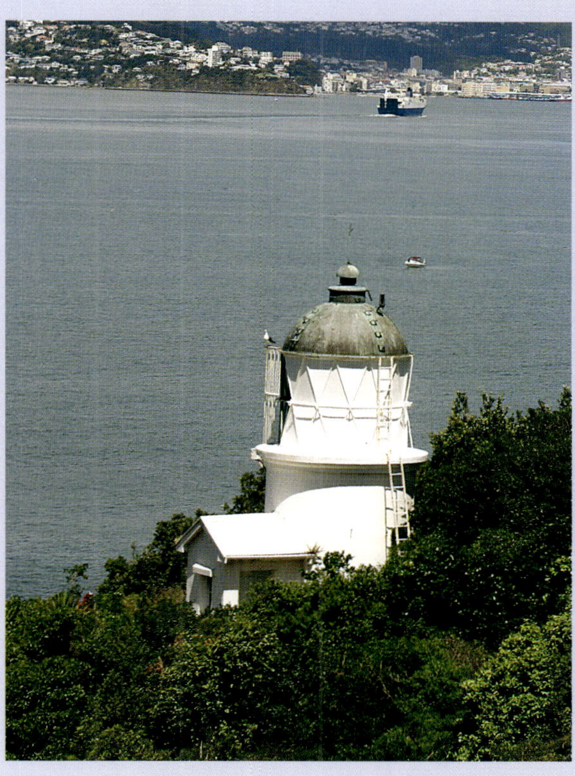

Once sitting on a bare headland, the lighthouse on Somes Island in the Wellington Harbour was the first harbour light in the country.

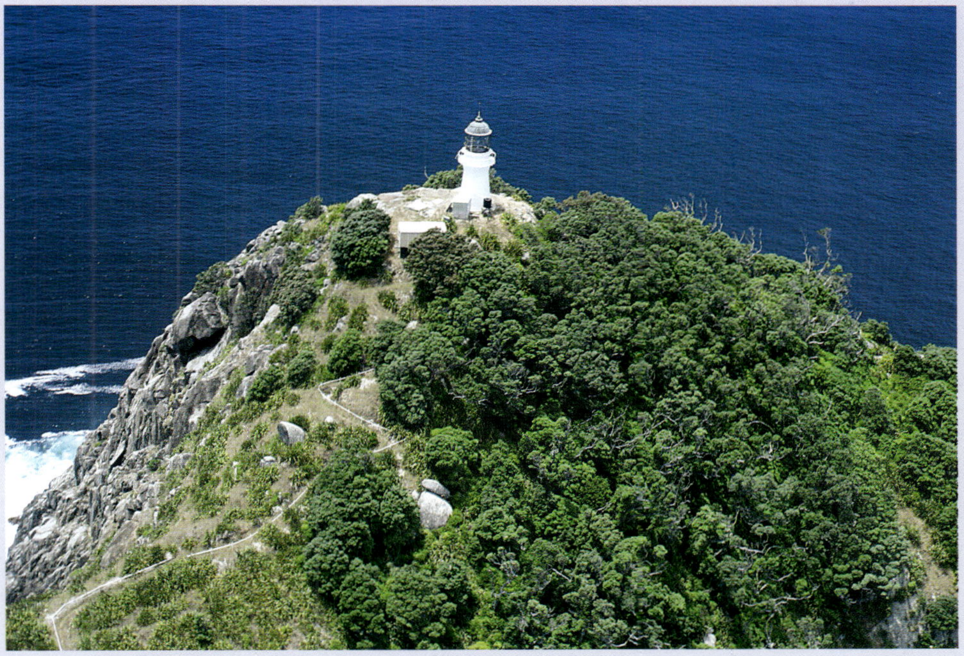

Life as a lighthouse keeper was lonely on islands such as Cuvier, situated well to the east of the Coromandel Peninsula.

Islands of the Earliest Landings

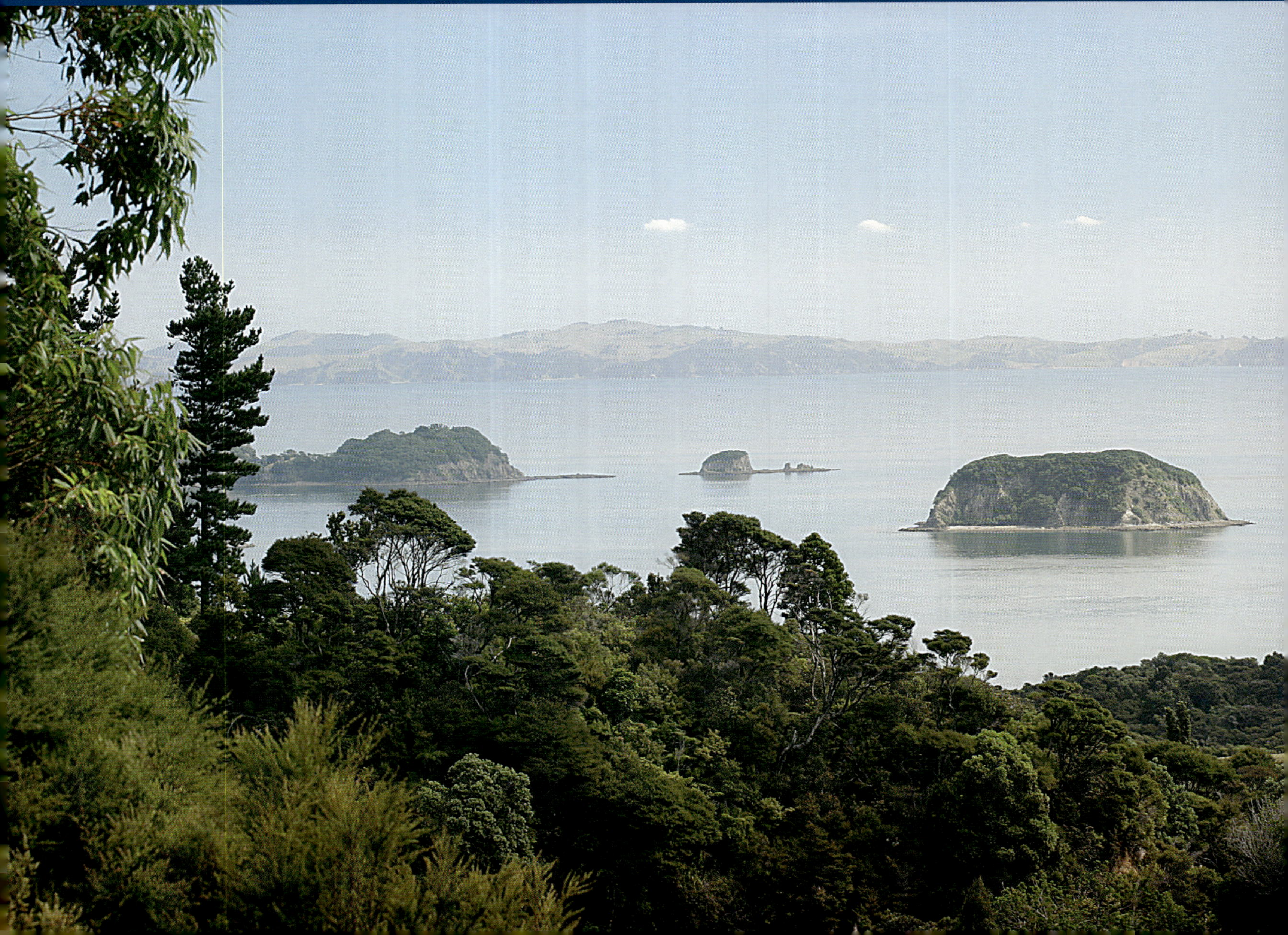

In days gone by islands often represented a place of refuge to people under threat. Nowadays it is more often the wildlife of our country that finds refuge on islands, while people enjoy their scenic aspect and the opportunities they offer for recreation. Around the Coromandel Peninsula there are islands enough for everybody, including wealthy families who have their own. Some are so small they are merely a stack of rocks; others, such as the Mercury Islands, are much more substantial. Some have a long history of human habitation, many now are devoted to conservation, several were formed by volcanic activity and others are the tops of drowned ranges. All offer spectacular scenery. Their beaches, sheltered anchorages and the marine life around their rocky shores make this a paradise for holidaymakers. In the summer months yachties and anglers converge on the Coromandel and the waters surrounding it.

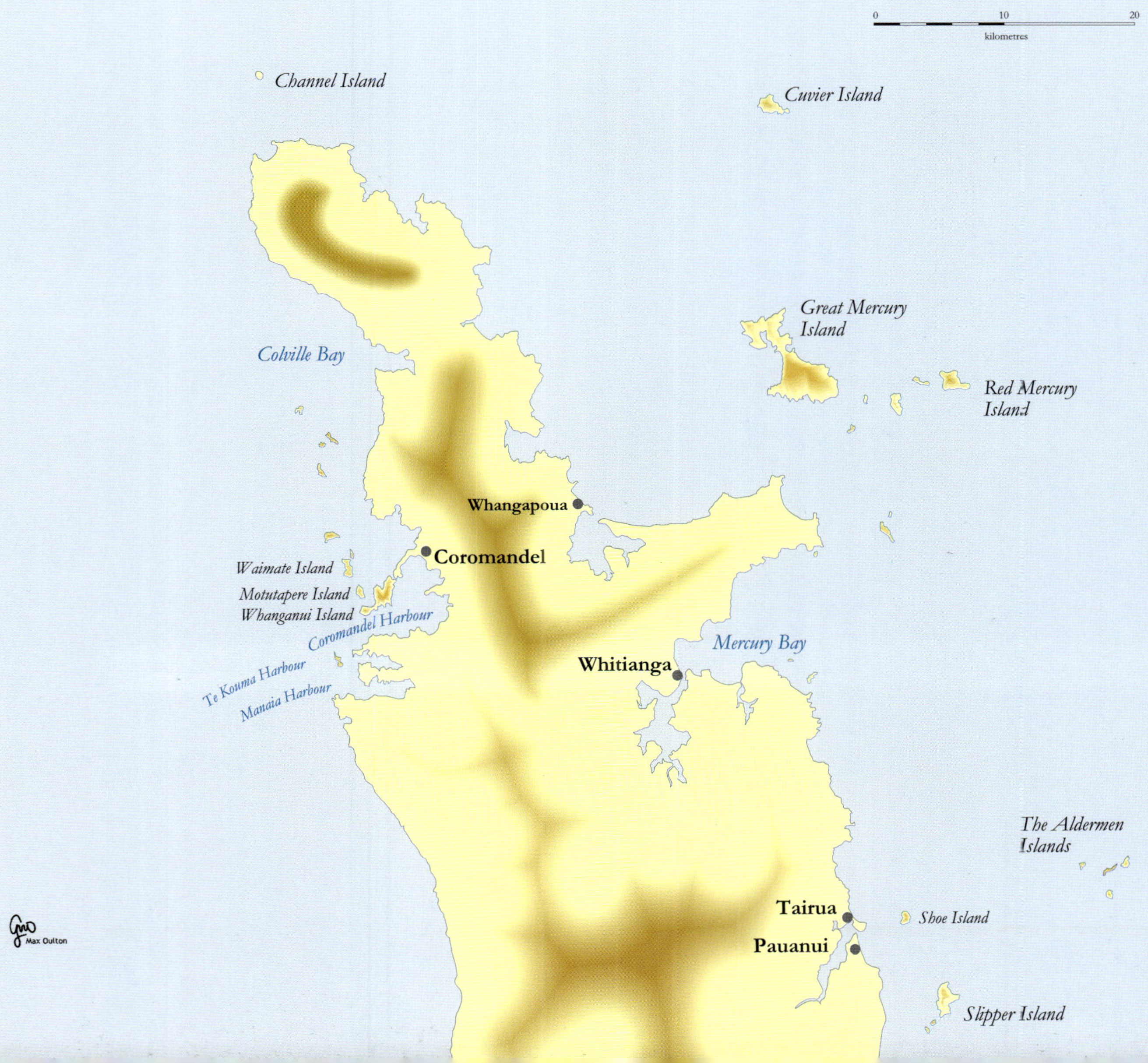

Western Coromandel

COROMANDEL HARBOUR, ON THE WEST COAST OF THE PENINSULA, WITH ITS FRINGE OF OUTLYING ISLANDS, IS A POPULAR DESTINATION FOR SAILORS. ON SUMMER WEEKENDS PLEASURE CRAFT REGULARLY ANCHOR IN THE BAY AT TE KOUMA ON THE MAINLAND.

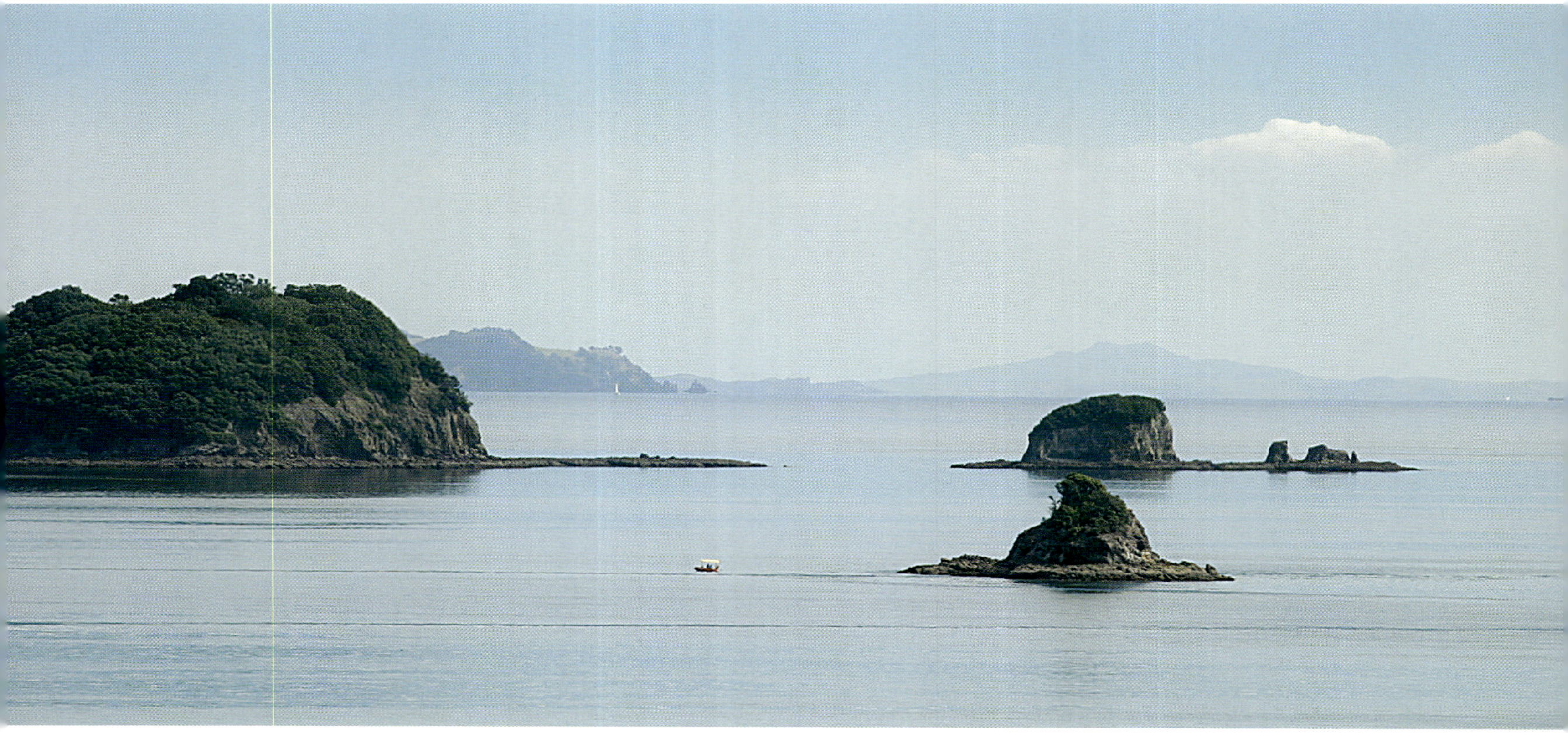

Islands abound off the western coast of Coromandel and this view near Papaaroha shows islands of the Motukaeao Group in the foreground with Thumb Point on Waiheke Island in the middle distance. The classic profile of Rangitoto Island is only just visible.

Previous page:
Further up the hill the view stretches to the 'bottom end' of Waiheke Island, giving a sense of the enormous potential for boating in this area.

But this was not always an area for leisure activities. In the early days of European settlement British seafarers coveted the plentiful Coromandel kauri for their ships' masts and entrepreneurial wayfarers prized the profits they could make from selling the cut timber.

One among them was American Bill Webster, known as Wepiha, the King of Waiau. He established a trading station during the 1830s at Herekino Bay on what is now known as Whanganui Island, at the mouth of the Coromandel Harbour. Webster was agent for a Sydney trading firm before setting up shop for himself. His merchandise included axes, blankets, muskets, tobacco, pigs and timber. Like many of the early Europeans in New Zealand he married a Maori woman and built up a profitable business, using his wife's connections.

Webster's village comprised a collection of weatherboard cottages, log cabins and whares. He ran an accommodation house of sorts and John Logan Campbell, who visited him in 1840, wrote about Webster's hostelry where a variety of guests sat down to eat at his table and paid for the privilege. Webster later extended his operations to Great Barrier Island and in the early days of Auckland his enterprise supplied much of the food and firewood for its citizens.

But his empire was short-lived. Webster purchased large tracts of land and engaged in land dealing between Maori and European interests. He overextended himself; frantic deals he put through before the Treaty was signed in 1840 were not validated and he returned to America an embittered man. It is not known whether his wife went with him.

Whanganui Island is still the preserve of entrepreneurs. It is owned and farmed by the

Spencer family. At 283 hectares, it is a relatively big island, with regenerating native forest and pine plantations. In an agreement with the Department of Conservation, captive-bred weka were released on the island in 1997. Only a short distance from the mainland, its main drawback as a conservation refuge is that a stoat or ferret can easily swim across the channel. Many of the first weka fell victim to stoats and Australasian harriers, attracted to the island by numerous rabbits. By 1998, however, calicivirus had decimated the rabbits, so fewer harriers were in the air and the stoats had been trapped. More weka were released and their numbers have been increasing.

In the vicinity of Whanganui Island are Motutapere Island, owned by DOC, along with Waimate Island, Motukopake Island and Motuoruhi. Further north off Amodeo Bay is the Motukawao Group, comprising several small inshore islands, all of which are home to many threatened bird species but the risk of stoat attacks is high also. DOC is constantly on the alert, laying traps whenever they suspect the presence of predators.

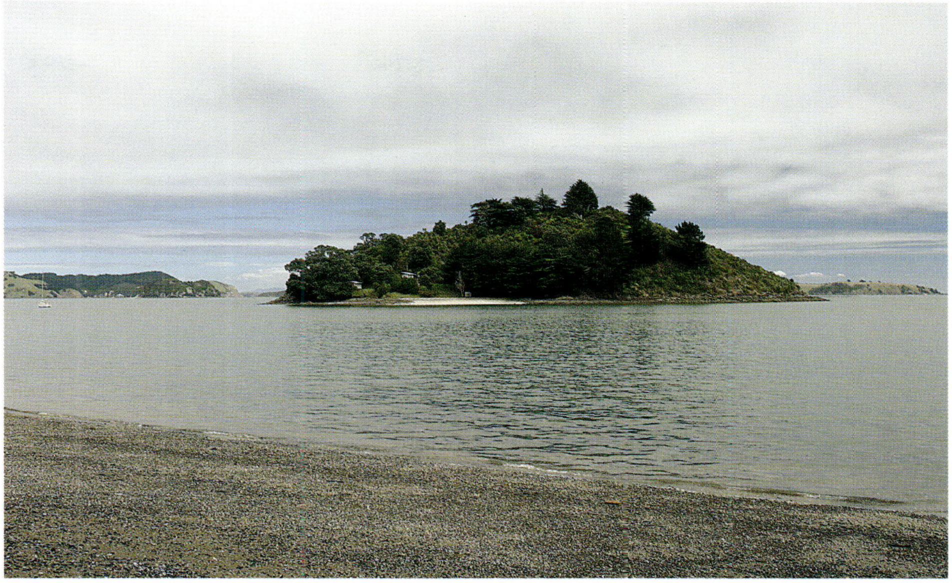

Two baches on Turkey Island off Oamaru (Koputauaki) Bay make an idyllic Coromandel holiday retreat.

These islands on the easternmost reaches of the Hauraki Gulf were formed when the low-lying plain that is now the Firth of Thames was inundated with rising sea levels in the distant past. The islands are the tops of ranges or mountains left above water level.

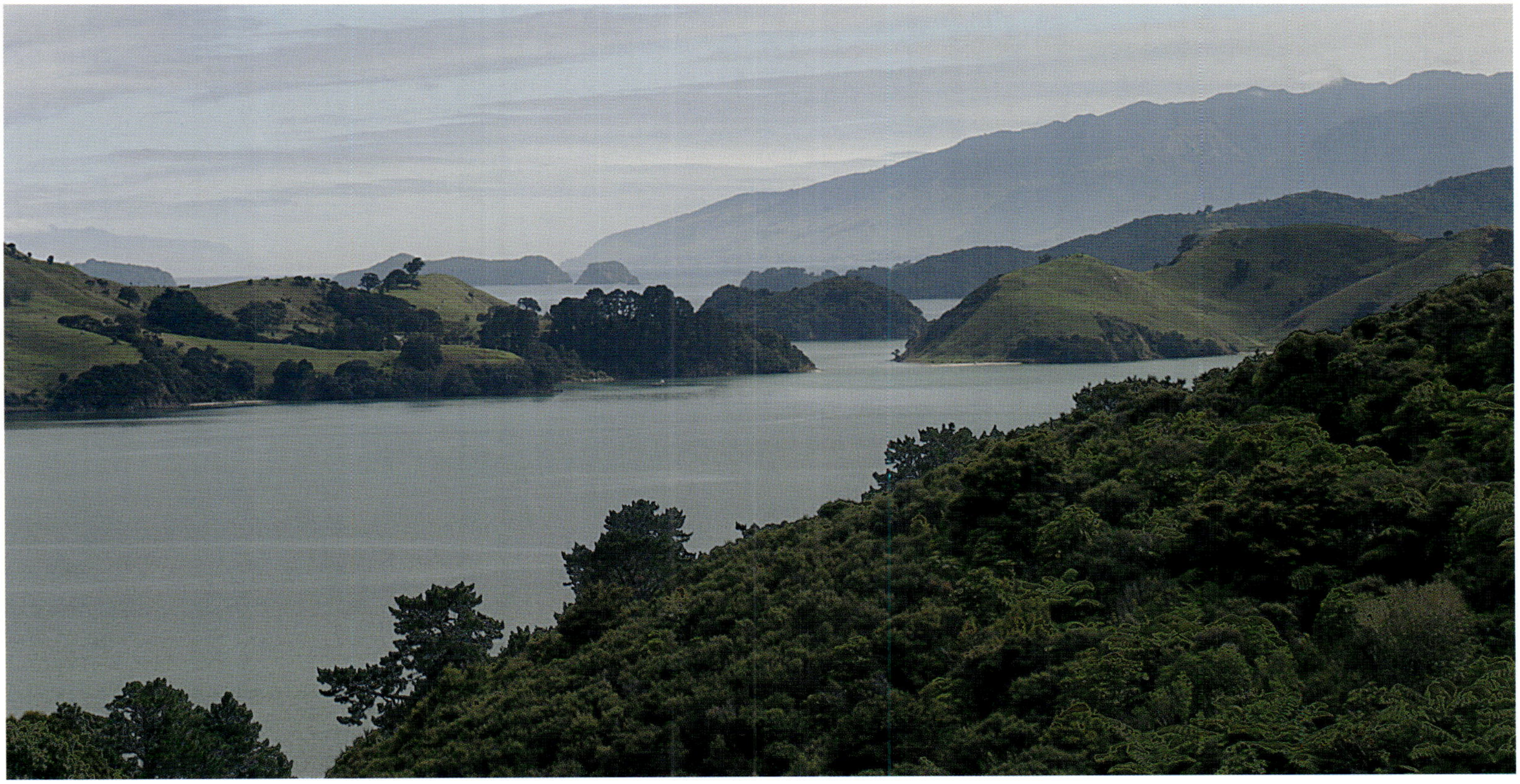

A peaceful roadside view near Te Kouma, looking northwest over Coromandel Harbour, which was the scene of much activity in the early days of European settlement. Whanganui Island, Little Passage and the tip of the Wyuna Peninsula are in the middle distance with Rat and Motukopake Islands behind.

Eastern Coromandel and Bay of Plenty

On the eastern side of the Coromandel Peninsula most of the islands scattered along the Pacific Coast from Cuvier Island east of Cape Colville, and south as far as White Island, were formed by volcanic action. The result is a series of rugged islets and larger islands, many of them covered in bush, often edged with dramatic cliffs that plunge into the ocean.

Marine life thrives around them. Seals frequent these waters in winter, dolphins are regular visitors and occasionally pods of orca swim by. It's an area to explore by boat and with scuba diving gear, for many of the islands have reserve status and landing is only permitted with authorisation.

Cuvier Island

Cuvier Island, named by the French explorer and navigator Dumont D'Urville when he came to New Zealand in 1827, is volcanic in origin. Its Maori name is Repanga and, according to legend, when the mighty waka (canoe) *Mataatua* came to New Zealand in the time of early Polynesian migrations, it was guided here by two magical birds. The waka stopped at Repanga, at the north-eastern entrance to the Hauraki Gulf, before sailing on. But the birds remained and it was they who so impressed D'Urville that he gave the island the name of a French naturalist of the time, Baron Cuvier. The birds were saddlebacks or tieke.

Nestled in the trees, the former lighthouse keepers' houses are just visible in Fairchild's Bay at Cuvier Island.

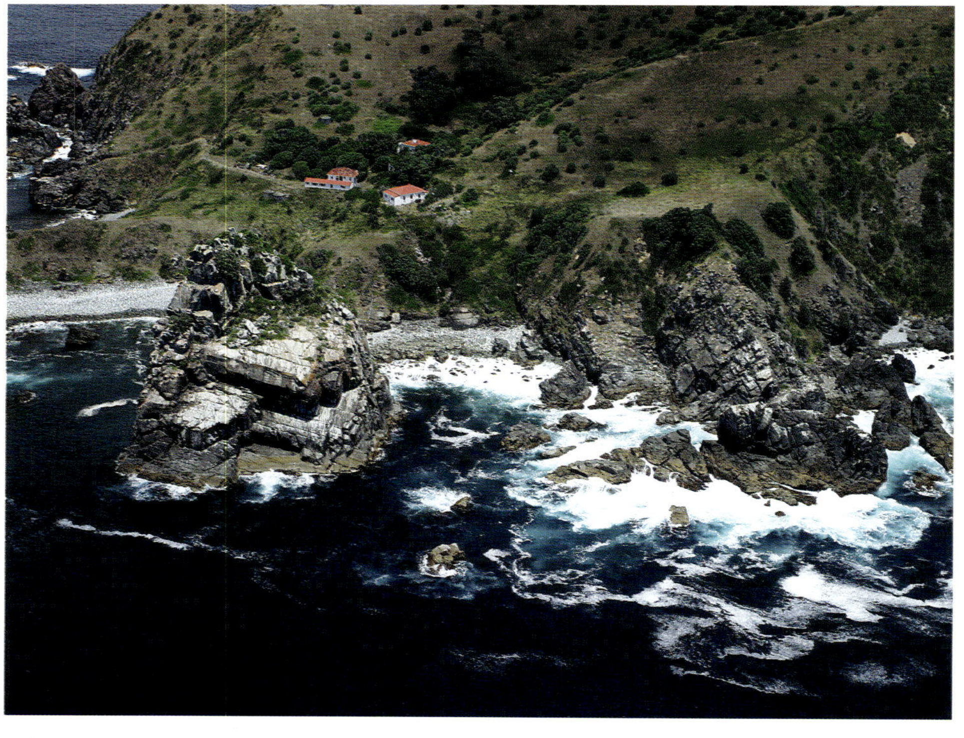

A lighthouse was built on Cuvier in 1860 and the keepers brought cats with them, and goats that destroyed the vegetation. Between the two, the native birds stood no chance and soon the saddlebacks disappeared along with any other birds that had made the island their home. Happily, the situation has been reversed. In the 1960s the New Zealand Wildlife Service got rid of the goats and set about eradicating the cats. The steep zigzag track that leads to the lighthouse is almost obscured now by regenerating bush.

Pohutukawa blooms in early summer and saddlebacks, relocated to the island, have bred so successfully that they have provided populations for other offshore islands. Stitchbirds are also doing well and a variety of seabirds have found a safe haven there too.

View Cuvier from the air and it is not difficult to imagine what a lonely outpost this must have been for the lighthouse keepers. In 1890, the keeper, in a report to the Marine Department, reminded his superior that 'posting letters by passing steamer are the only means of communication we have'. A carrier pigeon service from Auckland was never very successful and 10 years later the situation had scarcely improved. A newly posted Principal Keeper complained of not getting any mail in the three months since his arrival.

The last keepers left the island in 1982 and now access to Cuvier is restricted in order to protect the wildlife.

Mercury Islands

Further south of Cuvier and sitting to the north-east of Whangapoua Harbour are the Mercury Islands, comprising several smaller islands clustered around Great Mercury Island, the largest of the group. A copy of the Treaty of Waitangi was signed offshore here – the government boat carrying Major Thomas Bunbury dropped anchor, a couple of Maori chiefs arrived by canoe and promptly signed the document.

On Great Mercury Island, forestry and farming take up most of the land and the southern end is thick with pine forest, fringed with bush. A wild plunging cliff to the east contrasts with the rest of the island where beaches of the finest golden sand curve into turquoise water and cattle graze on rolling paddocks along the spine of the island. It offers a selection of sheltered anchorages but this is a private domain and landing is allowed only with permission, though there is fine diving and fishing around it.

It is on this island, some historians theorise, that the earliest Maori settlements may have occurred when the Polynesians travelled to this part of the Pacific and decided to stay.

The other islands in the group, a scattered handful of volcanic outcroppings, are all nature reserves and landing is restricted. They are home to many of New Zealand's endangered bird and reptile species including some Suter's skinks, reared at Victoria University and released on Korapuki Island.

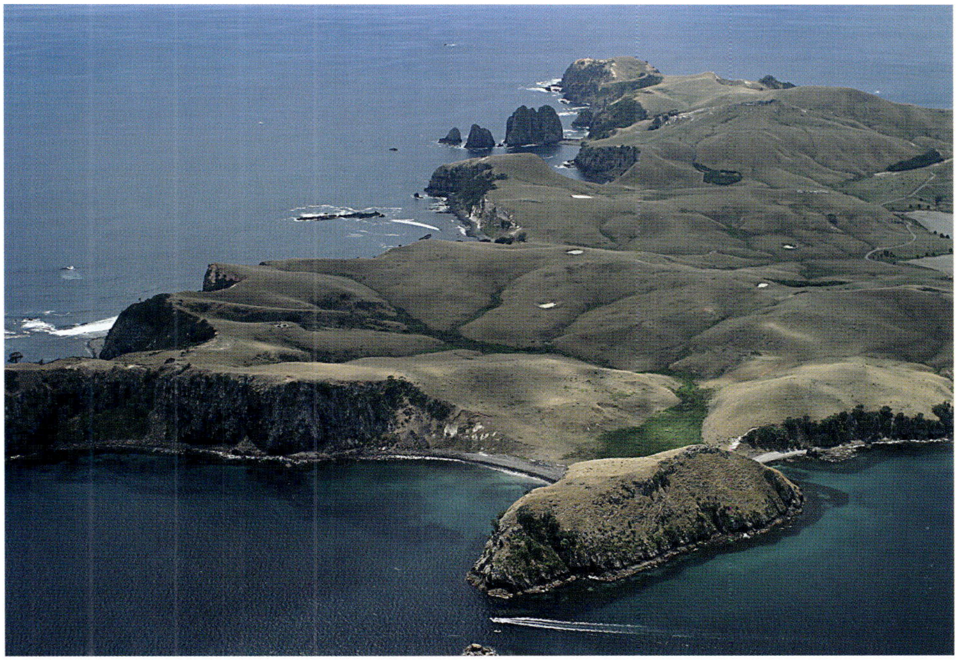

Privately owned Great Mercury Island is the largest of the Mercury group. Farming is the main activity but an area of the land is devoted to forestry.

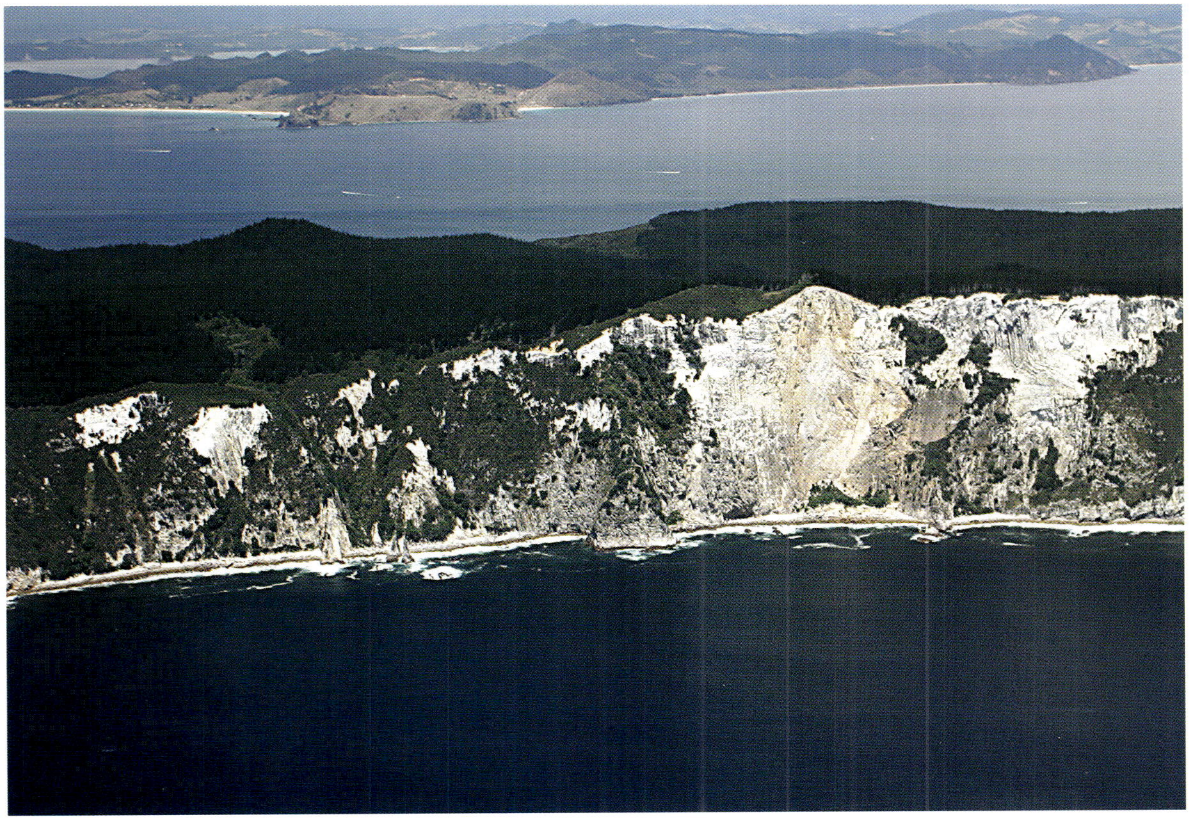

These imposing cliffs along the northern side of Great Mercury Island contrast sharply with the remainder of the island.

How to get there

Great Mercury Island is privately owned and there is no public access.
All the other islands in the Mercury group, Cuvier Island and the Aldermens, are conservation islands and access is by permit only.

Aldermen, Slipper and Shoe Islands

The Aldermen Islands, east of Pauanui, look like a dragon's broken black teeth jutting above the sea – the perfect rocky habitat for tuatara and skinks. Landing on the 'teeth' is not allowed but divers are free to explore the marine life, and the waters around the islands are popular fishing areas. Maybe the august aldermen of the London council would not have approved of Captain Cook's humorous naming of the islands in their image, though Mayor Island to the south looks dignified enough to suit the name he bestowed on it.

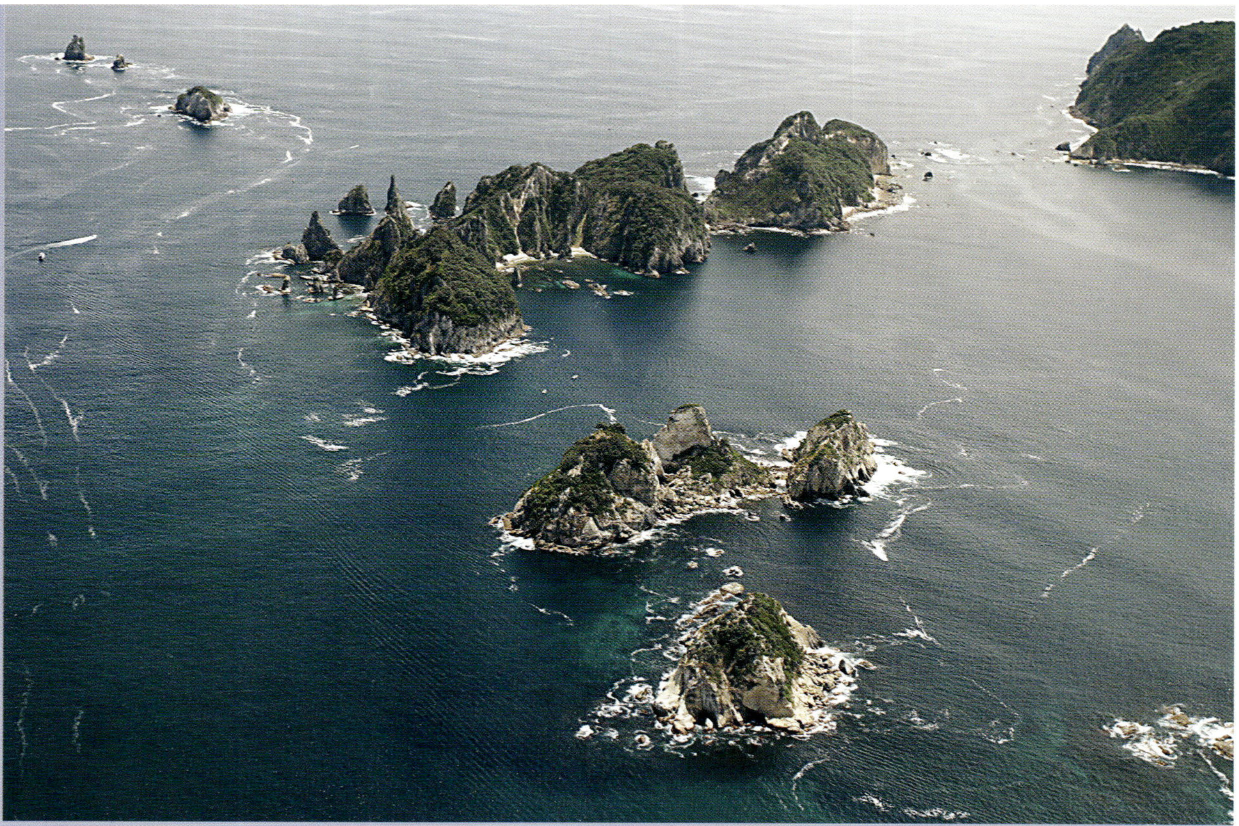

How to get there

Scheduled and charter dive trips to the vicinity of the Aldermens, Slipper, Shoe and Mayor Islands depart from Tairua.

Slipper Island is privately owned and features a resort. Lodge, chalet and camping accommodation is available. Transfer from Tairua or Pauanui.

Charter flights over the islands are available from the Tauranga Aero Club at Tauranga Airport or the Northland Districts Aero Club at Whangarei Airport. Activities include swimming, fishing and diving.

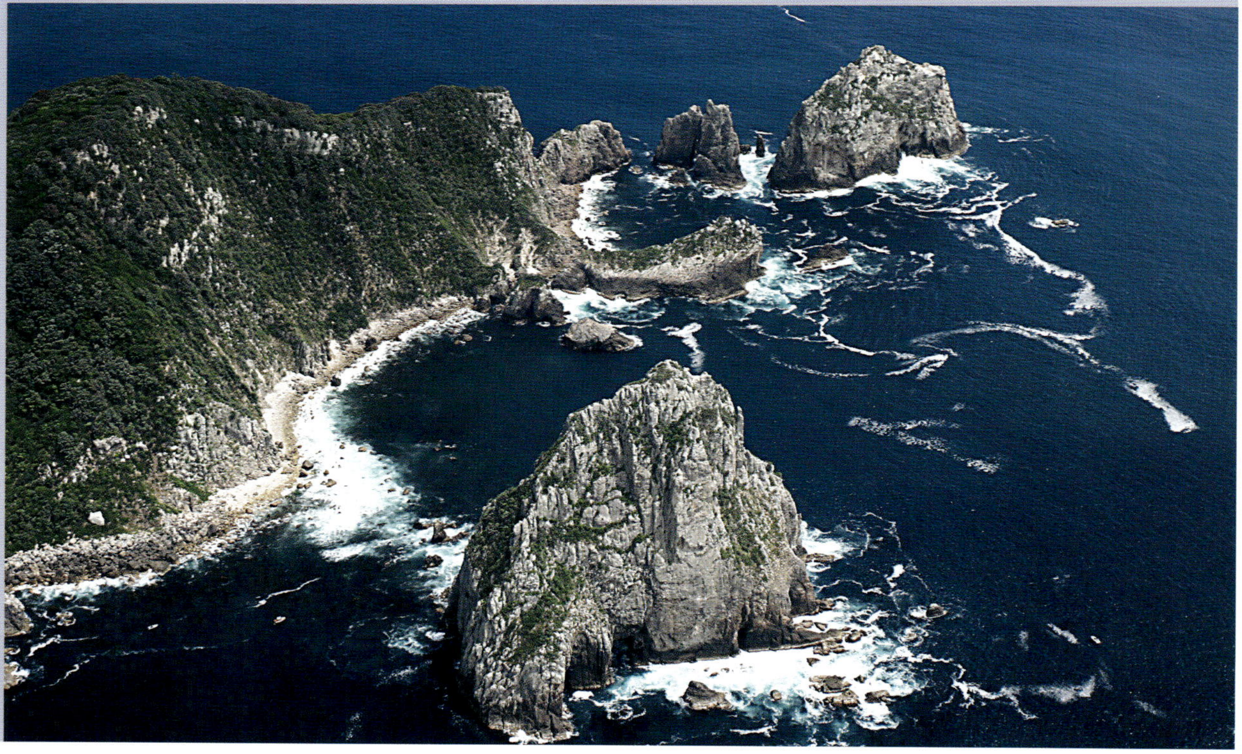

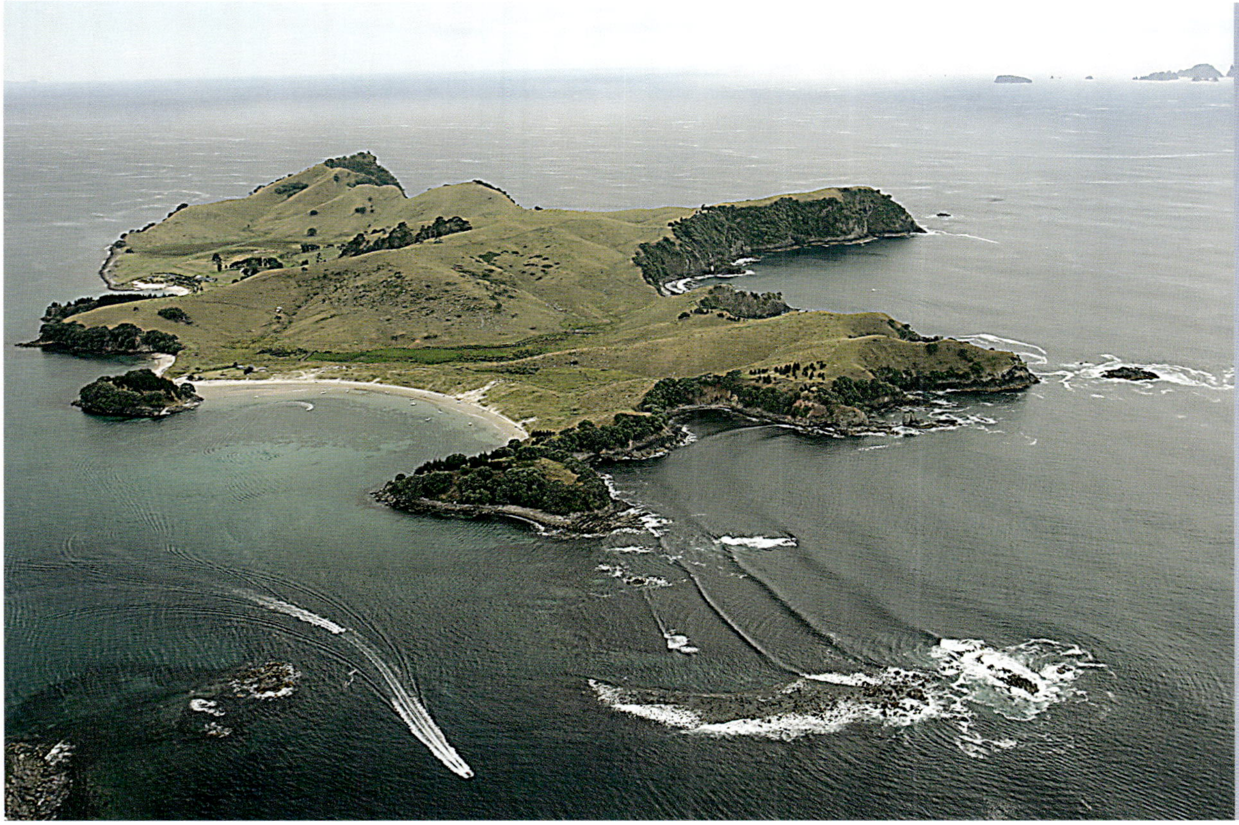

Slipper Island, slightly south of Pauanui, fringed with trees and curving beaches, is home to an exclusive resort.

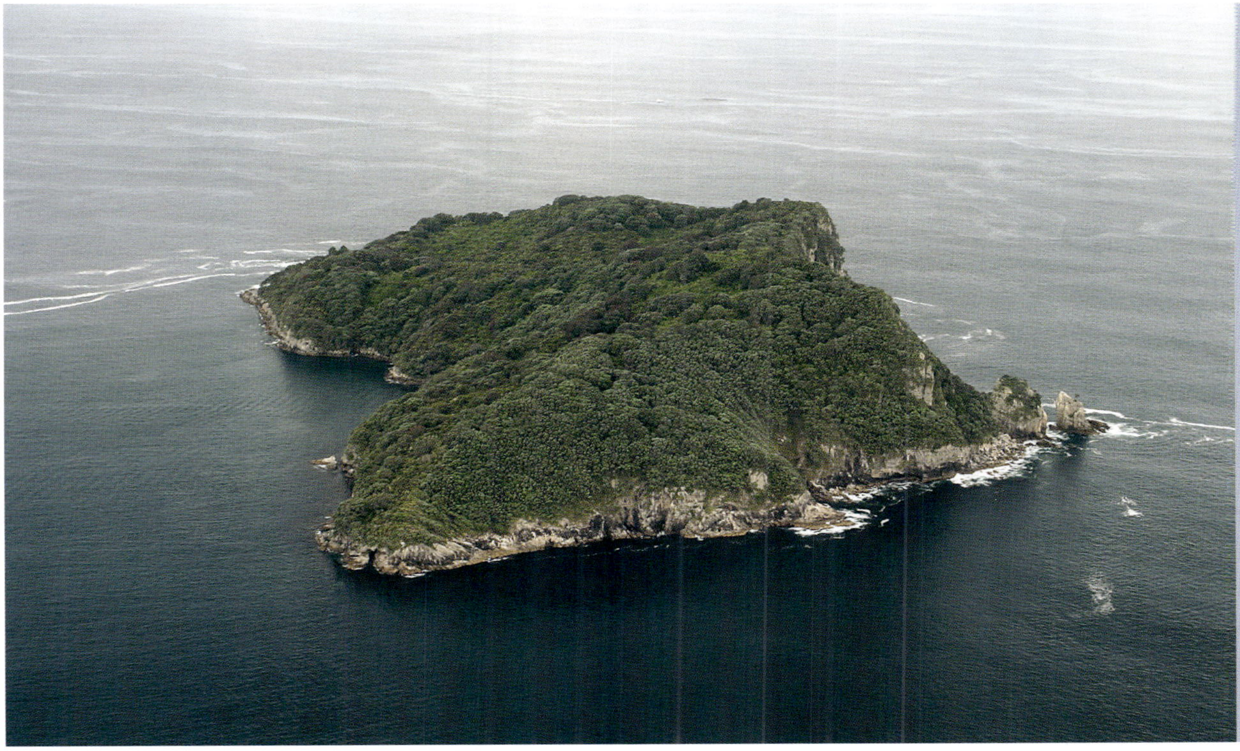

Shoe Island sits close inshore, visible from Pauanui and five kilometres north of the larger Slipper Island. Seen from the air the two islands sparkle in a navy-blue sea, like green jewels studded with pohutukawa rubies.

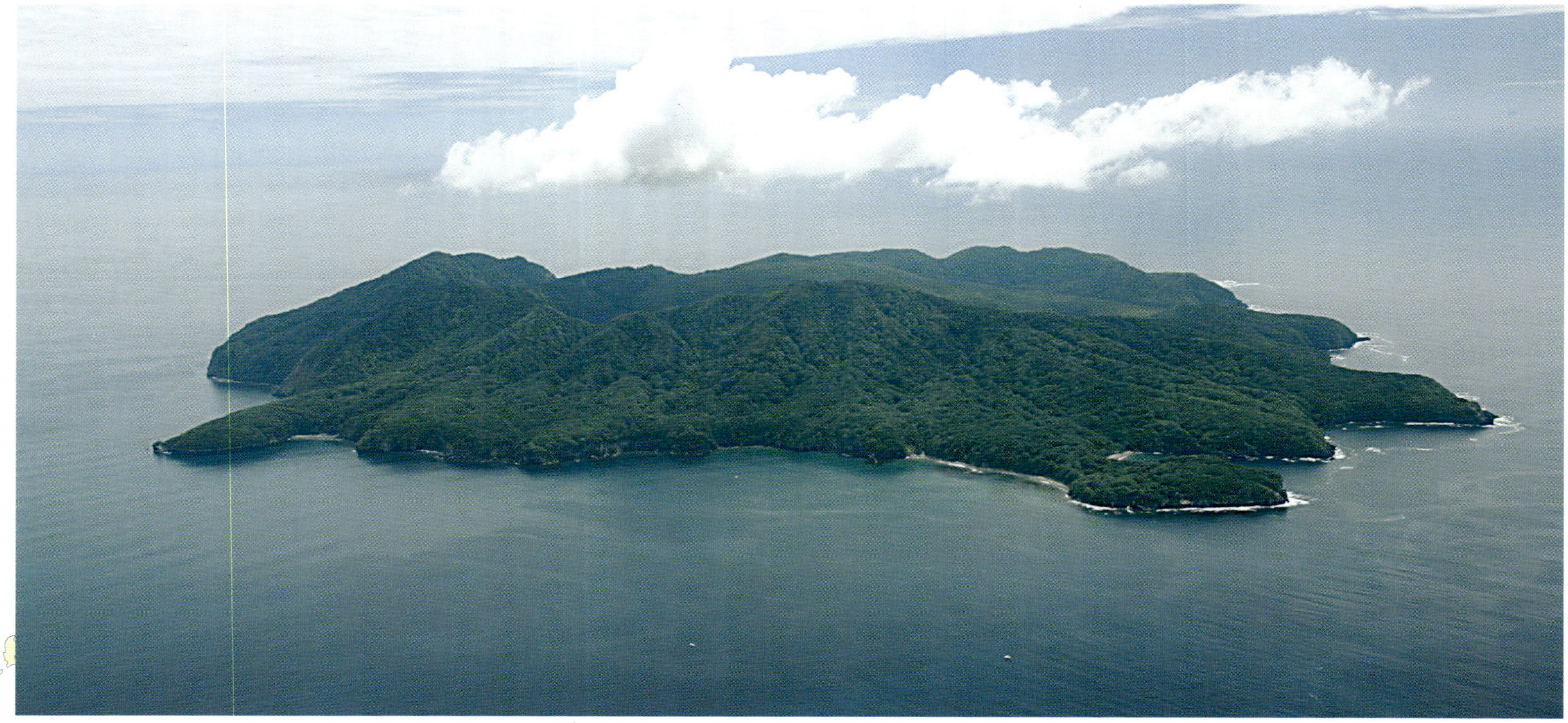

Bush-covered Mayor Island, or Tuhua as the Maori know it, is the summit of a volcano rising from the sea floor. In summer, crimson pohutukawa stud the bush with bright colour.

Mayor Island/Tuhua

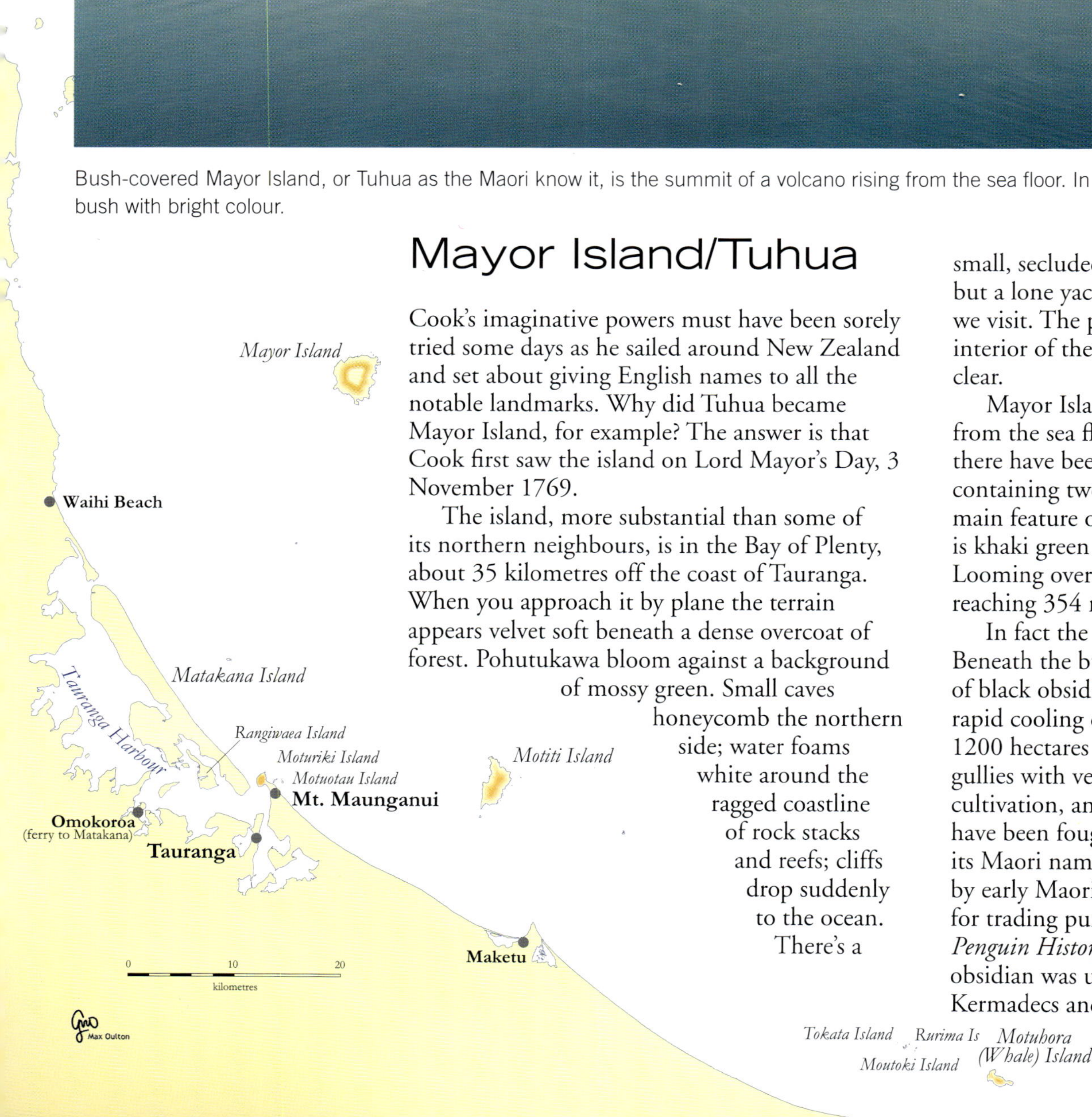

Cook's imaginative powers must have been sorely tried some days as he sailed around New Zealand and set about giving English names to all the notable landmarks. Why did Tuhua became Mayor Island, for example? The answer is that Cook first saw the island on Lord Mayor's Day, 3 November 1769.

The island, more substantial than some of its northern neighbours, is in the Bay of Plenty, about 35 kilometres off the coast of Tauranga. When you approach it by plane the terrain appears velvet soft beneath a dense overcoat of forest. Pohutukawa bloom against a background of mossy green. Small caves honeycomb the northern side; water foams white around the ragged coastline of rock stacks and reefs; cliffs drop suddenly to the ocean. There's a small, secluded bay where charter boats anchor but a lone yacht is the only vessel in sight the day we visit. The plane banks and swoops over the interior of the island. Suddenly its origins become clear.

Mayor Island is the summit of a volcano rising from the sea floor and over the last 130,000 years there have been numerous eruptions. A crater containing two lakes, both near sea level, is the main feature of the island. Lake Aroarotamahine is khaki green and Lake Te Paritu peaty brown. Looming over them is Opuahau, the highest peak, reaching 354 metres.

In fact the island is anything but velvet soft. Beneath the bush cover is volcanic rock composed of black obsidian, a natural glass formed by the rapid cooling of silica-rich lava. Its terrain of over 1200 hectares is a series of rock ridges and narrow gullies with very little flat land available for cultivation, and water is scarce. But many battles have been fought for control of the island. Tuhua, its Maori name, means obsidian, a resource prized by early Maori for making tools and weapons and for trading purposes. Michael King reports in *The Penguin History of New Zealand* that Mayor Island obsidian was uncovered on Raoul Island in the Kermadecs and dating processes indicate it had been there since the 13th century. Possession of the island obviously had enormous advantages and traces of ancient, heavily fortified sites can still be found.

The last permanent residents were the hapu of Whanau a Tauwhao. They moved to the mainland in 1901 but continued to visit the island frequently and planted crops to provision fishing parties who used the island as a big-game and commercial fishing base. The taking of obsidian has been banned in recent times.

Since 1953 Tuhua has been a wildlife refuge and is administered by the Mayor Island Board of Trustees. Both cats and rats have been eliminated and it has healthy populations of nectar-feeding bellbirds and tui. New Zealand pigeons, moreporks, fantails, kaka, grey warblers, silvereyes, kingfishers and tomtits also enjoy the island's environment and in summer the shining cuckoo is a frequent visitor.

A marine reserve at the northern end of the island displays a sea floor with fascinating caves, pinnacles, hot water vents and black obsidian beds. Many fish varieties inhabit the reserve, including subtropical species drawn in on the warm East Auckland Current. Add to this dense forests of paddleweed kelp and reefs festooned with corals, colourful anemones and sponges and it's not difficult to understand the area's attraction for divers.

For those who prefer walking to water-based activities, there are tracks on the island and it's possible to walk around the crater. Charter boats run regular trips to the island.

How to get there

Private charter boat trips are available out of Whangamata and Tauranga for fishing and diving in the proximity of Mayor Island.

Lake Aroarotamahine and Lake Te Paritu sit in the crater of Mayor Island.

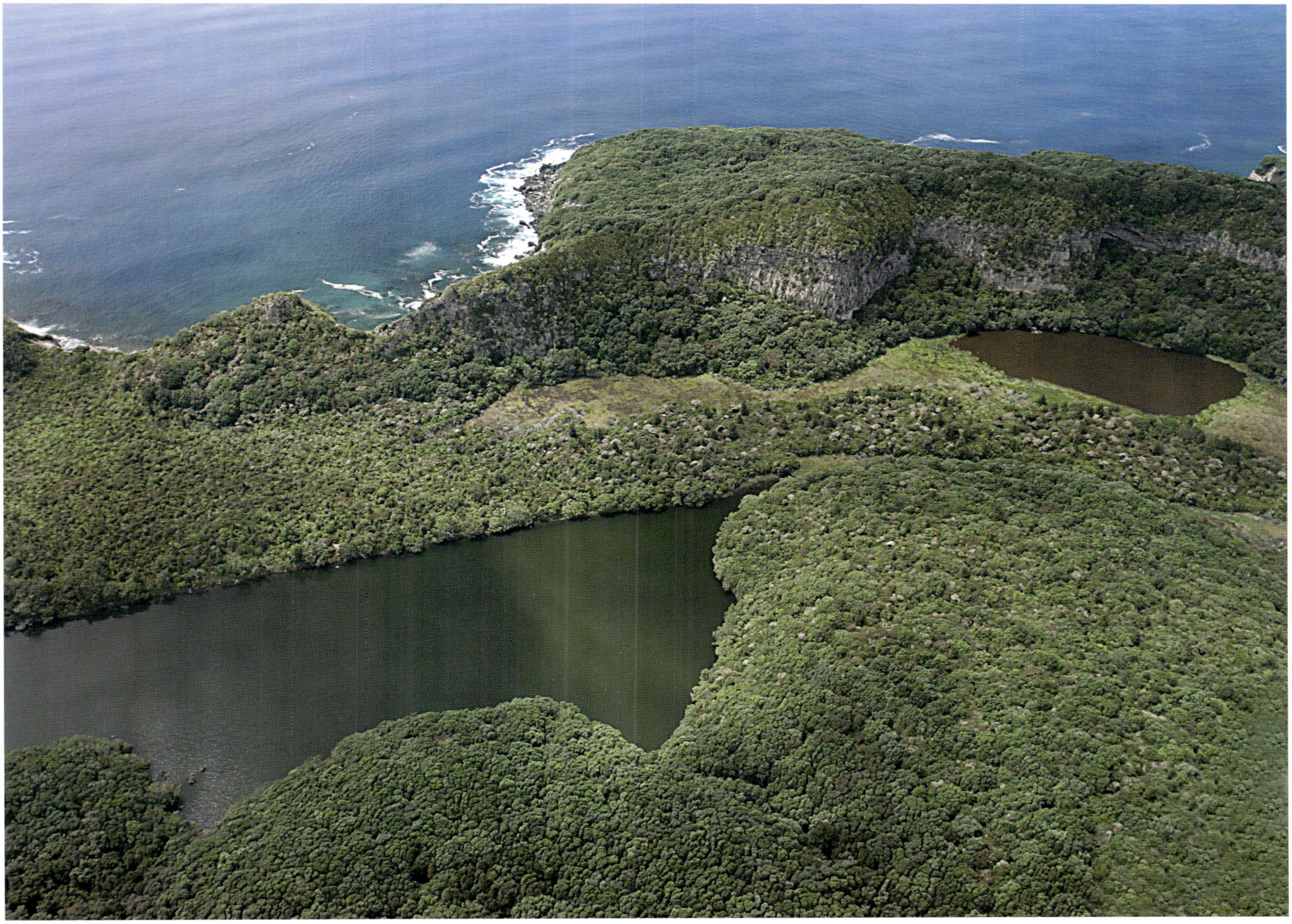

How to get there

Motiti is a privately owned island and visits can occasionally be arranged through Bay of Plenty Tourism.

Right: The rocky shoreline and numerous small bays around Motiti Island are ideal for snorkelling.

Motiti Island sits low in the water off the coast of Tauranga. Part of the island is farmed and avocados thrive in its mild climate.

Motiti

Motiti Island sits low in the water off the Bay of Plenty coast, a cultivated green patch of land not far from Mount Maunganui that looks as if it should still belong to the mainland. Frost-free and fertile, the island has an excellent climate for avocados and those produced there are among the earliest to ripen in the country. Long-term owners the Wills family also graze prime Simmental cattle.

As with many islands around the coast, Motiti in the past has been claimed by several Maori tribes. In 1867 a decision of the Native Land Court deeded the land to Te Patuwai on the northern end and Whanau a Tauwhao on the southern end, although the actual boundaries remained in dispute for some years. The southern part of the island was subsequently sold to Europeans and has been farmed ever since. Today there are about 50 permanent residents and people of the two tribes still live on the northern part, although the occupants are principally of Patuwai descent.

Excellent fishing and diving around the island attracts weekend boaties. Spearfishers work the reef off the island – big-game fishers go further afield. Enhancing the diving environment is the wreck of the Tug *Taioma*, scuttled in March 2000 on the south-eastern side of Motiti Island and already totally encrusted with marine life.

Dolphins and whales are a common sight – baleen whales are often seen around here with their calves, and dolphins frequently play alongside boats as they head out from the mainland to fish.

In the early days signal fires were lit on the island when communication was needed with the mainland. Carrier pigeons were the next mode of conveying messages and they were replaced by radio-telephone in the 1940s. Now the island is a mere 10-minute flight from Mount Maunganui and email makes communications fast and simple.

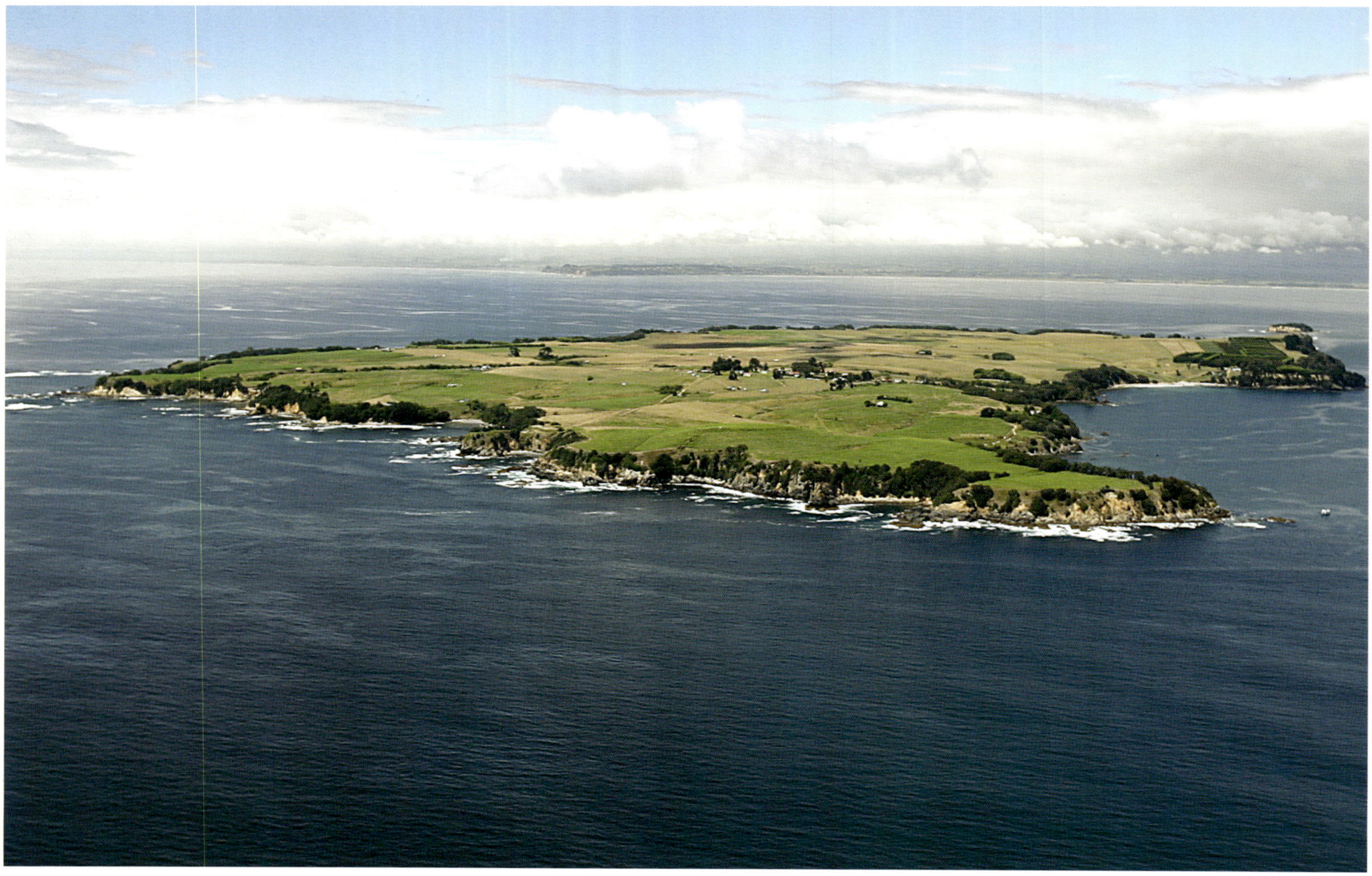

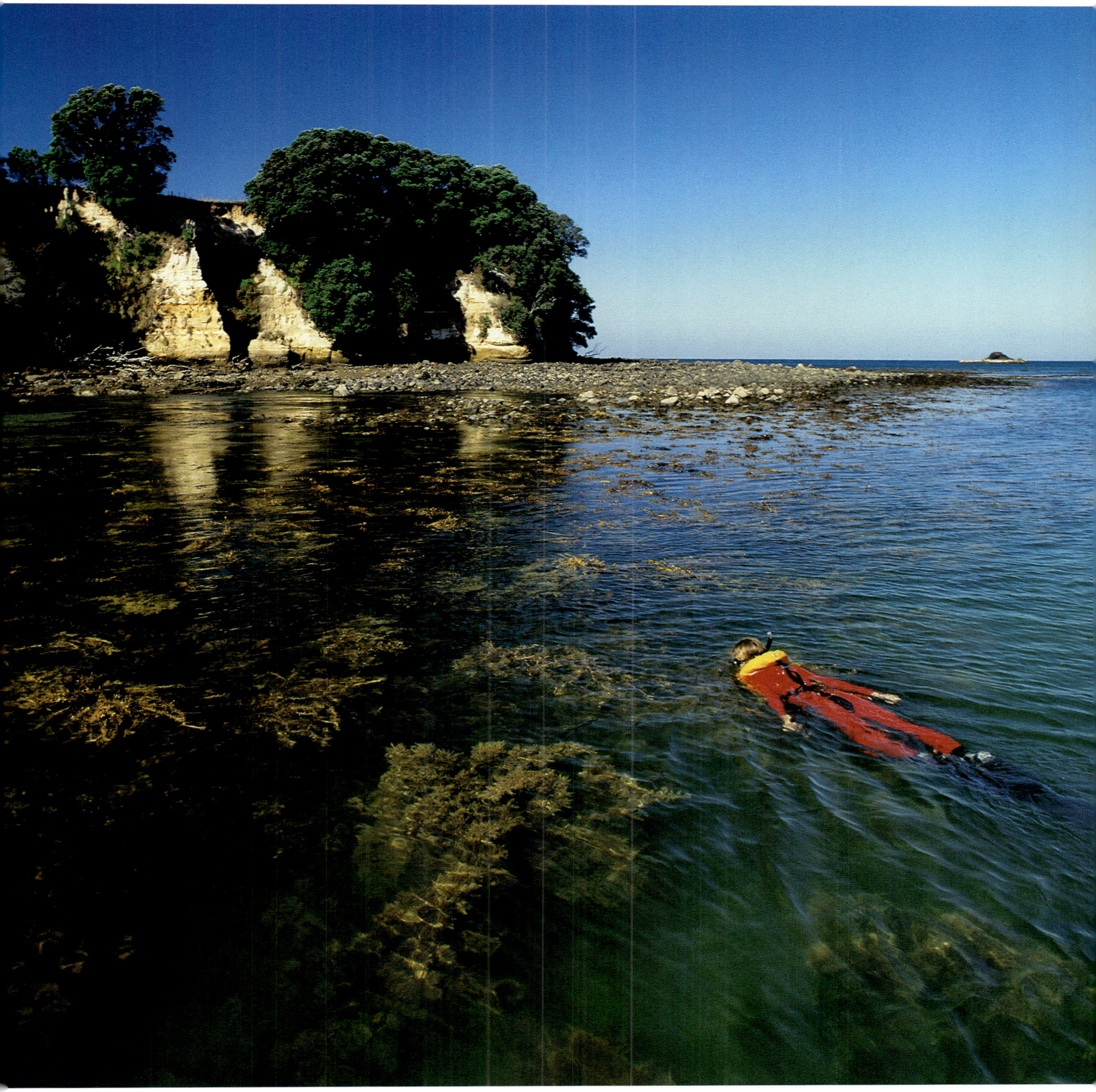

How to get there

Visits to Whale Island are restricted to permit holders, approved tour parties and school groups. Contact DOC office in Whakatane.

Moutohora/ Whale Island

'The petrels could be seen, just before dusk, gathering off the island. The birds were coming back to feed the young that were nearly due to fledge. As dusk fell, birds would skim over the island looking for their burrows, then they would crash through the trees and head for their nest site. Sometimes they liked to sit before going underground. Some birds get caught in the trees as they crash through, and die.'

Many people who have lived on or visited islands when the muttonbirds are coming home to their chicks relate stories similar to this diary entry on a website. This description relates to Moutohora or Whale Island in the Bay of Plenty and the writer is recording a success story. For this is another island that was cleared for cultivation and farmed through the 19th and early 20th centuries with its birdlife decimated as a consequence. Now Moutohora is covered with mainly pohutukawa, mahoe and kanuka forest and the birds are returning – in particular, the shearwaters, both the grey-faced petrel or northern muttonbird, and the sooty shearwater. Little blue penguins come ashore and the variable oystercatcher and the threatened New Zealand dotterel also breed on the island.

Initially known as Motuhora, the name was changed to Moutohora in the early 1990s as a result of submissions from Ngati Awa. Maori were the first settlers to burn the forest and clear land for cultivation on this island – remains of ancient pa sites and stone-faced walls around cultivation areas are still visible – and they also went there to collect the traditional harvest of titi, or muttonbird chicks. This custom continued until the early 1960s when it was apparent that their numbers were in serious decline.

Ngati Awa were the last Maori inhabitants of the island and although they sold the land in the 1860s they still take an active interest in the conservation activities on their ancestral island. It was declared a wildlife refuge in 1965, bought from its private owners by the Crown in 1984 and subsequently administered by the Department of Conservation. Regeneration from bush remnants, combined with an active planting programme, has restored much of the island to its former bush-covered glory.

Tuatara and saddlebacks both inhabited the island before humans destroyed their habitat. According to Maori legend, two tieke or saddlebacks accompanied the waka *Mataatua* from Repanga (Cuvier Island) to Moutohora in ancient times and the recent return of tieke raised on Repanga is of enormous symbolic significance, as well as conservation importance, as is the return of tuatara. Other forest birds thriving on the island include tui, captive-bred parakeet or kakariki, kaka, kiwi and the New Zealand falcon.

The next step in the conservation saga of Whale Island is the possible setting up of a marine reserve in its surrounding waters.

Cleared for farming in the 19th century, Whale Island in the Bay of Plenty is now covered in regenerating bush and is a home for both sea- and bush-birds.

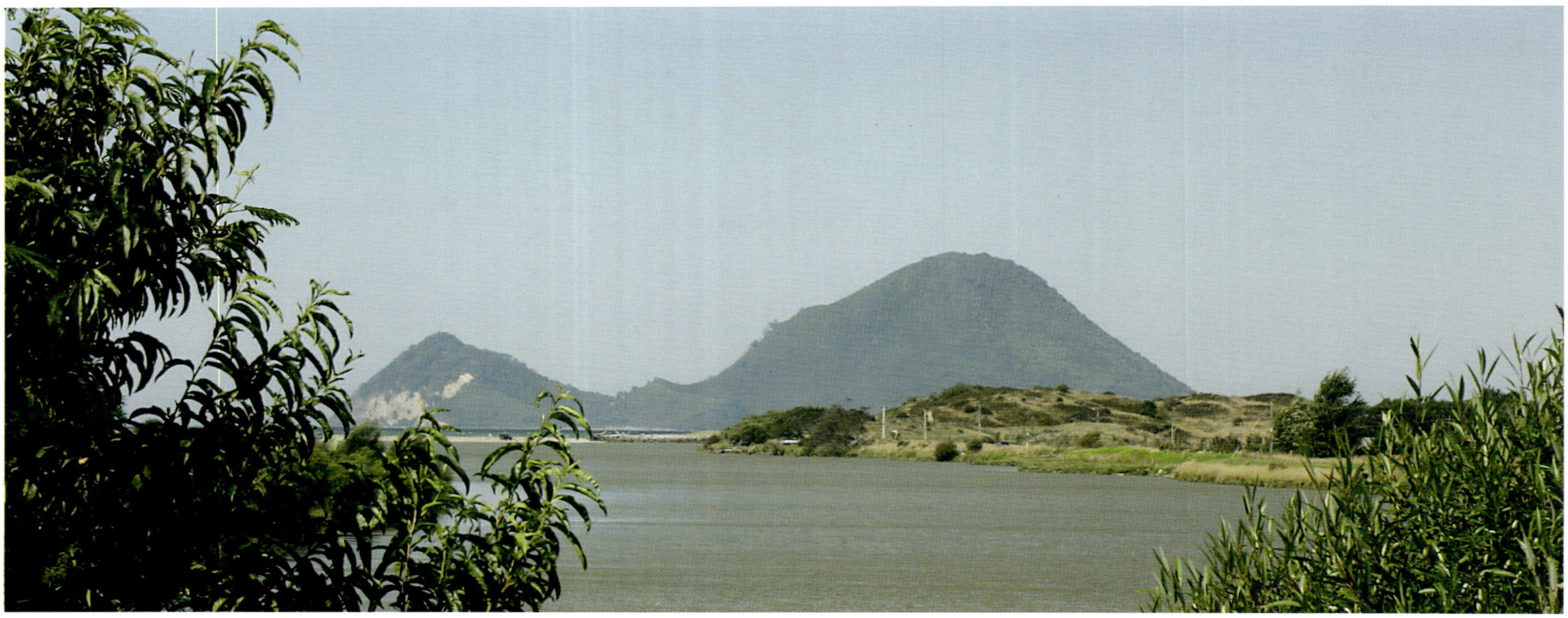

Matakana

The story of Matakana Island is a very different one. Like its neighbours, it was first owned by Maori. Unlike them, however, it has been retained in Maori ownership and the community of 300 living on the island – all members of a large extended family – has a long history there. No bush-covered remnant of volcanic action, Matakana is a long, thin sliver of low-lying land, sprawling across the entrance to the extensive tidal bay that is home to Tauranga and its satellite towns.

Two ferries traverse the narrow stretch of water that separates Matakana from the mainland and people from Tauranga cross to the island to work every day, just as islanders make the journey the reverse way. It's a short trip and often a chance for a neighbourhood chat. Vehicles pull up to the ferry ramp in Omokoroa or at Sulphur Point. There might be a four-wheel drive waiting, filled with the week's groceries or a utility vehicle with a team of sheep dogs in the back or a car packed with family returning home.

The ferry pulls in and the vehicles rumble aboard. Five minutes later it heads back to sea. It hasn't gone far before there's a shout and a woman runs down the ramp. The ferry stops, turns around and returns to shore.

'I had to park my car,' the latecomer explains, catching her breath. 'It's cheaper to leave it here. My uncle,' and she points to a guy on the ferry, 'he should have told them I was coming.' She scrambles aboard amid laughter and jokes and this time the ferry makes the run to Matakana without incident.

In the past, families left the island to find work. Now a few of them are returning. Forestry and a sawmill provide jobs. Homestays are available offering the experience of daily living in a Maori community. Visitors to the Bay of Plenty can take horse-drawn wagon trips around the island.

'Where else would I live?' asks one islander.

'It's the best place,' another confirms. 'I came back a couple of years ago. Came to my grandfather's tangi. Got a job the next day and stayed. It's peaceful – no traffic to worry about. When I feel like it I can jump on my horse. Go hunting. Go fishing, man.'

On the ocean side a long, deserted beach of white sand runs the length of the island. Surf creams on to it from a jade green sea. The rollers here are better than Mount Maunganui across the water, according to some surfers.

At the southern end of Matakana, separated from the farmland on the main island by tidal mudflats, is another island, Rangiwaea. There used to be three productive dairy farms there, but when the cream run stopped and milk tankers became the mode of transport, the farms became uneconomic. Families moved away. The land reverted to wilderness where wild horses and cattle grazed.

In 1980, a group of elders from the island community decided to take action. They organised the amalgamation of all the individual land blocks and set about making the land productive once again. With help from the Maori Affairs Department they planted pine trees and orchards. A wharf was built so machinery could be brought by barge from the mainland and produce could make the return journey. A few years later, electricity was connected. Now maize is grown as a commercial crop, and kiwifruit and avocados are exported. The manager of the enterprise, and son of one of the founders, is looking at ways to expand the business further.

How to get there

A daily ferry service operates between Matakana and Sulphur Point in Tauranga and the wharf at Omokoroa. Holiday accommodation is available on the island and tours of the island can be arranged.
Contact the Tauranga Visitor Information Centre.

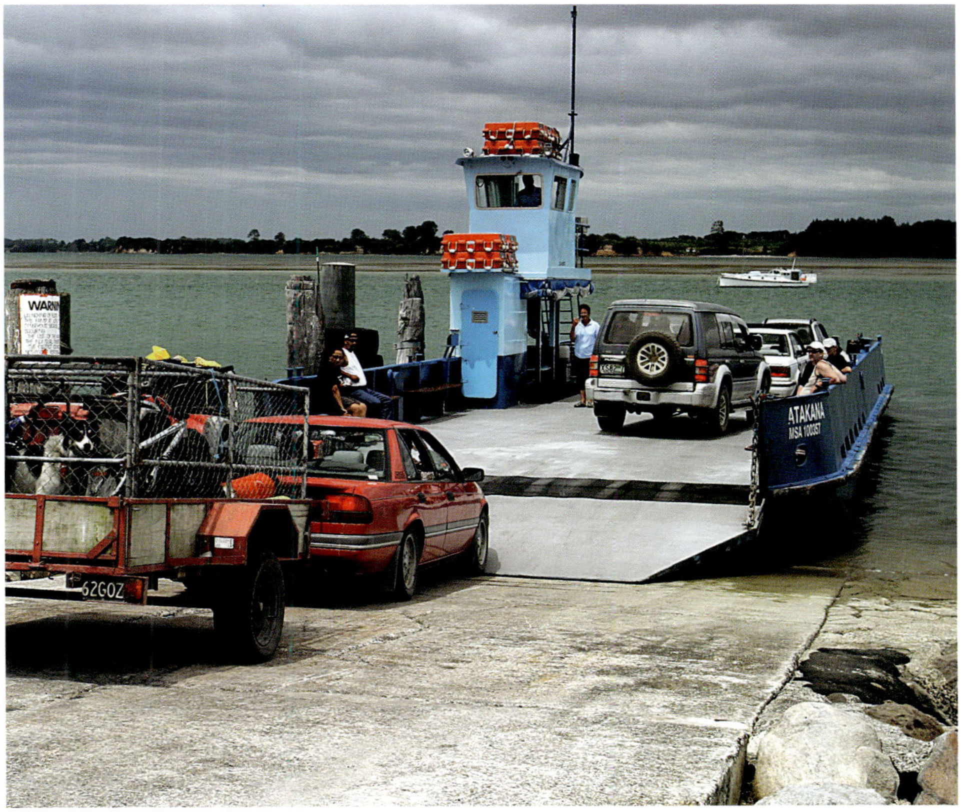

Residents of Matakana Island board the ferry at Omokoroa, ready for the short ride home.

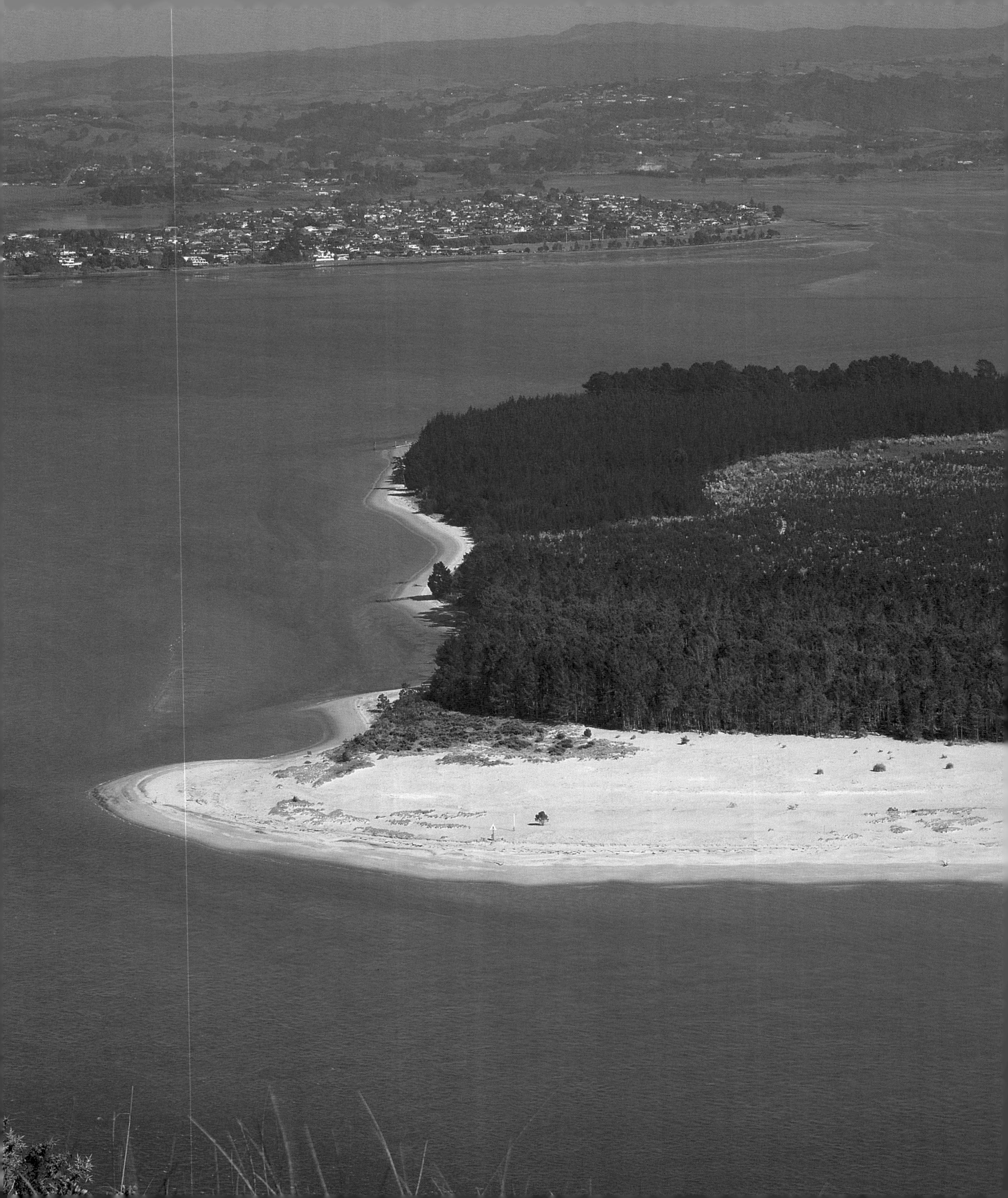

A project that has caught the imagination of the islanders is the Dotterel Recovery Programme. In 1992 the Department of Conservation chose Matakana as one of the breeding sites in the country for the endangered New Zealand dotterel. The department took charge of predator control around nesting sites and set about arousing the interest of the local community by involving the children. They visited the school, taking the children to view the nesting sites and the chicks in their natural habitat. The result is a group of children who have become ardent champions of the birds along with the Matakana Island Trust, which also supports the dotterel programme. Recently a guardian was appointed to keep watch over the birds and educate people about their needs. Their habitat, so near and yet so far from the bustling expansion of Tauranga, seems assured of protection at present.

White Island/Whakaari

It's an early start, the day I go to White Island. 'Bring your bathing suits,' the tour company woman reminded me when she rang the previous night to confirm that it was all go. Arriving at their office at Whakatane at 8 am, I find a crowd of travellers receiving last-minute details of the day's tour.

Nothing is left to chance. Everyone signs a document absolving the company of responsibility should the island erupt while we are on it, annihilating us in a cloud of smoke and ashes. We climb on board the superbly appointed vessel, a crew member goes through the safety drill and on a shining morning we surge through the breakwater that guards the entrance to Whakatane Harbour. Whale Island comes into view. White Island sits far out on the horizon.

Left: Low-lying, long and largely covered in pine forest, Matakana Island runs the length of Tauranga Harbour and protects its many bays and beaches.

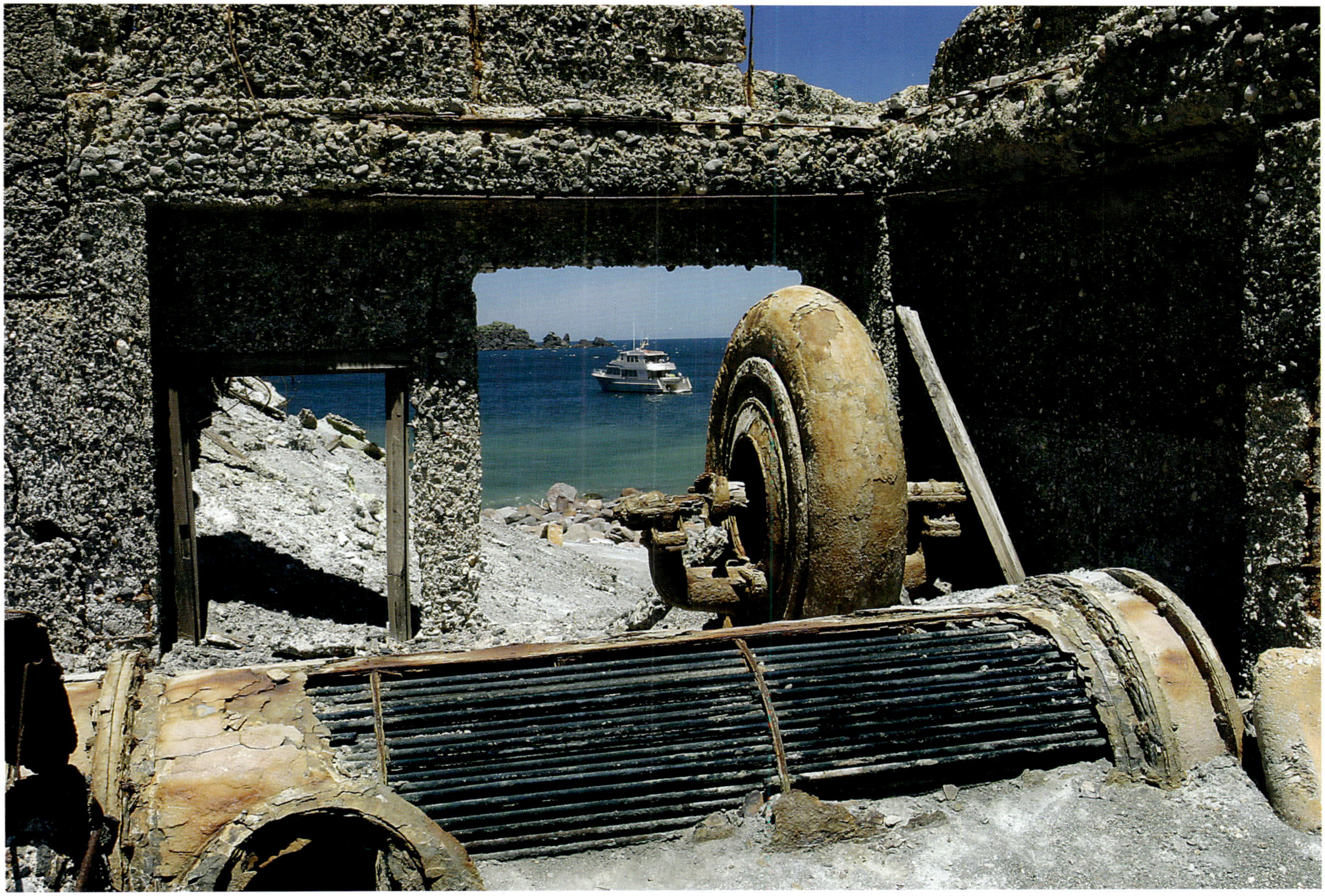

Mute they may be, but these remains of the sulphur works destroyed in 1914 by an eruption on White Island speak eloquently of desolation.

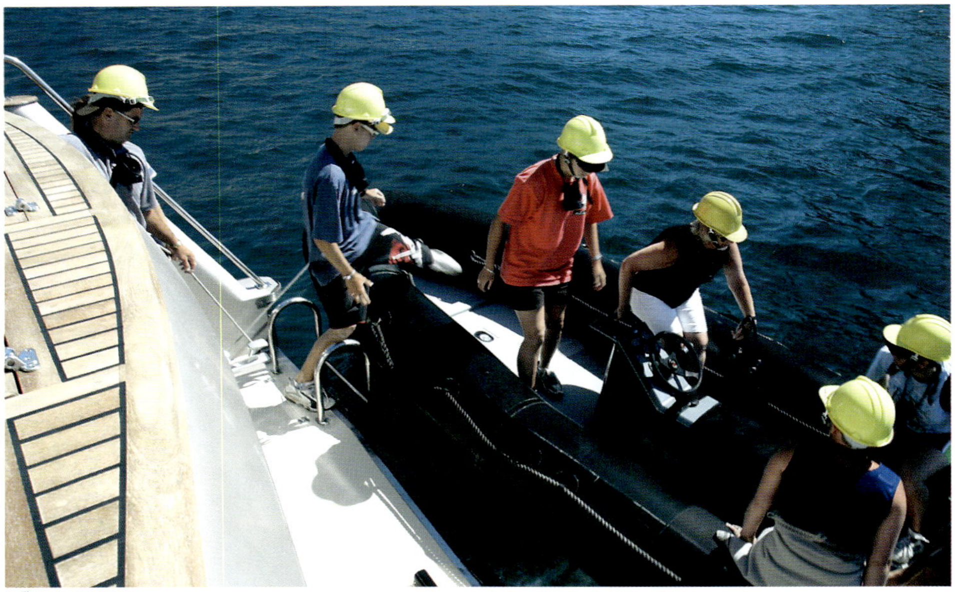

Intrepid visitors prepare to go ashore on White Island.

Right: Residents of Whakatane are accustomed to seeing White Island steaming on a calm day.

To visit White Island is to visit a landscape like no other in New Zealand and probably few others in the world. This is a living, breathing volcano – the realm of a fiery dragon trapped beneath the sea – that could erupt at any time. It did erupt in 2000. In March of that year, the dragon heaved and shook itself and smoked with fury. Three small vents developed in the northern wall of the main crater complex and merged. Clouds of ash poured out and continued for months. Peter Tait, the skipper of our boat, who visits the island on more than 200 days a year, likened walking in it to trudging through powder snow.

On 27 July the island erupted. Red ash turned everything dark, rusty red. Mud and scoria lay 20 centimetres deep, the lake was displaced and a new explosion crater was formed.

This volcano, one of the most fascinating and accessible on earth, is New Zealand's only live marine volcano. As with an iceberg, approximately two-thirds sits under the sea and the main vent is actually below sea level although it is shielded by high crater walls and sealed off from seawater. Built up by continuing activity over the last 150,000 years, the island is in a constant state of change. There are earth tremors every day. The dragon turns and a new fumarole or boiling mud pool appears.

The Institute of Geological and Nuclear Sciences monitors the island's activity. Scientists visit regularly and a solar-assisted seismometer constantly feeds seismic data back to the mainland. A camera in situ transmits photos on an hourly basis. International scientists collaborate with those from IGNS to study the island's eruption history, its volcanic processes and its geothermal system.

Sulphur deposits like this, formed by constantly steaming fumaroles, are common on the island.

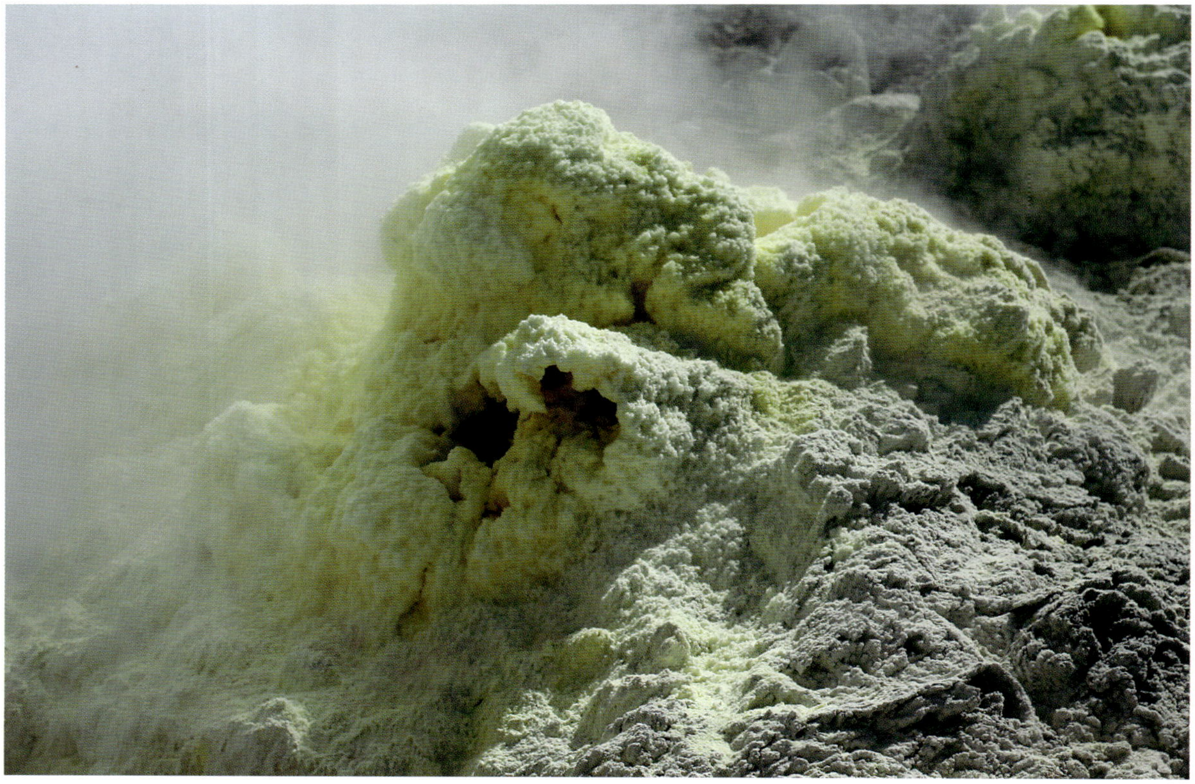

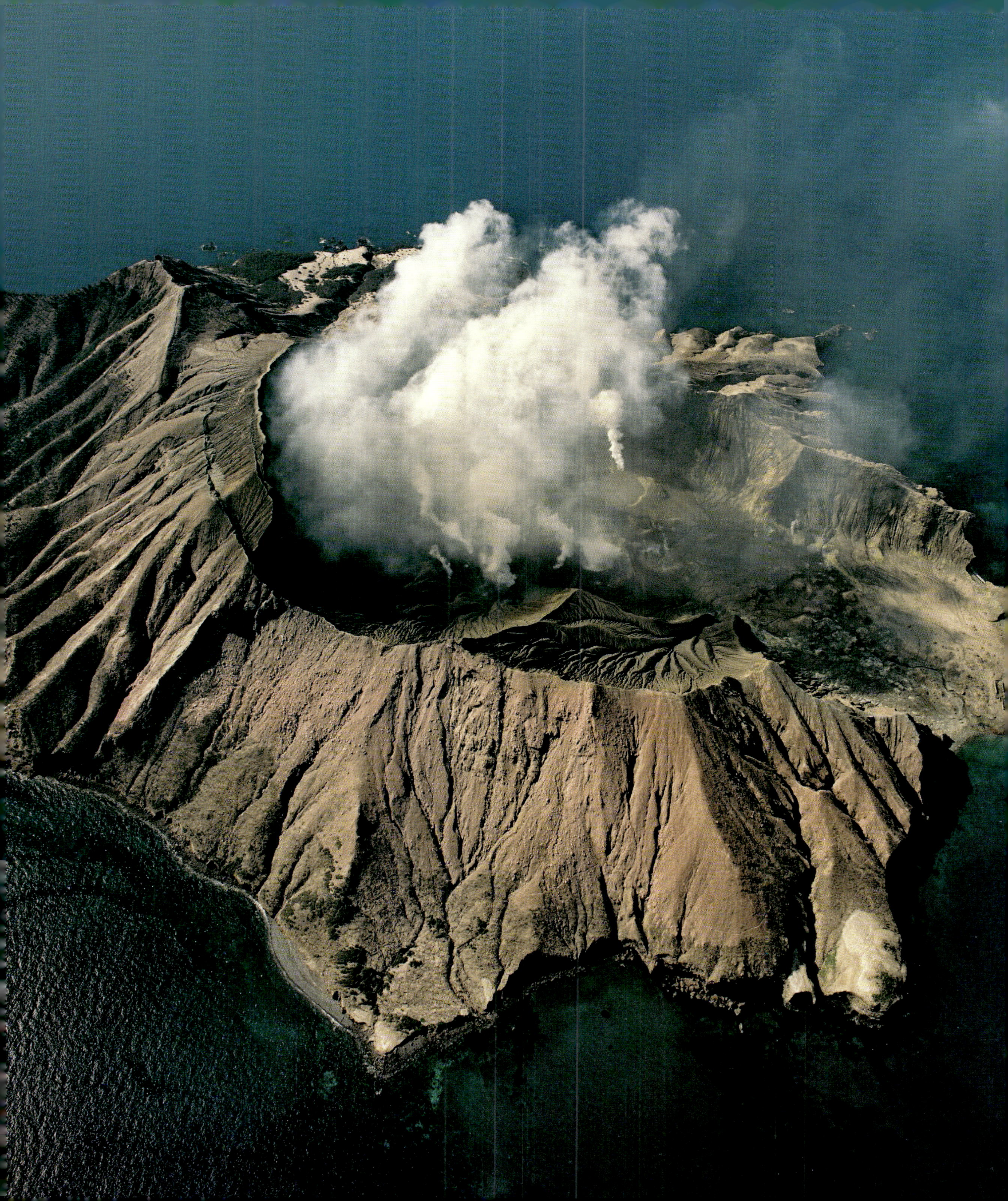

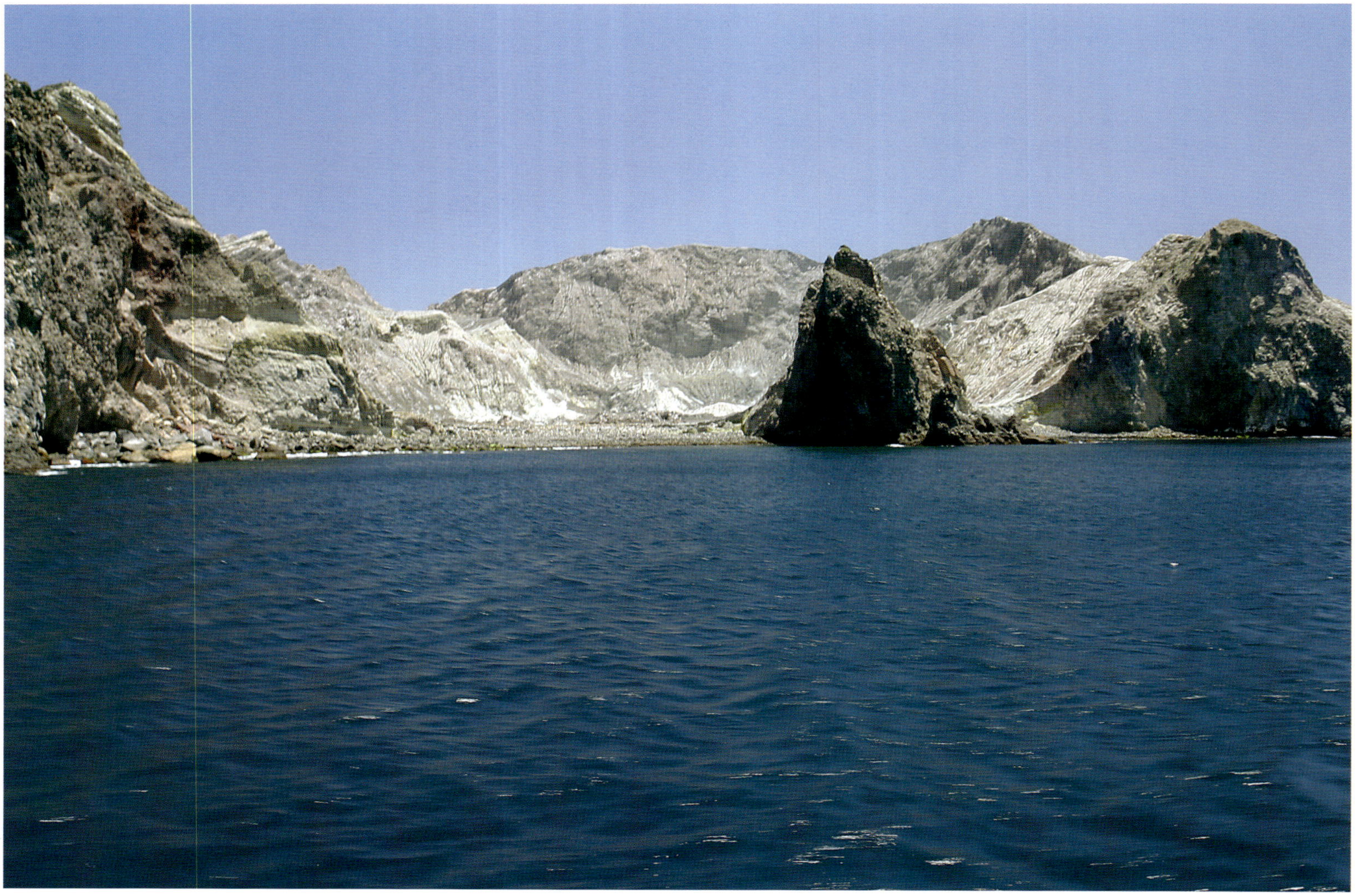

The harsh environment means there is very little vegetation on White Island.

The day I visit, the island is quiet, but residents in the Bay of Plenty are accustomed to seeing a plume of steam and gases rising from the island on clear days.

The first European visitors were slow to realise that White Island or Whakaari ('To Make Visible') was a volcano. Cook called it White Island but made no mention of its volcanic nature. In 1827 the Frenchman Dumont D'Urville, following in Cook's path, sighted the island but assumed the smoke came from the inhabitants' cooking fires.

By the 1870s more accurate information about the island had been collected and men eyed the sulphur deposits with a view to profit. Several attempts were made to extract the mineral – the first in 1885 – but none was conspicuously successful and the enterprise set up in 1913 ended in disaster. A new plant was built on the flat area below the crater, where tourists now come ashore. It consisted of several buildings, including living accommodation, a railway, boilers and other machinery. The following year 10 men were brought from the mainland to live and work on the island. In September 1914 a huge section of the volcanic cone collapsed into the crater, unleashing a lahar of rock and boiling mud that tore through the new installation, destroying everything in its wake and killing all the men. Two weeks after the tragedy, wreckage and debris was washed up on the shore of the mainland.

In spite of this disaster, the sulphur works were re-opened in 1927, with new plant and equipment. The men's camp, in contrast to the earlier operation, was set up around the corner from the crater area and they either rowed to work or, on bad days, walked along a steep path on the crater wall. Supply ships called in every week, weather permitting. In winter the men could go two or three weeks without fresh supplies. Sulphur mining finally stopped on White Island in 1934.

Several people have owned the island since it passed into European hands in the 1830s. One former owner described his purchase as 'the most desolate, God-forsaken and, at the same time, the most awe-inspiring place on the face of the earth'.

George Buttle bought the island in 1936 and when asked why, replied that he 'rather liked the idea of owning a volcano'. The island is still in Buttle family ownership and it became a private scenic reserve in 1953. Tourist operators require permits to take visitors ashore.

As the boat approached the island the crew leapt into action. Everyone was issued with a hard hat and a gas mask with instructions on how to use it. Candies were handed out, to suck in case the stench of sulphur was overpowering. Group by group we stepped into an inflatable tender and were motored ashore.

Desolation is everywhere. Gaunt metal ruins – shocking yet weirdly attractive – are a fitting preview to the bleak moonscape ahead. Dusty scoria scrunches underfoot. Citrus-coloured sulphur crystals glitter in the harsh light. An acrid tang hangs in the air. Changing shades of amber, red and mauve streak the crater walls and in its maw, a lake of milky jade is wreathed in steam. The dragon causes mighty chaos.

For two hours we wander on the island on a strictly monitored route, in awe at the unimaginable savagery wrought by nature at its most violent.

Incredible to think that gannets build nests in ash on the outer wall of the crater and line them with seaweed, that pohutukawa exist on the lower slopes and grey-faced petrels make their home here. Not so incredible to learn that divers come here, as fascinated by the underwater volcanic environment as we are by the landforms.

We return to the boat, dive overboard for a swim and hope that the island doesn't choose this moment to erupt.

How to get there

Private boat charter companies in Whakatane offer daily excursion trips to White Island.
Diving charters are also available.
Helicopter charter companies operate out of Whakatane and Rotorua airports.

A moonscape with a weird beauty all its own.

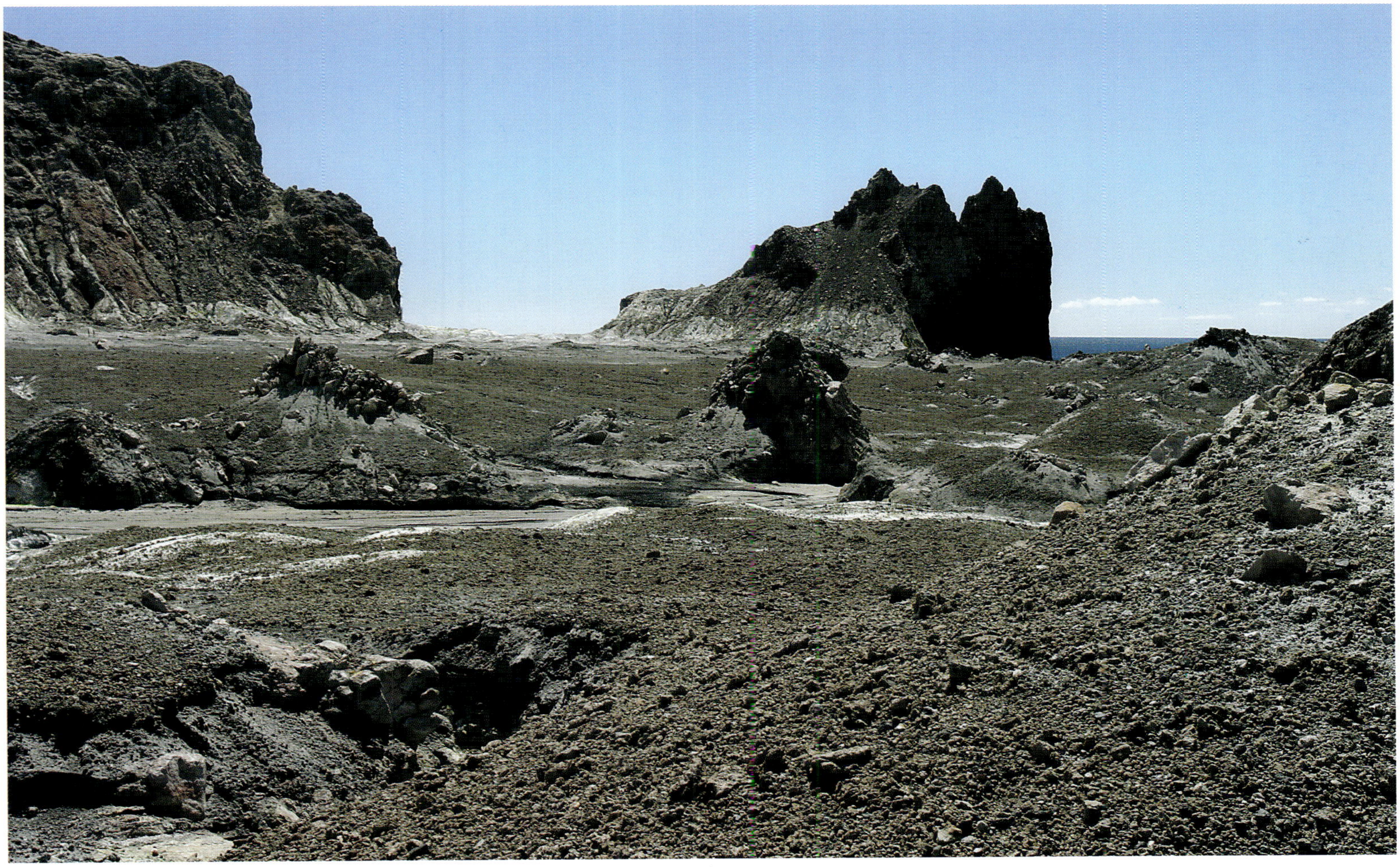

Islands of Refuge

In Europe, in bygone days, people under threat of attack retreated behind castle walls. In New Zealand, they retreated to islands. Some, such as Matiu or Somes Island in Wellington Harbour are relatively substantial and people lived there for extended periods. Others are so small and so rocky that it is incredible to think that anyone actually found shelter there and managed to survive.

Such an island is Tapu te Ranga. So small is the island that it rarely appears on maps, though it has given its name to the Wellington suburb of Island Bay. When a southerly roars in from Cook Strait and rain lashes the coast it almost disappears from sight beneath huge crashing waves.

A hill near its centre has been levelled off, as if to form a lookout platform. On the rocky flat, south of the hill, a small well of brackish water evidently existed. Burnt oven stones, remains of a stone wall, a shell midden and store pit have all been discovered on the island.

Piecing together its story, historians believe that Tapu te Ranga became the refuge briefly of the high-born wife and family of Ngati Ira's chief Whanake in about 1828. Warring tribes from the Waikato came south and when Ngati Ira were attacked the chieftainess Tamairangi and her children were taken to the island where they took shelter in a pa there.

More rugged than Tapu te Ranga are the Sugar Loaf Islands, eroded remnants of early volcanic activity. They shelter the port of New Plymouth and also sheltered local refugees under attack from marauding Waikato tribes in the early decades of the 19th century. The biggest island is a symmetrical cone – the shape evidently of sugar lumps in Captain Cook's day, hence the name. Steep, rocky, inhospitable, it is difficult to imagine anyone living on it, but nearly all the islands in the group, as well as Paritutu onshore – steeper even than Sugar Loaf – show signs of human habitation.

On Saddleback Island, Motumahanga, there is a fortified rock wall. The terrain has been terraced and a kumara pit, neatly lined with rocks, has been uncovered – no doubt used in the past to stockpile food in times of trouble.

Now the islands are a refuge for wildlife. There are colonies of Cook's scurvy grass on Saddleback and terns and seagulls breed there. Black-backed gulls, black shags, blue penguins, gannets and fluttering shearwaters all frequent the islands and a marine park established in 1986 surrounds the Sugar Loaf Islands.

We still need refuges, though warring tribes no longer stalk the country. Today it's the turmoil of city living we often need to escape, and Quarantine Island in Otago Harbour is a modern-day refuge. Once its inhabitants were forcibly confined to its shores. Nowadays, visitors come to Quarantine Island by choice, to find peace and spiritual renewal.

In 1958, James Matheson, a Presbyterian

Sugar Loaf Island, close to New Plymouth, was a temporary place of refuge in the past for Maori under attack.

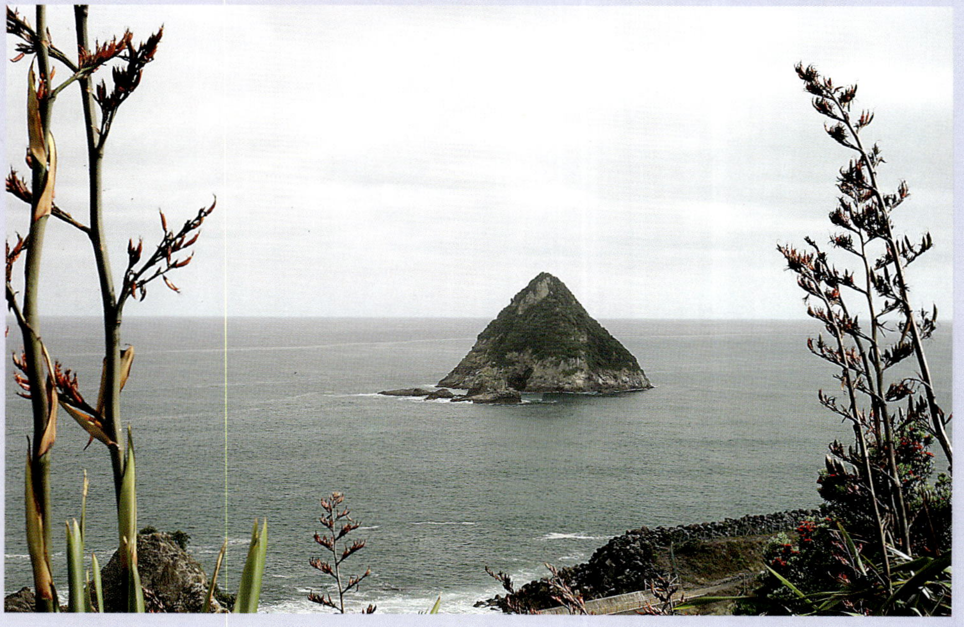

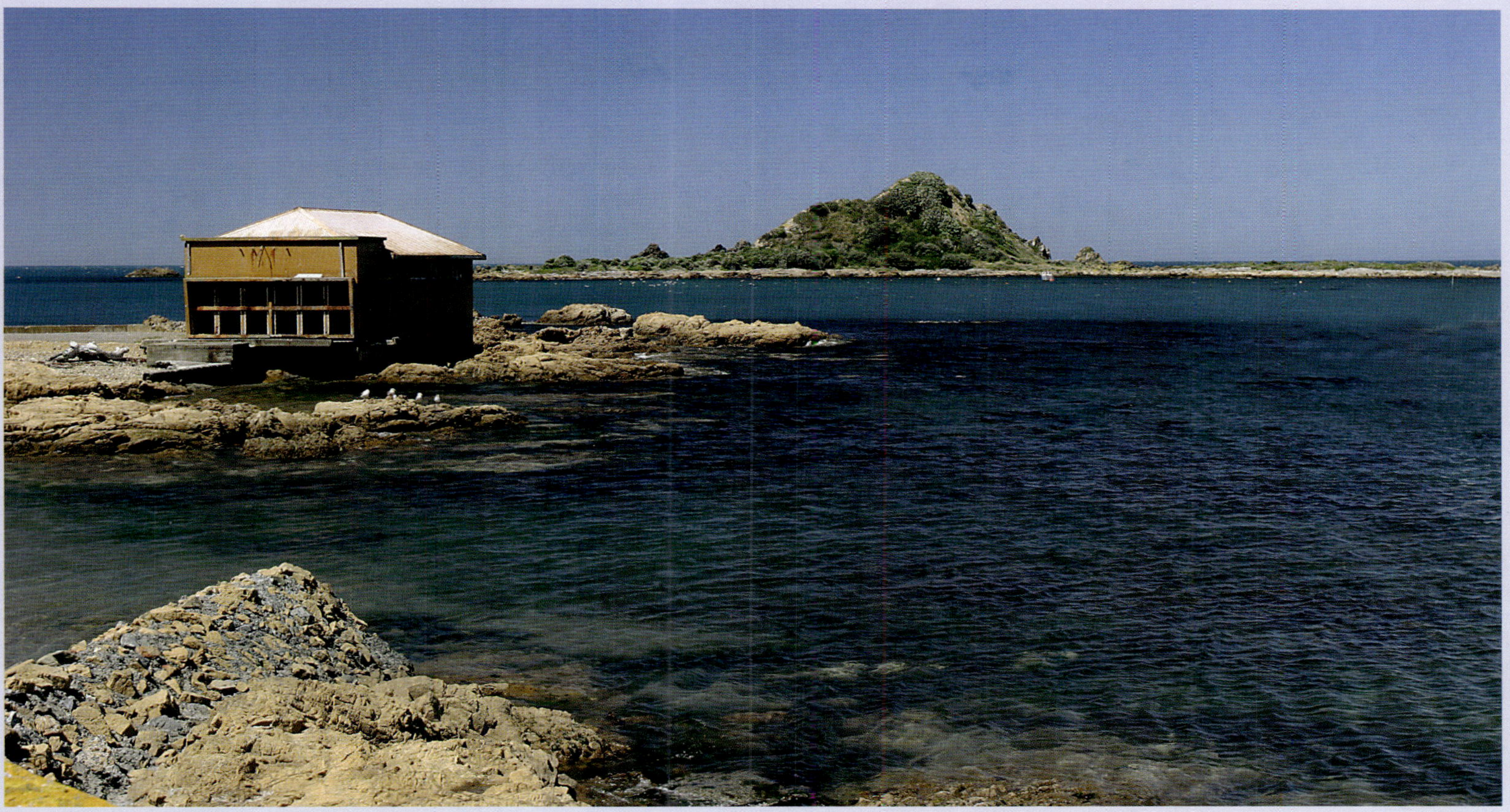

minister from Dunedin's Knox Church, and Dean Hurst from St Paul's Anglican Cathedral, formed the St Martins Community and bought the lease of half the island, which is a recreation reserve.

They aimed to establish a temporary refuge for religious groups and others who needed a quiet place for contemplation, work or study. Inspiration for the idea was drawn from a community established on the Isle of Iona, in Scotland, that promoted the concept of physical work as a complement to worship, based on the principles espoused by St Martin of Tours. It was to be Christian in origin but open to people of all beliefs.

In 1984, Dave Wilson, who had run a drop-in centre in the city, went to live and work permanently on the island. Now Douglas Black and his partner Kathy Morrison live there full-time. 'We originally came for a three-month spell, to help out after Dave died. But we came back to live a few months later. It's a special place. There's a tradition of offering comfort. Anyone is welcome to come here,' Douglas says. 'The community is not about money and it's very important that it doesn't become commercially driven.'

On the Saturday I visited, Kathy came to the mainland to pick me up in their boat. Already several people had arrived at the island to spend the weekend working. They were preparing for the arrival of a Buddhist group who spend a week on the island every year. Their days are spent in silence, hands involved in manual labour, minds turning on spiritual matters.

The chapel is a fitting symbol of what the island community is about. 'To work is to pray. To pray is to work' is the inscription at the entrance. The inspiration for its design is the upturned prow of a boat, the first shelter for anyone coming ashore on an island. Inside is a plaque to the memory of Dave Wilson. Engraved on it are the words 'He cared for many and was a friend to all'.

I wander over the island, visit the old cemetery, gaze down the harbour at the distant view of Dunedin and look out to the end of the island where the Department of Conservation and the Dunedin branch of Forest and Bird are working to re-establish the native bush cover. The only noise is the distant bleating of sheep and the occasional seabird's cry. Peace has replaced past anguish.

When aircraft approach Wellington Airport from the south, passengers can see the tiny island of Tapu Te Ranga just off-shore from Island Bay.

Inside the chapel on Quarantine Island, Otago Harbour.

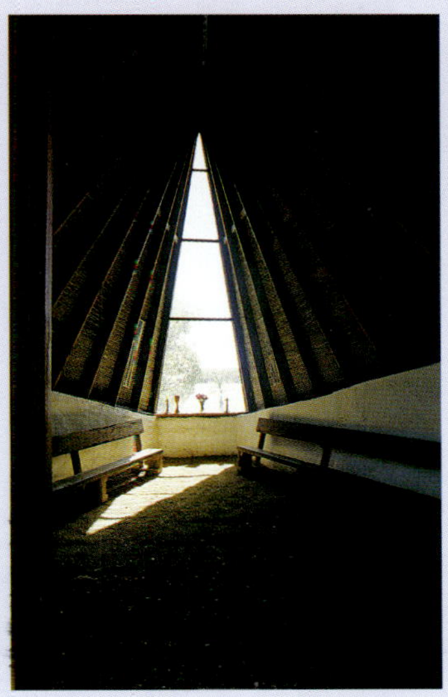

The Islands of Kupe and Te Rauparaha

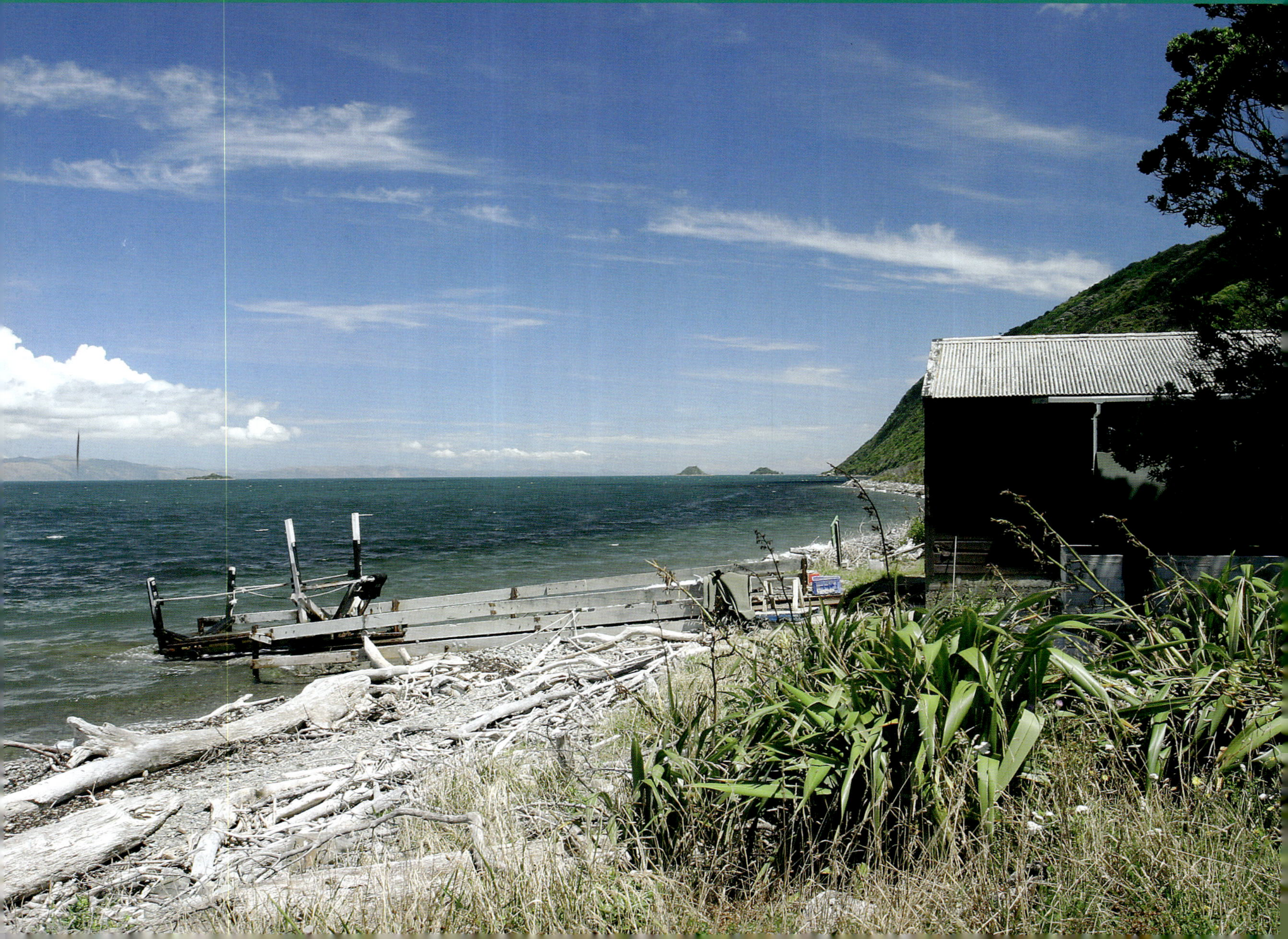

It is easy to think of New Zealand as a long, thin country – North Island at the top, South Island below it, Stewart Island at the bottom. But in reality the top of the South Island and the bottom of the North Island sit side by side. There are extensive areas of the Marlborough Sounds that lie north of Wellington and in pre-European times as well as in the early days of colonial settlement there was much to-ing and fro-ing between the two areas. Canoes, trading vessels and whaling boats made frequent journeys between the islands, big and small. Kapiti Island was a convenient base for much of this commerce as Te Rauparaha and his Ngati Toa tribe found when they staged their raids on the South Island.Maori legend gives Kupe the credit for creating Kapiti and Mana islands as well as Arapawa Island in Marlborough. After discovering Wellington Harbour and naming the two islands there Matiu and Makaro after two of his daughters, he sailed north, severing Kapiti and Mana from the mainland with his paddle as he went. The steep cliffs on the western side of the islands is where his blade sliced.

Kapiti

When you drive north from Wellington it takes almost an hour before Kapiti Island appears sitting out from the west coast, protecting Waikanae from the worst of the winds that sweep in from the south-west. Wellingtonians have taken Kapiti to their hearts. They see it as their slice of primal New Zealand.

Left: Boatshed and ramp on Kapiti Island.

Getting there takes planning – you must book ahead – and a certain amount of courage. There's no wharf at Paraparaumu where the boat departs. It sits on a trailer and visitors climb aboard on dry land. The day I make the crossing the passengers are a mix of New Zealanders and overseas tourists. Several are bird watchers. One couple are regular visitors. Most of the men are bearded. Everyone is issued with life-jackets and a tractor tows us into the waves. Thankfully the weather is calm. Arriving at the island 20 minutes later, the skipper hauls out a gangplank, we clamber onto the stony beach and Kapiti rears above us, a steep bush-covered island with no hint of its early history apparent at first meeting.

Geological evidence suggests that Kapiti may once have been joined to the South Island, part of a land bridge that connected what are now the two main islands of New Zealand, and that it became an island only because it is the summit of a mountain range, drowned in the course of global warming over the last 12,000 years.

Kapiti's history has been determined both by its strategic position at the northern entrance to Cook Strait and its terrain. As one of its former caretakers stated, once on the island, the only way is up – and the tracks are surprisingly steep. Only at the northern end is there more than a border of flat land and much of this is occupied by Okupe Lagoon. Precipitous, almost impregnable cliffs fringe all but the eastern side and these, combined with the fact that there is abundant fresh water and food, make it a natural stronghold.

The first inhabitants probably arrived about 800 years ago and in pre-European times various tribes occupied the island in waves of conquest and displacement.

In the 1820s the pattern of conflict changed. Te Rauparaha of Ngati Toa moved south from his ancestral lands at Kawhia. A natural leader, his tactical skill in warfare was augmented by the acquisition of muskets from the newly arrived Europeans. In 1822 he attacked Kapiti, conquered the inhabitants and set about establishing his base on the island, successfully fighting off a counterattack two years later. He and his people built and occupied three pa on the eastern shore and, using Kapiti as his base, Te Rauparaha carried

How to get there

Visitor numbers are restricted to 50 per day on Kapiti, day trips only. Permits are required from the Wellington DOC office. During holiday periods and weekends it is often necessary to book well ahead.
The boat leaves from Paraparaumu Beach.

Precipitous cliffs on its western side make it easy to understand why Te Rauparaha chose Kapiti Island as the base for his raids further south.

out yearly raids on the South Island, establishing a strong presence for his tribe in the Marlborough Sounds.

Te Rauparaha maintained his 'empire' on the island for more than 20 years. But after he died in 1849 Maori gradually drifted away to the mainland and for the first time in centuries Kapiti was almost deserted.

For the next few decades farming was the main occupation on the island.

By the 1870s, several far-sighted naturalists had come to realise that, with the clearing of trees to provide pastureland, many of New Zealand's native birds were becoming rare. The concept of island sanctuaries for the protection of birds and plants was promoted. Public and parliamentary opinion was slow to grasp the idea but, in 1897, The Kapiti Island Public Reserve Act was passed, following legislation that had conferred similar status on Little Barrier Island and Resolution Island. It established Kapiti as 'a preserve for the flora and fauna of New Zealand'.

Few would argue that the purpose of the Act was commendable but Maori who still owned land on the island were very reluctant to sell what they regarded as their birthright, the sustaining force of their people. The settlement, which included compensation for their loss, laid the ground for future claims under the Waitangi Tribunal. However, Wi Parata and his brother Hemi Matenga were able to retain their family land. During the 1920s their heirs, the Webbers, lived permanently at Waiorua Bay, actively farming and developing the property as a resort that attracted summer visitors, especially tramping club members. With the death of Hona Webber and the outbreak of the Second World War, his widow moved to the Waikanae area, without relinquishing Waiorua, but profitable farming became more and more difficult to sustain and finally ceased in the 1950s.

Despite government efforts to induce them to sell, the Webbers have held on tenaciously to their land. A dialogue is continuing between the

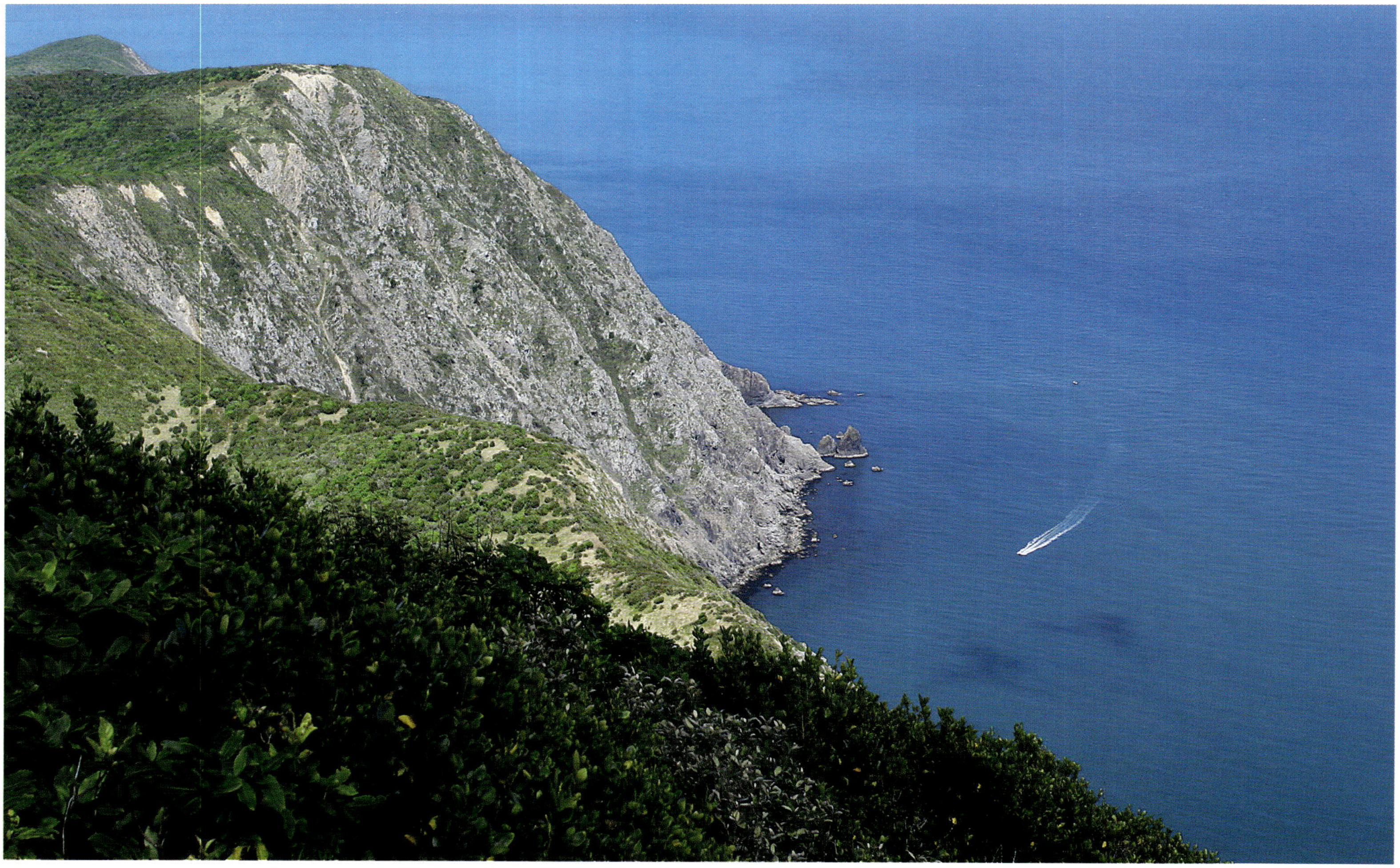

Department of Conservation and the various members of the extended family over a management plan that will protect the sanctuary and satisfy the needs of the family to exploit their birthright.

Once the island had been declared a reserve further action was slow until a permanent caretaker was appointed in 1906. Shortly thereafter, Leonard Cockayne, botanist and pioneering ecologist, was employed to visit the island and carry out a botanical survey. His recommendations, which included restricting visitors to the island, destroying foreign animals and introducing native birds from all over the country, are the principles on which the island has been developed and is still administered today.

It wasn't until the 1980s that an attempt was made to totally eradicate introduced pests; a mind-boggling exercise when the terrain is taken into account. Chris Maclean in his book *Kapiti* gives a revealing description of tracking down and killing the final possum. Between February 1980 and November 1986, 22,500 possums were taken from Kapiti, the first-ever successful possum eradication.

Rats were eradicated in 1996 and only a few months after their disappearance, lizards and insects were already becoming more noticeable in the undergrowth. Ongoing vigilance will always be necessary – it would take only one pregnant rat to land on the island and all the years of hard work, all the bird populations that have gradually become established, would be put at risk.

The first birds were transferred to Kapiti about the time the first caretaker was appointed. One of these species, the little spotted kiwi, is now extinct on the mainland but is alive and well on this island. So successful has its survival rate been that descendants of the first arrivals have started colonies on other island sanctuaries.

From 1981, a series of experimental transfers were made of endangered saddlebacks (tieke) and stitchbirds (hihi). Both are thriving, though special feeding stations have been set up to allow the honey-eating stitchbirds to feed in the face of competition from the more aggressive tui and bellbirds.

Weka are common and usually appear out of the bush as soon as visitors arrive. During summer kaka frequent the Rangatira area, occasionally even landing on visitors' heads or shoulders, and New Zealand pigeons fly overhead. Other birds, such as parakeets (kakariki), the North Island robin and tomtit are more often seen in the bush. Recently, takahe and kokako, two birds that appeared bound for extinction not so long ago, have been introduced.

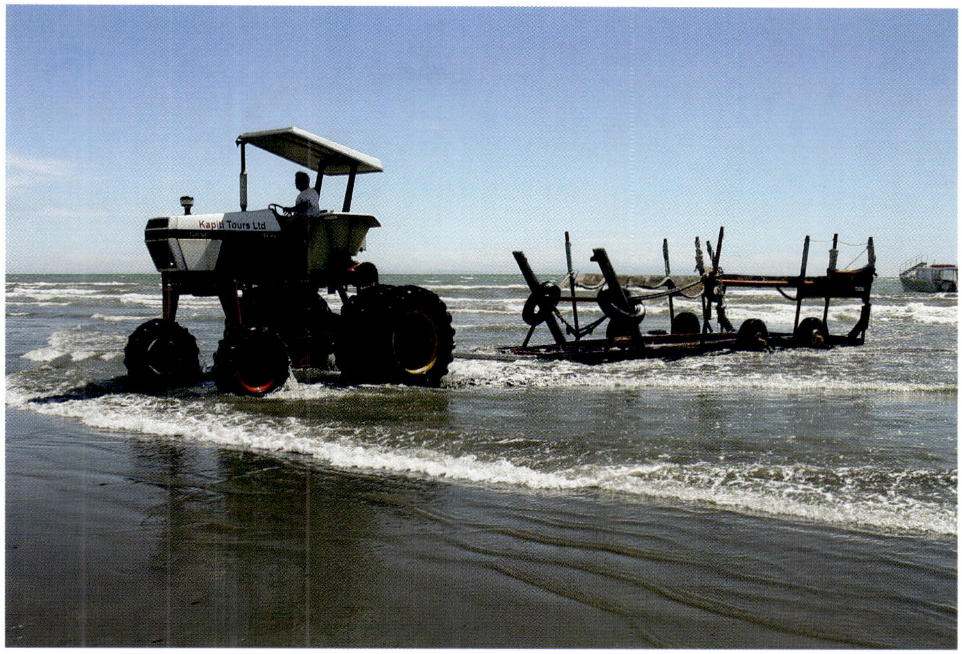

A specially adapted tractor tows the Kapiti boat and its trailer into and out of the waves.

Of course the birds would not survive without the bush. Most of the huge rata and podocarps such as matai and miro that were common before the arrival of humans have disappeared. The main forests on the island today are kohekohe, tawa and kanuka. Some plants, such as karo (*Pittosporum crassifolium*), have been introduced to provide food for the nectar-eating birds; karaka (*Corynocarpus laevigatus*) is a summer fruiting species and many birds eat the orange-yellow flesh of the berries, though only New Zealand pigeons actually pass the seeds and help to increase the tree's distribution.

Visitors may only spend a day on the island. After an introductory talk by the Department of Conservation ranger I decide to hike to the top. It's a steep climb and there's no respite. I'm glad possum eradication was not my job!

The lower reaches of the bush are punctuated by young nikau palms. In the upper areas of forest where cloud often sits, the bush cover becomes denser. The nikau disappear and tall tawa trees predominate. Suddenly I arrive at the summit and the contrasts are startling. Sudden cliffs drop away at my feet, forbiddingly steep, the manuka on their flanks sculpted by the wind to a hard-wearing cloak. North and south the blue-green hills roll away to the sea in a series of ranges, their

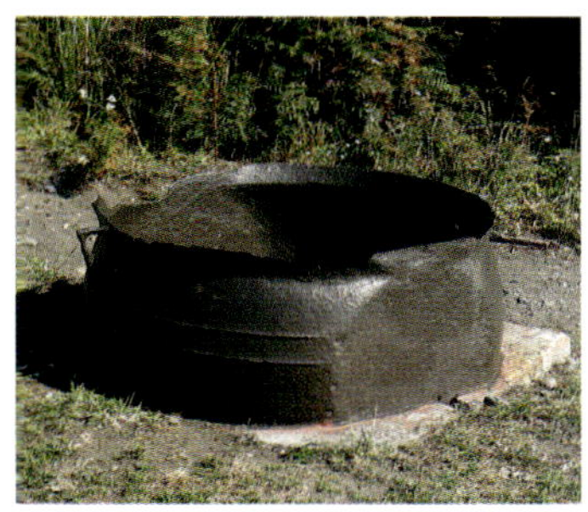

Various relics of the whaling days used to be common on Kapiti. This trypot was one of two that were rescued before they disintegrated and is now on view in front of the visitors' centre.

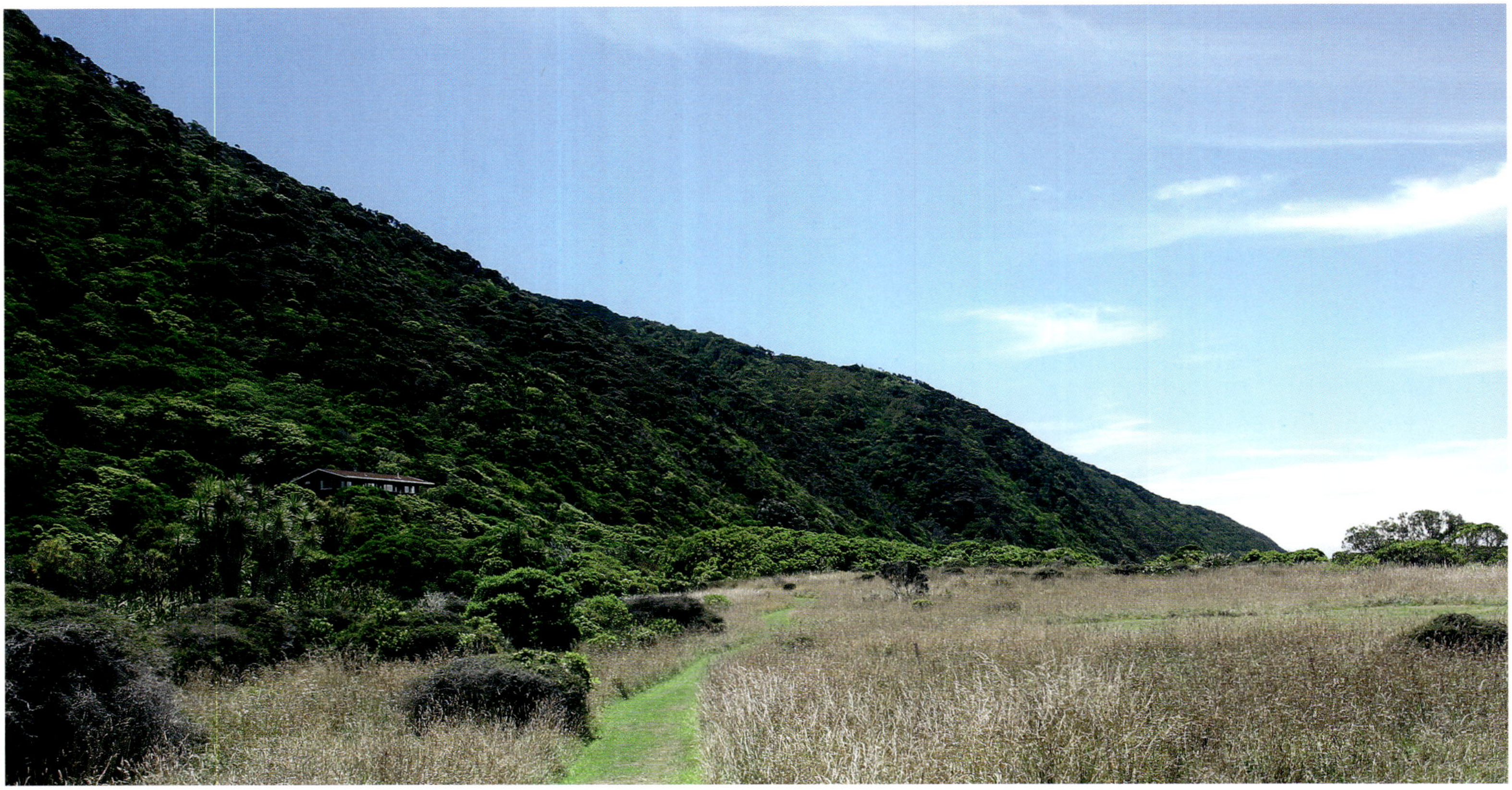

The ranger's house on Kapiti, situated in the bush not far from the beach, gets a superb view towards Paraparaumu.

outlines softened by an unbroken mantle of bush. On the way down I stop to watch stitchbirds feeding. Fantails dart through the bush and somewhere, far off, there's the screech of a kaka. The turbulent times of Te Rauparaha are difficult to imagine. It is little wonder that thousands of visitors come to Kapiti every year to spend time wandering in a secluded world where birds are the main inhabitants and the trees are left to grow at their own pace.

Mana

STEEP-SIDED AND FLAT TOPPED, MANA ISLAND FROM A DISTANCE LOOKS LIKE A LONE TABLE SET IN THE SEA. ITS HISTORY REFLECTS THAT OF KAPITI, ITS NORTHERN NEIGHBOUR, AND IT IS CLEAR FROM ARTEFACTS FOUND IN ARCHAEOLOGICAL DIGS ON THE ISLAND THAT MAORI LIVED THERE AT LEAST AS EARLY AS THE 14TH CENTURY.

When Te Rauparaha swept down from the north and made his base on Kapiti he and his nephew Te Rangihaeata also spent time living on Mana, building a small village and cultivating kumara. At around the same time a few Europeans arrived on Mana, attracted by the whaling possibilities in the area. In the 1830s John Bell, one of the first settlers, started clearing vegetation to create one of the earliest pastoral farms in the country. He brought merino sheep to the island, and cattle, which he used to supply meat and milk to the whaling station. The woolshed he built still stands and is one of the historic sites on the island that the Department of Conservation is committed to preserving. Fleeces from Mana Island provided the wool for one of the first shipments to be exported.

In 1863, a lighthouse was built on the northern end of Mana, but its tenure there was short-lived. Two shipwrecks on the coast were blamed on confusion between the Mana light and the light at Pencarrow and in 1877 the lighthouse was removed and re-erected at Cape Egmont.

The original land sales on Mana led to complications and finally in 1865 the Crown purchased the island, leasing it out with a right of renewal, which could be withdrawn if the island was needed for defence. Two generations of the Vella family, of Italian descent, farmed successfully on the island for nearly 70 years and became well-known in the area, establishing a homestead on the mainland as well as on Mana.

In the 1970s Mana Island was taken over by the Lands and Survey Department in conjunction with the Ministry of Agriculture and Fisheries. It was used by Ruakura Research Station for research on sheep, but after scrapie disease broke out the entire flock on the island had to be destroyed and strict quarantine was maintained for the following five years.

Since 1987 Mana has been maintained as a scientific reserve. A massive replanting project was started with the aim of restoring bush and forest cover with the hope that the island would become a refuge for a number of endangered species.

Rats, fortunately, have never been a problem on Mana but the island was overrun by mice, which were finally eradicated in 1990.

As part of the Department of Conservation's 20-year planting programme, community groups such as Forest and Bird and Friends of Mana Island, established in 1998, have become involved in the restoration. Project Crimson has provided funding to plant northern rata, which had almost disappeared from the island. So far the project has been a resounding success.

Reforestation has been confined to the valleys and slopes, leaving sufficient grassland area to support takahe that were first transferred to Mana from Fiordland when DOC took over its administration. A colony is now thriving there. Other birds released on the island include the North Island robin, diving petrels, the brown teal and, most recently, fairy prions from Stephens Island.

Mana is also a refuge for the Cook Strait giant weta, McGregor's skink and the goldstripe gecko.

How to get there

Visitors are welcome to visit Mana Island between 8am and 5pm.

Boats must remain offshore after visitors have landed but dinghies can be drawn up on the beach.

Fishing and excursion trips available by charter out of the town of Mana.

Appearing flat as a table top from the coast, Mana Island nevertheless offers spectacular views from its several walking tracks. A historic lighthouse and a concrete gannet colony are also of interest to visitors.

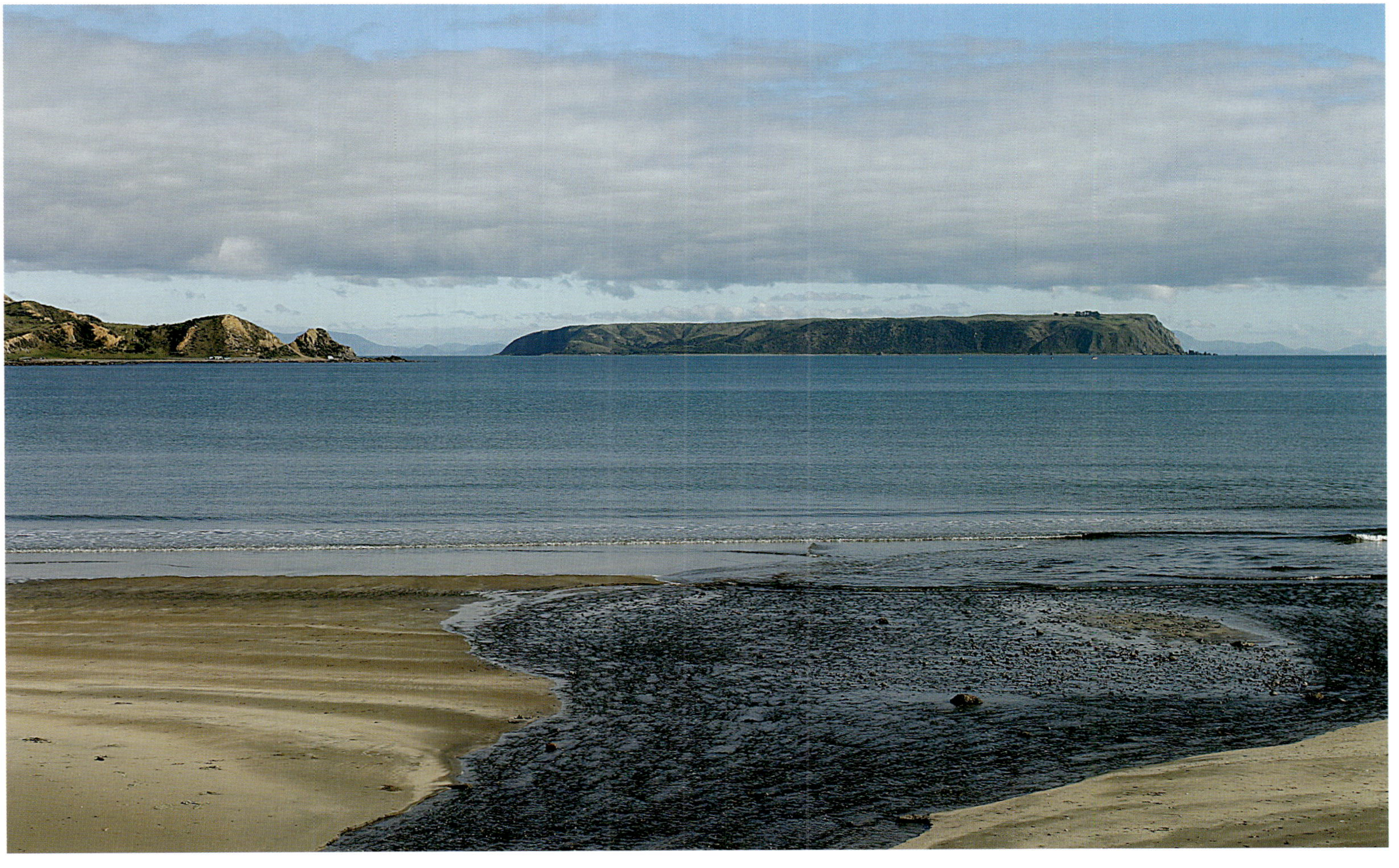

Matiu/Somes Island

JANUARY. SUN GLITTERS ON THE SEA. VISITORS THRONG THE CAPITAL. IN DOWNTOWN I JOIN THE CROWD OF JOSTLING TOURISTS WAITING FOR THE FERRY THAT TAKES DAY-TRIPPERS TO MATIU OR SOMES ISLAND, THIS MOST CONTROVERSIAL OF ISLANDS SITTING IN WELLINGTON HARBOUR.

Right: Once situated on a bare headland, the lighthouse, with Wellington's western suburbs behind, is now surrounded by regenerating bush.

Visitors to Matiu/Somes arrive at this wharf and all bags are inspected for rats before people are free to wander where they will.

Controversial because for periods in its history people were forced to spend time there against their will and at other times Wellingtonians campaigned in vain for the right to visit the island in their leisure time. Only since 1995 has the general public enjoyed free access.

As we step ashore from the ferry, seabirds scream overhead and a huge black-backed gull surveys the scene from atop an old macrocarpa. The caretaker arrives on his quad bike and in a few minutes we have disembarked and assembled in the Kiore Whare for a bag inspection. Matiu/Somes is pest-free and it's in everyone's interests that it remain so. A quick introduction to the history of the island and then we're free to spend the day wandering at will.

Few islands in New Zealand have been witness to so much despair, so much frustration and so much hope as Matiu/Somes. The despair is evident as we climb the track towards the top and reach the cairn erected to commemorate all those who are buried on the island, victims of infectious diseases. The names and dates tell the story.

Mary Elizabeth Maude Rudman, aged two years, died 20 July 1872, victim no doubt of the smallpox infecting the immigrant ship *England* that sailed into Wellington Harbour that year reporting a contagious disease on board. Mysterious new illnesses occurring nowadays help us to understand the dread such outbreaks must have induced. In 1872, panic seized the population. The crew and passengers, forbidden to land on the mainland, were unceremoniously disembarked at Somes Island. Makeshift housing was provided for them and a purpose-built quarantine station was constructed.

1874 was the worst year for quarantined immigrants. Ten children died – of measles, whooping cough, mumps, pneumonia or typhoid – and were buried on the island. The names of the settlers who died are recorded on the cairn along with those of six sailors who died in 1919, victims of the deadly 'flu pandemic that swept the world at the end of World War I.

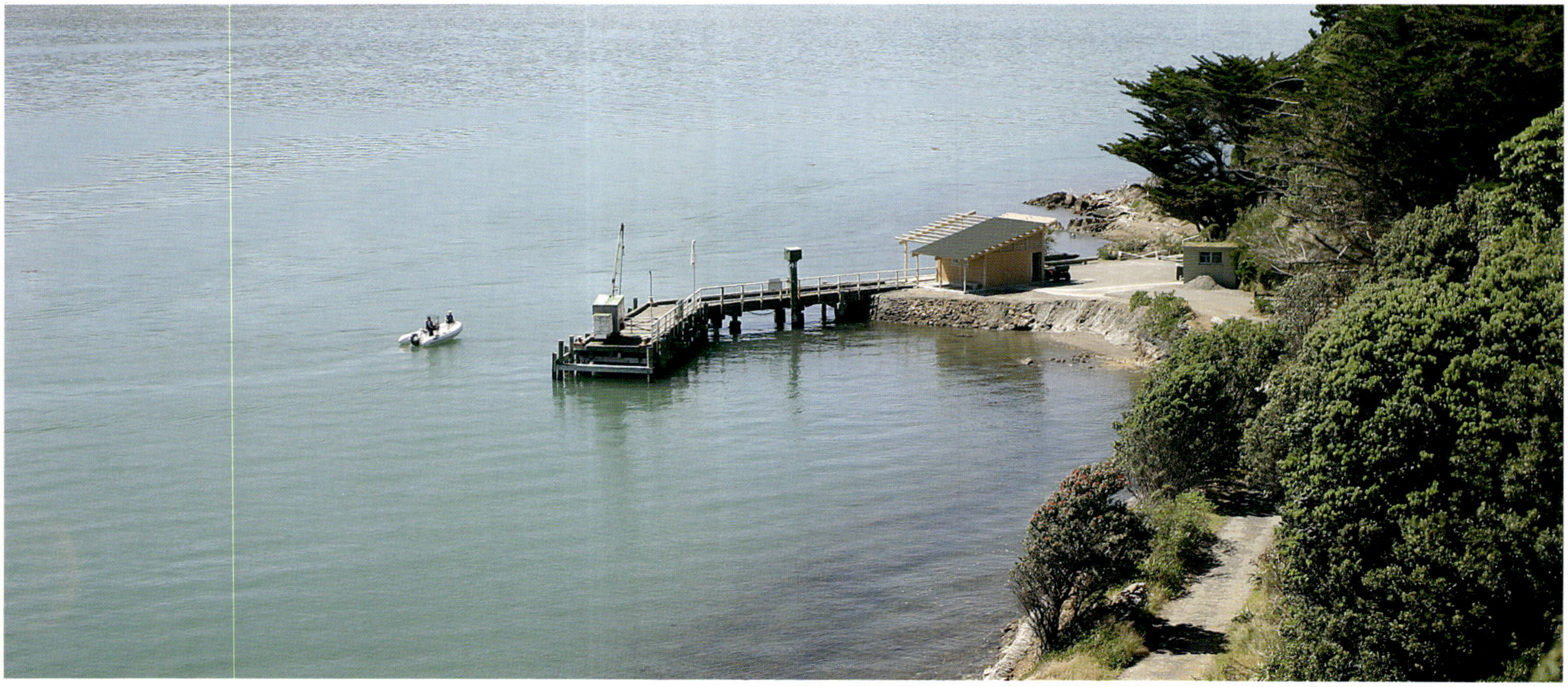

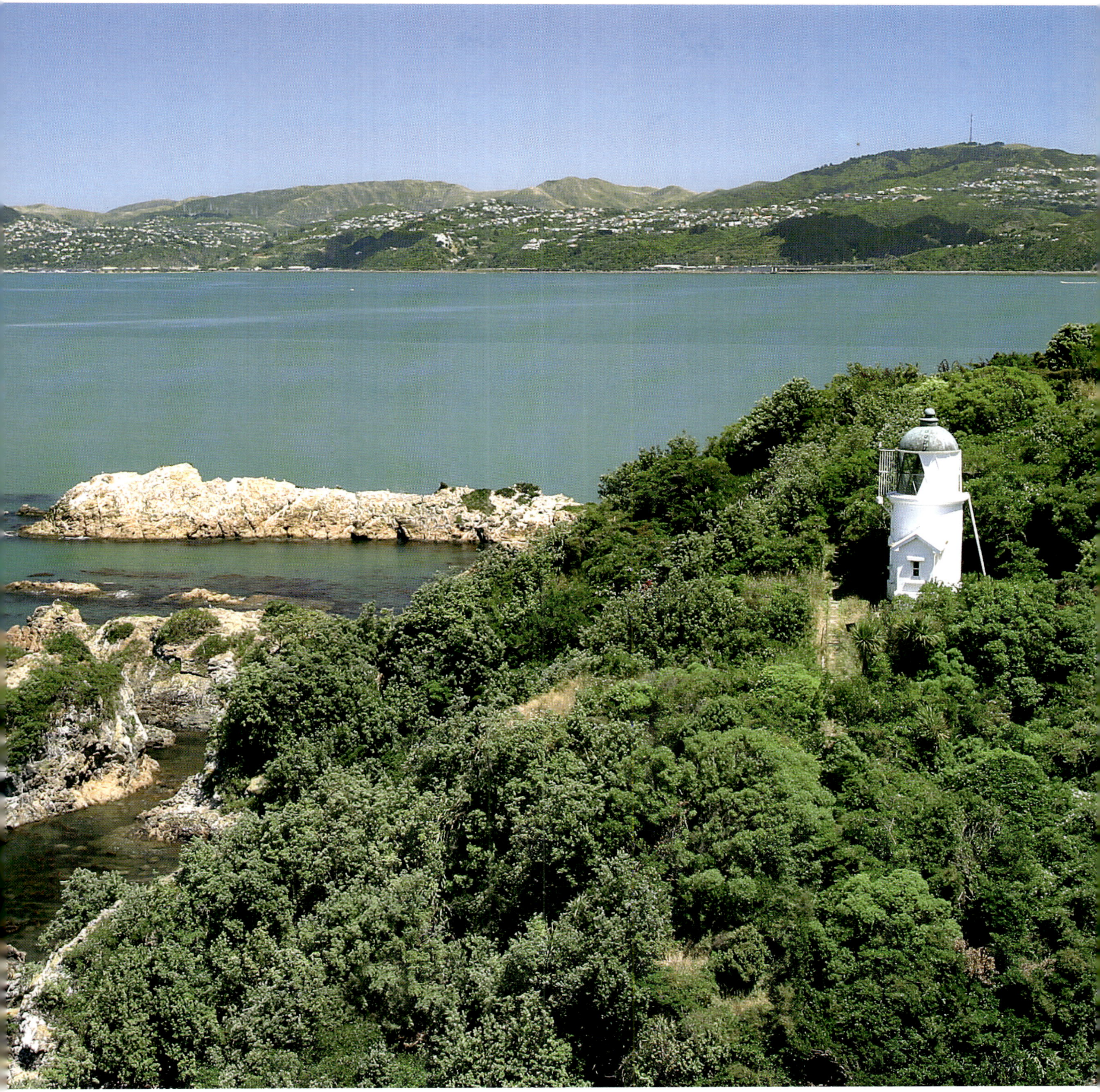

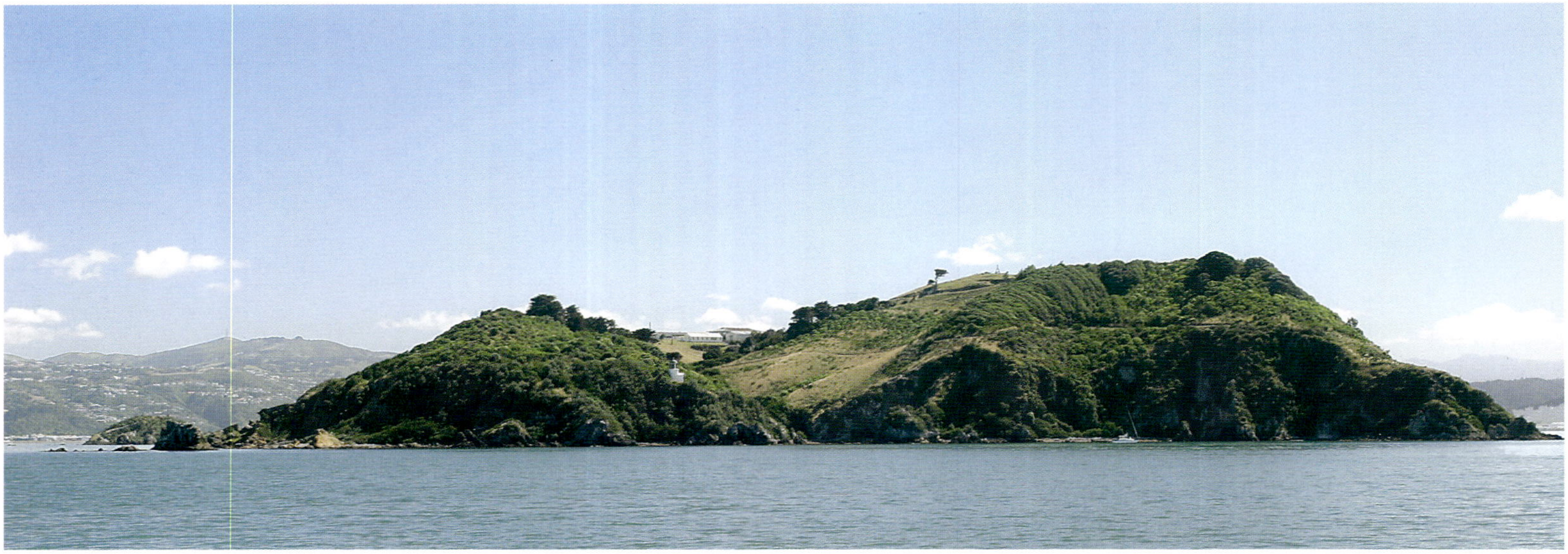

For decades, Somes Island was off-limits to visitors and Wellingtonians could only enjoy 'their' island from afar. Now it is an open reserve and visitors can go there to picnic, to wander round the island and spend a day by the sea.

The burial dates stop there and the human quarantine station officially closed in 1920, though animal quarantine facilities remained until 1995.

The island was also used as an internment camp, the first time during the First World War when aliens were detained – mainly German or Austrian immigrants who were suspected of sympathy with the Kaiser's cause. Twenty years later, the saga was repeated. I spend time reading the inscription on a memorial cairn: *In remembrance of the Italian internees who were held on Somes Island during World War II.* There are 38 names engraved on a tablet.

No doubt those confined in the past took little joy from the magnificent panoramic views at the top of the island, but today's visitors are entranced. The city glitters in the distance, numerous small craft criss-cross the harbour and the *Inter-islander* sweeps past on its journey south.

Navigation within the harbour limits in the early days of the colony was fraught with difficulty. Concerns for the safety of ships led to the construction of a lighthouse in 1865 at the southern end of the island, along with cottages to house the keepers and their families. Replaced in 1900, the light still operates today but the keepers and their cottages have long gone.

Between the world wars Wellingtonians were eager to visit the island. They wanted to see for themselves where the internees had been kept. But any efforts to open it to day-trippers were quashed, both then and after the Second World War. Its use as an animal quarantine station, along with the risk of fire, was used as an excuse to keep the island isolated in spite of imaginative suggestions for an alternative use. In 1979, someone proposed setting up a gambling casino. The next year another optimist envisaged a mink farm. Neither eventuated.

In 1981, as a result of ongoing lobbying to regain recreational and conservation access to the island, the Lower Hutt branch of Forest and Bird began the first of many planting expeditions. In 1989 rats were finally eradicated from Somes and in 1995 the Department of Conservation took over the island. It was officially renamed Matiu/Somes and declared a scientific and scenic reserve.

Now as you walk around the island, native trees and shrubs abound. Pohutukawa, karo, ngaio, kauri, rimu, kowhai and totara are all thriving. At a lower level olearia, flax, taupata and aciphyllas flourish. In a period of 30 years, Forest and Bird has planted more than 100,000 trees and their aim is to cover the island in forest. Black-backed gulls, variable oystercatchers, the blue reef heron and blue penguins are all regular visitors and recently tuatara have come back to the island, transferred from The Brothers with the help of DOC.

Makaro or Ward Island is a tiny, one-hectare island in the middle of Wellington Harbour, a popular picnicking place for cruising yachties. Designated a recreation reserve, it has a breeding population of white-fronted terns and is home to blue penguins, variable oystercatchers and black-backed gulls. Relatively intact because no one has ever lived on it, the island is also rat free and provides a good habitat for a range of lizards. But it is relatively unknown compared to its big sister, Matiu/Somes.

How to get there

Matiu/Somes Island is open to the public every day between 8.30am and 5pm and there is a regular daily ferry service from the Wellington waterfront. No public accommodation is available on the island.

Islands of the Marlborough Sounds

LOOK AT A MAP OF THE MARLBOROUGH SOUNDS, AT THE TOP OF THE SOUTH ISLAND, AND THE COUNTRY APPEARS FRAYED. JAGGED SLIVERS OF LAND STRAGGLE INTO THE SEA. PUZZLE-SHAPED PIECES DETACH THEMSELVES FROM THE MAINLAND AND FORM RAGGED ISLANDS. WATERWAYS RESEMBLE FALSE LEADS IN A MAZE.

It takes a searching eye to detect whether they eventuate in open sea or if they merely trace a complicated path between tentacles of crumpled land.

Arrive in the Sounds by ferry from Wellington and the view is much more straightforward. Long fingers of sea reach between bush-covered hills, the water is turquoise in its intensity near the shore and the ship slips past a series of bays populated here and there with scattered houses, tantalising in their tranquillity or lonely in their inaccessibility.

For centuries various craft have plied these waters. Firstly it was canoes, their people seeking

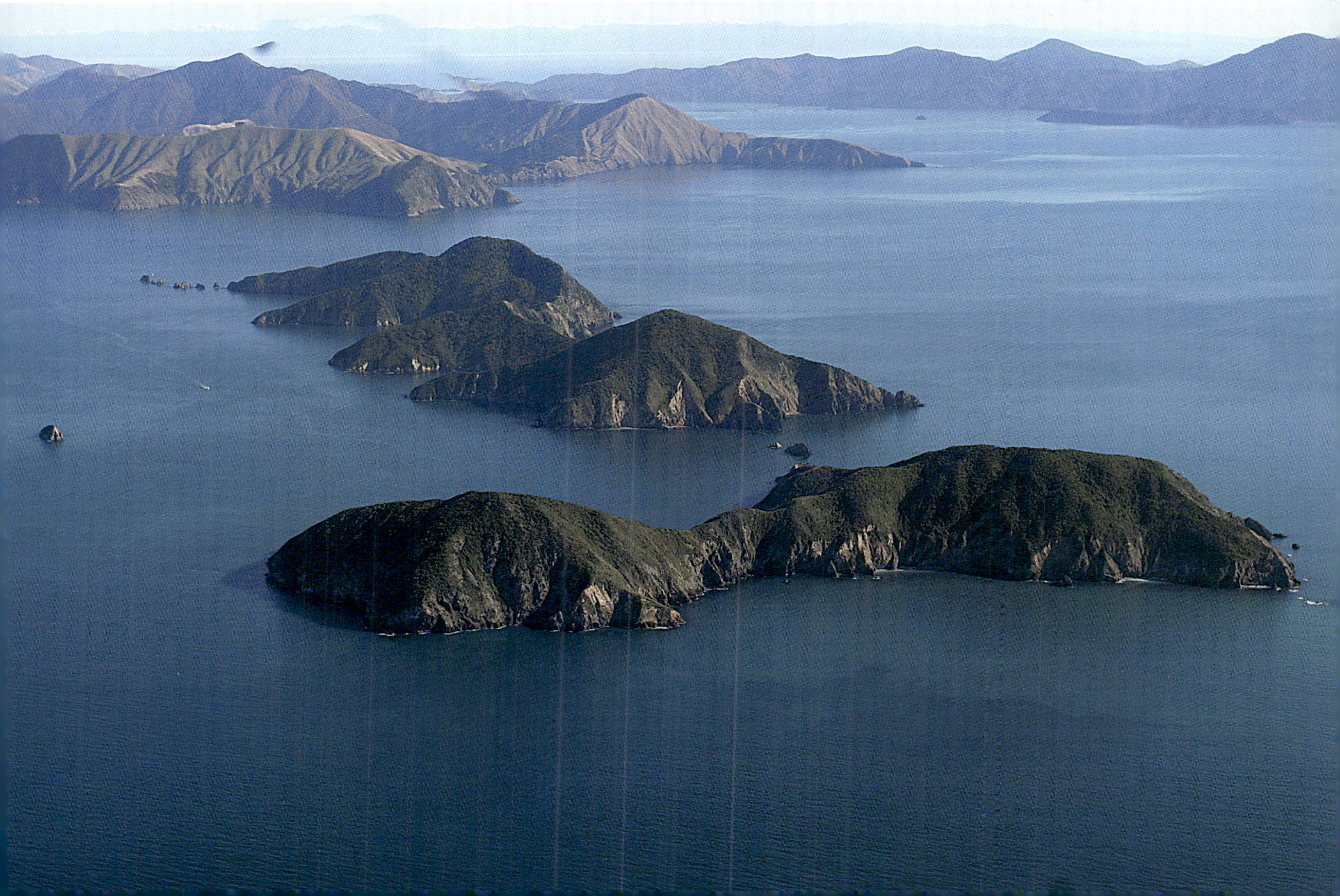

The Chetwode Islands, at the northern extremity of the Marlborough Sounds.

Right: An idyllic scene from the sheltered waterways of Queen Charlotte Sound.

refuge, seasonal food, stone for tools or plunder. Later came the sailing vessels of Europeans, their goals similar. More recently the Sounds have become a recreational playground and only with a boat and unlimited time is it possible to thoroughly explore this disorderly complex of waterways.

Maori legend provides a picturesque rationale for the shape of the broken land mass, describing it as the shattered prow of an intricately carved canoe in which the gods came down from heaven.

The scientific explanation is less romantic if more believable. Unlike many others in New Zealand these islands are not the results of volcanic activity. The Sounds have been formed by two processes. Firstly, the region is being tilted north, causing an uplift of the mountains of the Richmond and Bryant Ranges in the south and the sinking of the earth's crust from the Sounds region and north to Wanganui – a result of New Zealand being situated over the boundary of the Pacific and Australian tectonic plates. Imagine the action of a giant see-saw.

The second process is sea level changes, connected with climate warming over the 12,000 years since the last Ice Age. These changes have led to the drowning of large valleys, and the many islands are the highest points remaining above the waterline.

Visitors come here to enjoy the scenery and the activities associated with a unique system of waterways. A number of the islands are administered by the Department of Conservation and have been restored to their natural state. They offer a glimpse of primeval New Zealand before farming wrought enormous changes on the landscape and wildlife. Others support a lifestyle achieved by hard labour and a desire for independence. A few provide relaxed hideaways for their private owners or resort holidays for those who can afford to pay.

On a day when the sun shimmers on the water and you step ashore to the song of a bellbird and the humming of cicadas, it's not hard to imagine that this is a corner of paradise. It provided a welcome refuge when Abel Tasman spent five days sheltering on the eastern coast of D'Urville Island in 1642. Captain Cook must have also found it a tranquil paradise when he sailed into the Sounds in January 1770, seeking fresh food for his crew and a place to carry out much-needed maintenance on the *Endeavour*. He spent 23 days there, moored in a bay on the mainland, which he named Ship Cove, and returned four more times in subsequent trips to New Zealand. Ship Cove was where he discovered the plant we call Cook's scurvy grass (*Lepidium oleraceum*), which provided much-needed vitamin C for his sailors. In fact it looks nothing like grass but is a member of the cabbage family, sometimes described as a cress, and has become a rare plant mostly confined now to offshore islands.

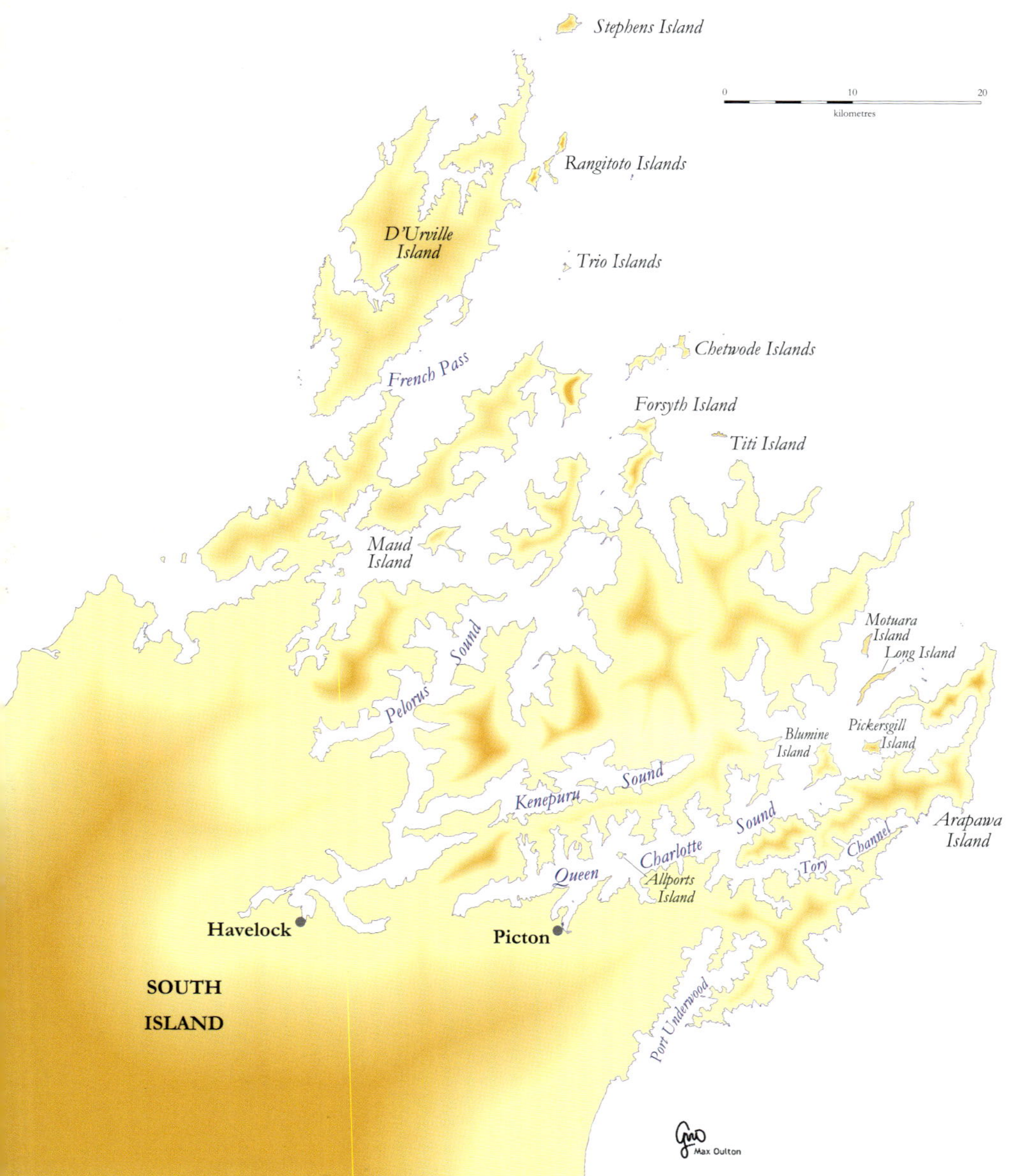

Motuara

Two weeks after his arrival Cook climbed to the top of Motuara Island, opposite Ship Cove and, raising the British flag, claimed the South Island for 'king and country', as he was wont to do. The long inlet of water that gave him shelter, named Totaranui by Maori of the area, he rechristened

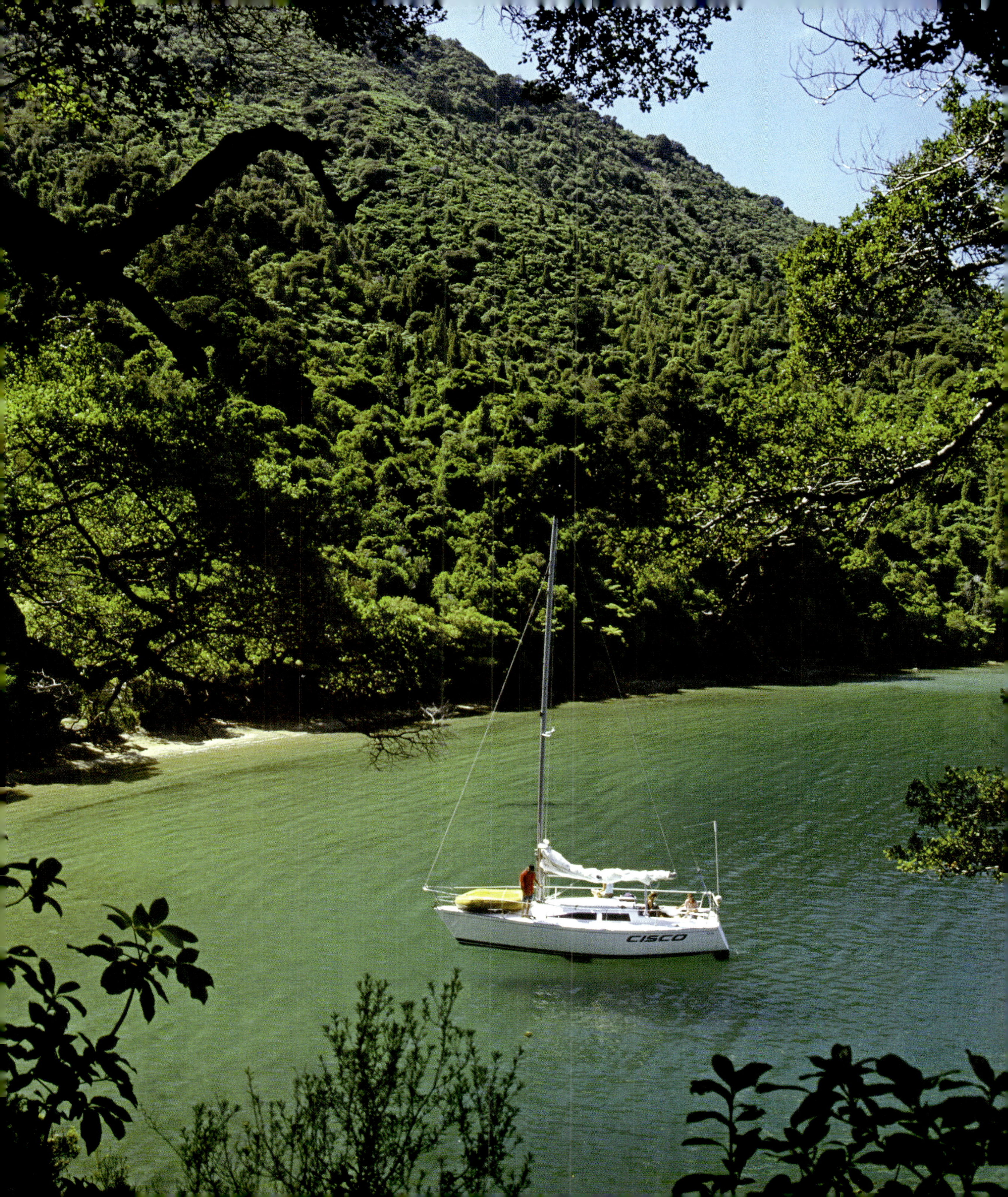
CISCO

How to get there

Motuara, Long and Blumine Islands are open to the public but all are conservation islands and certain regulations apply.
The Brothers, Stephens and Maud Islands are DOC conservation reserves. Public access is not normally allowed and landings are by permit only.
A private company in French Pass has a concession to land on the Chetwode Islands; otherwise landing is by permit only.
The Rangitotos are privately owned.

Queen Charlotte Sound after the wife of the King of England, George III. A plaque at the top of the island commemorates his visit. It is thought that Cook created gardens there to grow fresh vegetables and also established a hospital for his men. In 1777 Cook released rabbits on the island, the first to be introduced to the country, though they were reputed to die on the island. Sixty years later the Maori chief Te Rauparaha evidently used Motuara as a staging post for his raids on South Island tribes.

Cleared of native bush in the early years of the 20th century, Motuara was farmed briefly, and sheep and angora goats roamed at will. In 1926 it was declared a scientific reserve; the native forest has re-established and since the final eradication of pests in 1991 it has become an important conservation island. Now saddlebacks fossick in the bush, New Zealand pigeons crash through the trees and fantails, South Island bush robins and bellbirds are permanent residents. If Cook were to return today, he would no doubt find an island very similar to that which he knew so long ago, though the Maori tribe who reputedly lived on Hipa Island at the southern tip of Motuara has long since moved on.

Tread quietly if you take the track to the top of the island: DOC has built nesting boxes for little blue penguins along the way and they need peace to raise their young.

'Island in the Path' is a rough translation of the island's Maori name and it is the first of several small islands that boats pass in Queen Charlotte Sound as they head towards Picton.

Long, Pickersgill and Blumine Islands

Further into Queen Charlotte Sound are three more small islands – Long, Pickersgill, and Blumine – each of them important for the role they play in conservation. Long Island is a thin, elongated strip of land, also visited by Captain Cook. He reported seeing a small fortified pa on the island, but by the time of European colonisation all trace of it had disappeared. The native forest cover was burned off and for a brief period the island was farmed, with limited success. Now free from introduced predators and designated a scenic reserve, it is covered in second-generation native bush and supports a wide range of sea and forest birds, including the little spotted kiwi. A marine reserve surrounds Long Island and the neighbouring Kokomohua Islands.

During the Second World War, several large gun emplacements were built in the Marlborough Sounds with the intention of protecting an anchorage for the United States navy in Queen Charlotte Sound. In the event, the navy did not

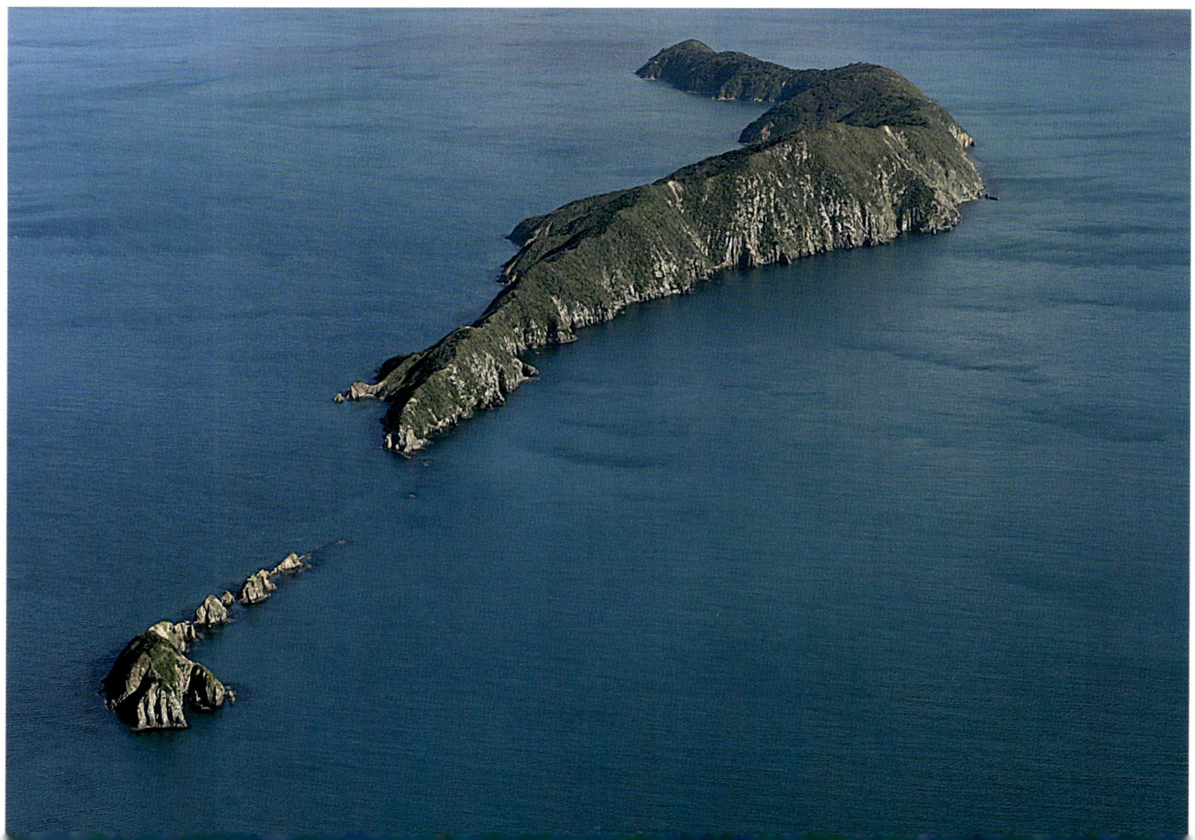

Long Island in Queen Charlotte Sound.

Exploring the Marlborough Sounds

Picton is the principal tourist centre for the Marlborough Sounds.
Charter boats, water taxis and kayaks are available for a wide variety of tours within the Sounds. Enquire at the Picton Visitor Information Centre for details.
Charter flights with Sounds Air are available from Picton Airport or Wellington Airport.

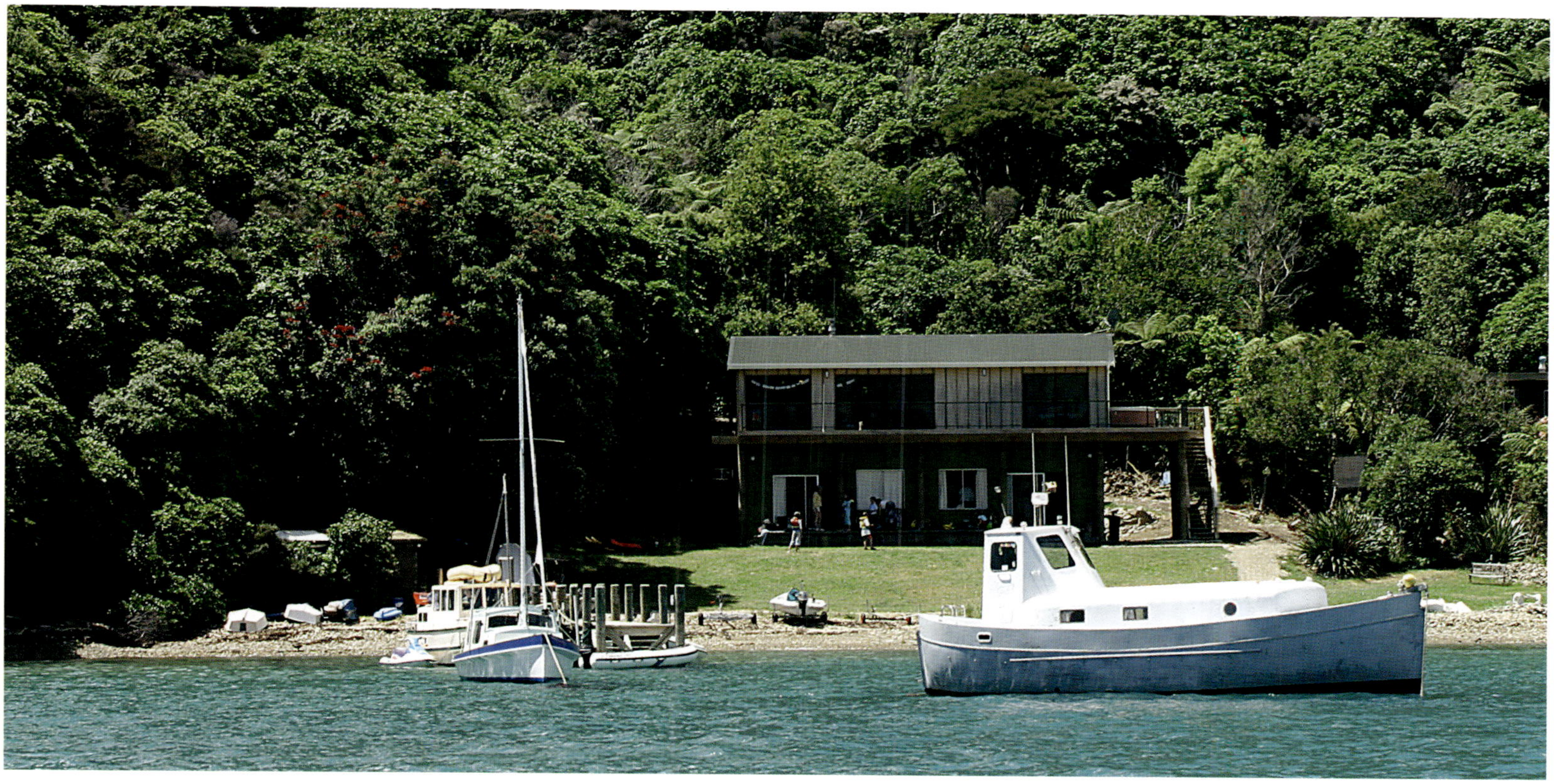

Arapawa Island in Tory Channel is a popular place for holiday homes.

use the secluded anchorage but two guns, with an observation post and magazine, were built on Blumine Island and some remains are still visible. Construction was not easy. The steep terrain and soft rock caused building delays, but they were completed in 1942 followed by further emplacements on Maud Island, at Whekenui on Arapawa Island and at several mainland locations.

Moioio and Allports Islands

Closer to Picton are Moioio Island and Allports Island. Moioio is a bush-covered, tiny hump of land rising out of the sea not far from the mainland. In 1839 a pa existed here with about 150 inhabitants. Long used as a burial site for Maori it is sacred ground, regarded as tapu by people who grew up in the area. When the fast ferries between Wellington and Picton first came into service, the little island's coastline was one of the areas affected by the strong wash left in the wake of the boats and it was feared that important archaeological sites would be lost by erosion. The problem was solved with an agreement that the ferries must slow their speed within a certain distance of the shore.

Native bush flows down to the water's edge of Allports Island, another scenic reserve. It forms part of the mineral belt running out to Cape Jackson at the mouth of Queen Charlotte Sound and there is evidence on the island of early mining operations. Now people go there to enjoy its bush walks or for the diving opportunities around the island. It provides safe moorings, there are good scallop beds in the vicinity, kingfish frequent the waters and occasionally dolphins and even orca swim by.

Arapawa

Much bigger than all these is Arapawa Island at the entrance to Cook Strait. Queen Charlotte Sound laps its western side and Tory Channel borders the east of the island. It was from a hill on Arapawa Island that Captain Cook first saw the sea passage from the Pacific Ocean to the Tasman Sea, subsequently named Cook Strait, and realised the dream of a vast southern continent cherished by geographers of the time was not a possibility. Maori tribes lived here for varying periods before Cook arrived and Ngai Tahu probably visited to get supplies of greenstone and perhaps argillite for the making of tools.

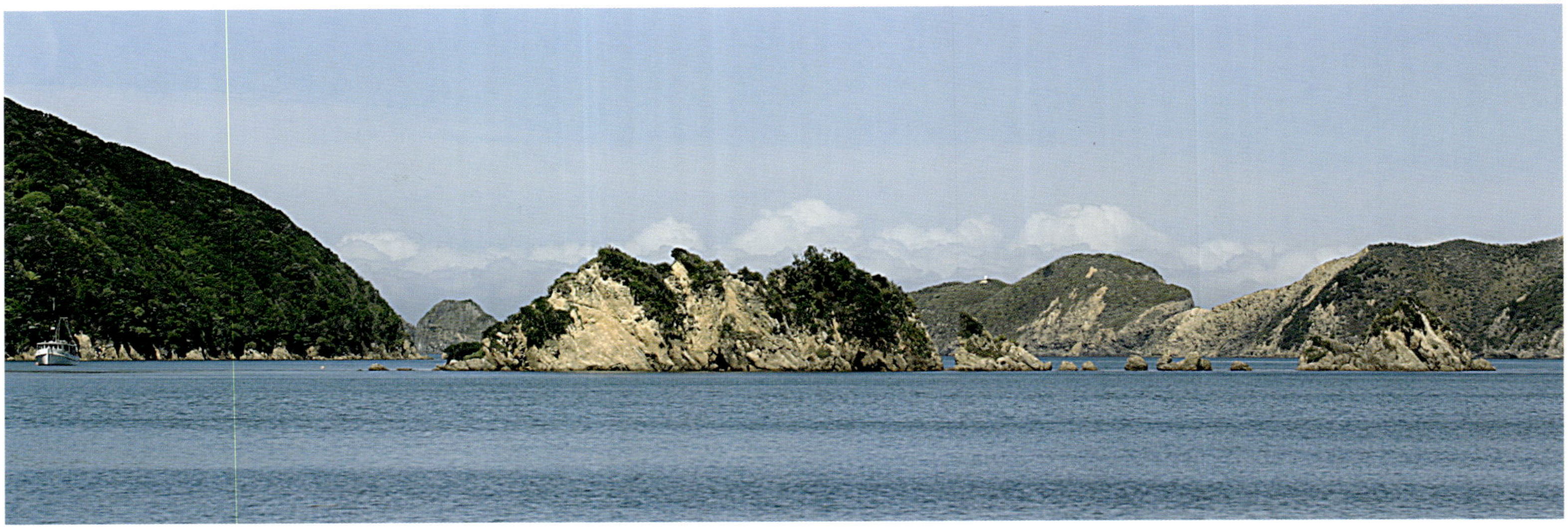

Rocky islets off Arapawa Island.

Europeans first came to the island in the 19th century, lured by the prospect of harvesting whales. In the mid-1820s Jacky Guard, a former convict from Australia, set up the first shore whaling station in New Zealand at Te Awaiti, on the eastern shores of Arapawa. By this time the Maori chief Te Rauparaha had launched his attacks on the South Island and had established his authority over Arapawa. Other whalers followed Guard, several of them intermarried with local Maori and their trading interests became intertwined. Testimony from an old Maori lady in a Treaty of Waitangi hearing talks about her aunties going out in small boats in Tory Channel, as teenagers, and shooting off the harpoons when a whale was within reach.

Another whaling station was set up in 1911 by Joseph Perano, not far from Te Awaiti. He also farmed at Whekenui and built a second whaling station at nearby Fishing Bay. This operation turned the area into a small community, with sometimes more than 30 whalers and their families living there throughout the winter season.

During the Second World War, the community grew temporarily, when 60 soldiers occupied an army camp on the Whekenui farm. When Joseph died in 1954 the whaling business was continued by his two sons until it closed in 1964 and whaling in New Zealand waters finally ceased. The houses built by the Perano family are now operated as tourist accommodation.

When Europeans settled on Arapawa they cleared areas of land for farming. Some families like the Heberleys, whose forbear James 'Worser' Heberley whaled around Marlborough for more than 10 years from 1830, combined fishing and whaling with farming for a living. Heather Heberley, married to a fifth-generation Arapawa Islander, in her autobiography *Weather Permitting* paints a vivid picture of island life in the days when access to the mainland was infrequent and depended on fair weather and availability of boat transport. Supplies came from Picton, sometimes only once a month. Women had to be able to conjure up a meal for multitudes on a coal range, deal with sickness and accidents and live with loneliness. When electricity was finally connected to the island in 1982, life in the home became less onerous.

Modern communications have also eased daily living. But husbands still have to be practical, competent fix-it men, equally able to muster a mob of sheep, handle a fishing boat in rough seas or mend a broken farm bike. Fierce storms are not uncommon in the Sounds and local fishers become involved in search and rescue operations, at times putting themselves and their boats at risk.

There was no school on the island in the past and there still isn't. The mother of the family has to be teacher and nurse, boat skipper and mechanic, cook and plumber.

'But it's a great life. We like the independence. Over the years we've made our own fun and the kids loved it,' Janice Saunders told us. She came to Arapawa with her husband Harold in the 1960s. 'We came here to escape,' he said, 'but bureaucracy caught up with us even here. It's ruined farming – we're no longer allowed to burn off,' and Harold pointed up the hill behind the house he built to where the cleared land is reverting to gorse, manuka and pine seedlings.

Harold combined farming with boat building, a trade he learnt from his father. He has built 'about 70' boats, all of them in wood. Now he's almost retired, but he led us to a huge shed where he is working on four more boats, all at various stages of construction.

In the bay next door there's a paua (abalone) farm. Denis and Nita McManaway started with 60 wild paua seven years ago. Now they maintain 50,000–60,000 at any one time in large tanks in a boat-building shed that once belonged to Harold Saunders. The industry is still small and they sell cocktail-sized paua to New Zealand restaurants.

Arapawa Island is known for breeds of pigs, sheep and goats found only on the island. The goats are among the few survivors of the Old English milk goat and it is thought these animals were brought to New Zealand by Captain Cook and left to breed on the island as a future source of food for explorers. The goats came under siege in the 1970s when the Forest Service decided they were destroying native vegetation and proceeded to shoot all the sheep and goats they could round up. A battle ensued between the Forest Service and supporters of the animals, led by Betty and Walt Rowe, an American couple who had settled on Arapawa and championed the survival of the rare breeds of animals they found there. DNA testing has proved their unique genetic make-up, and to ensure their long-term survival Arapawa goats have been sent to sanctuaries and wildlife parks in New Zealand and overseas.

The origin of the sheep is less defined but it is certain that they have lived on Arapawa since at least halfway through the 19th century, and they are considered rare and endangered. One theory suggests they are a Middle Eastern breed, introduced by whalers. Others consider they may have originated from Australian merinos. Certainly the sheep look prehistoric, with long, curving horns and very fine wool.

The Rowes set up a wildlife park on Arapawa, donating 300 acres of their land, and representative numbers of the sheep, goats and

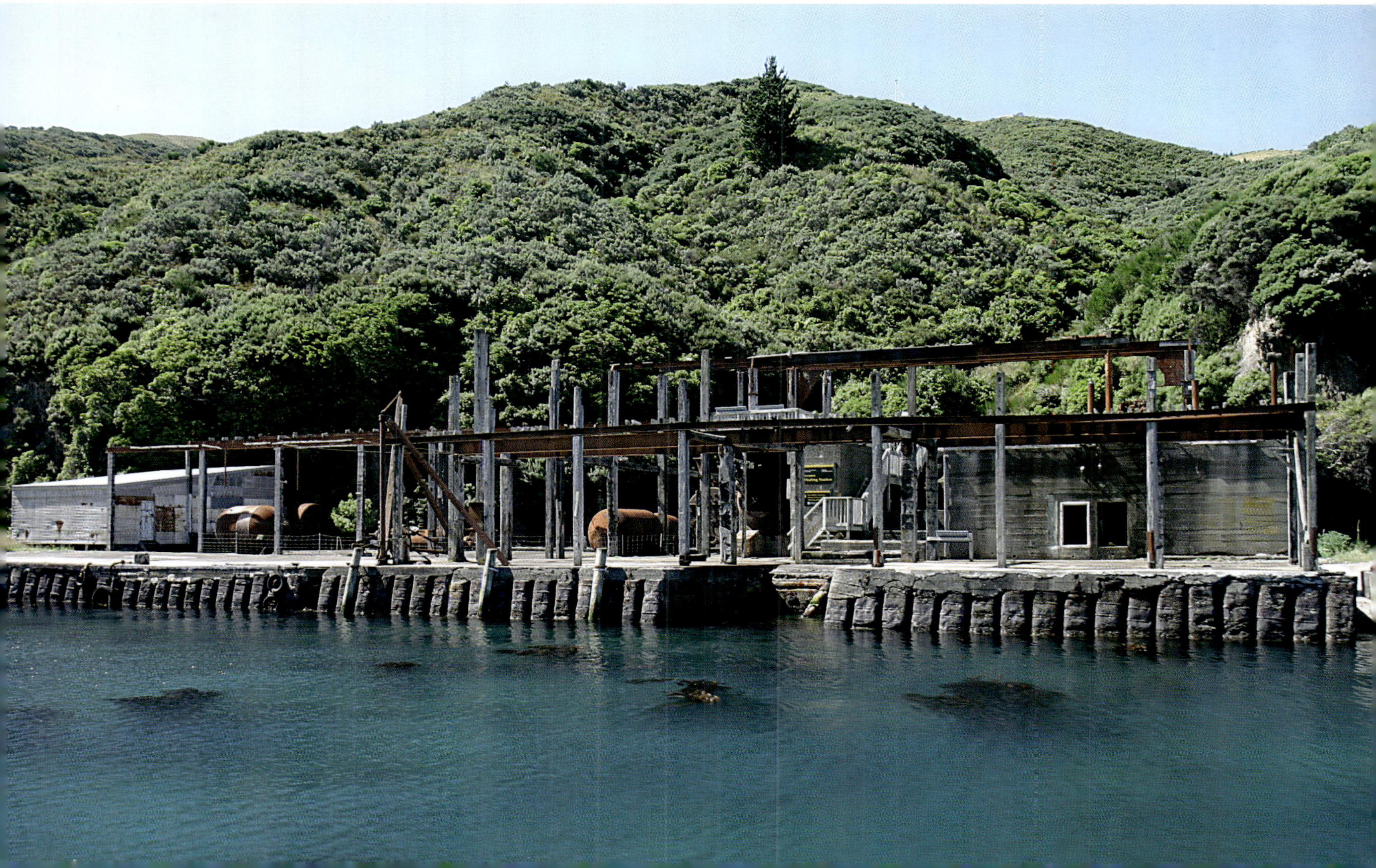

Remains of the Perano whaling station at Fishing Bay on Arapawa Island.

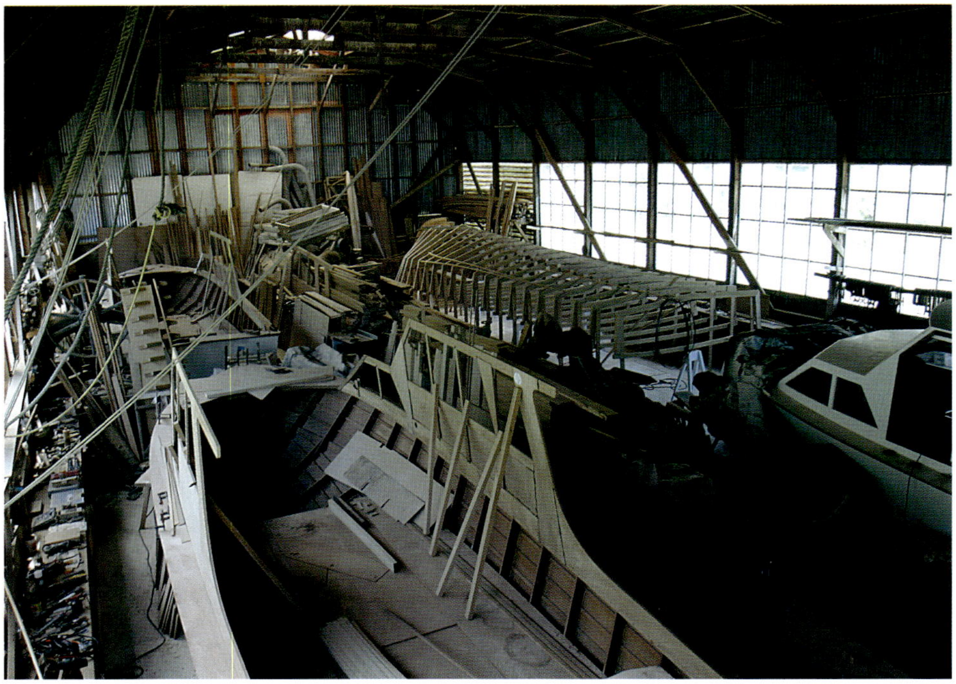

Harold Saunders' boat building shed on Arapawa Island where he has made 'about 70' boats since he first arrived there in the 1960s.

pigs live there, with care being taken to prevent their mating with other breeds.

Nowadays Arapawa is a mixture of open farmland, scrub, Department of Conservation reserve and privately owned pine forests. Holiday houses dot the shores of the island and it is a peaceful haven for many visitors to the Marlborough Sounds.

D'Urville Island

D'Urville Island on the western edge of the Marlborough Sounds sees fewer visitors. Mainly covered in bush interspersed with patches of pine forest, most of the island is now owned by the Department of Conservation, although farms still exist on the island, some in foreign ownership.

A wilderness resort offers overnight accommodation and visitors come here to tramp, mountain bike, admire the waterfalls, visit glow-worm grottoes or enjoy a quiet fishing holiday. Although there are walking tracks around the island offering outstanding views of the Sounds, there are few roads and holiday house owners leave their cars at the mainland township of French Pass. Boaties have been attracted to the island for decades and Port Hardy in the north provides a sheltered anchorage in stormy weather.

Although Abel Tasman sheltered here in 1642, the island was named after French explorer Dumont D'Urville, who sailed into the Marlborough Sounds in 1827 in the *Astrolabe* and successfully negotiated the treacherous strip of water known as French Pass that separates D'Urville from the mainland. The *Astrolabe*'s officers named the island in their captain's honour.

'Lofty and mountainous, the coast was steep and gloomy,' was D'Urville's comment on the island that bears his name.

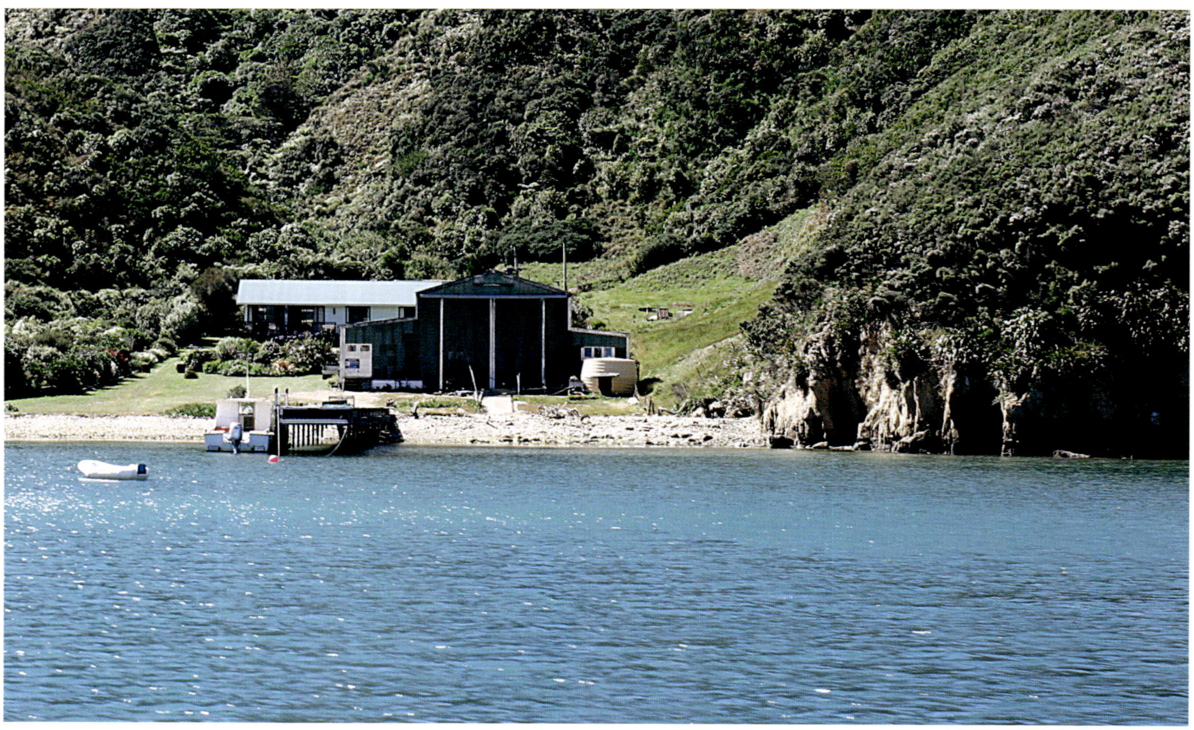

A paua processing plant on Arapawa Island.

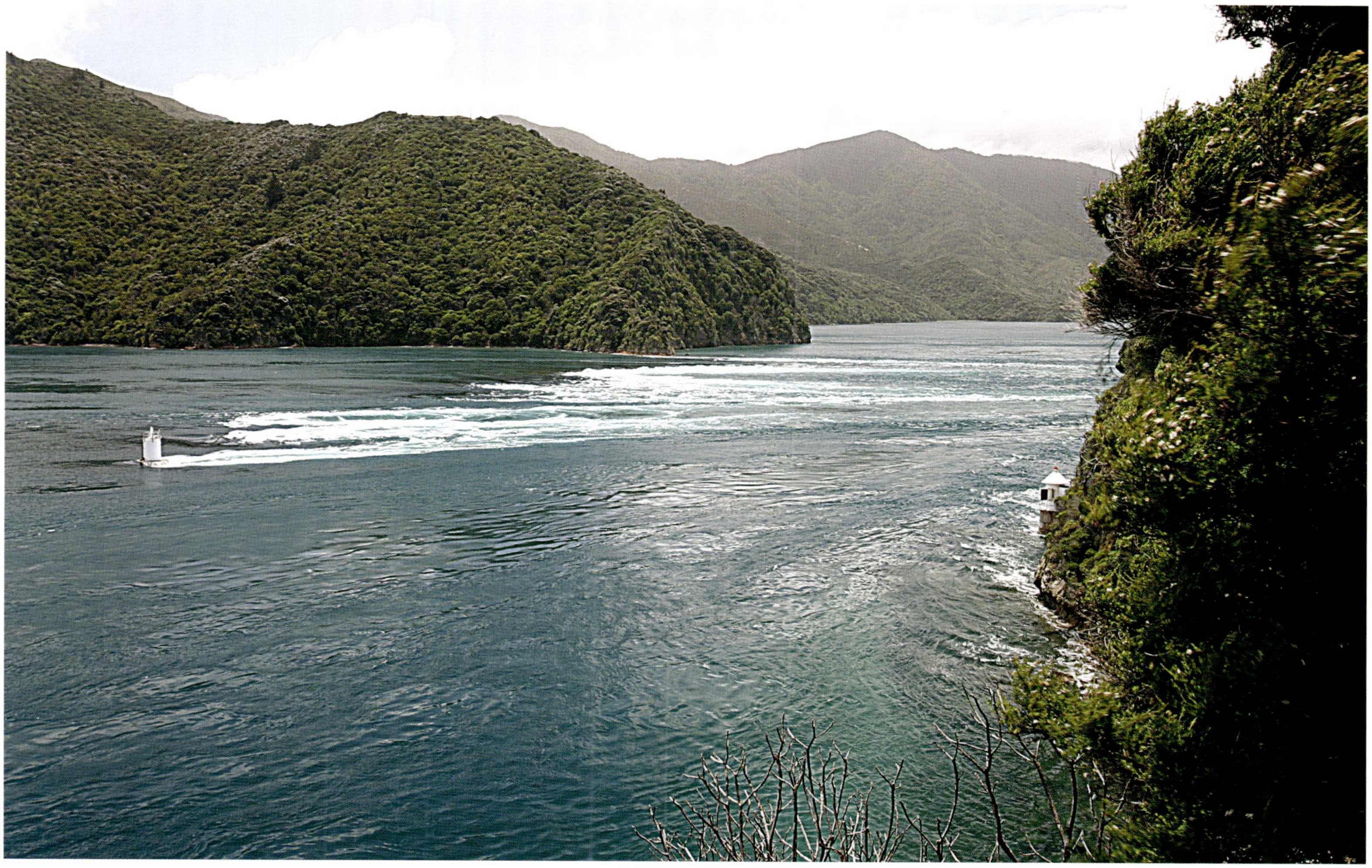

French Pass, the treacherous strip of water that separates D'Urville Island from the mainland.

D'Urville's wife is also remembered in the naming of Adele Island, further to the west in Tasman Bay off the coast of Abel Tasman Park.

Ariana Rene remembers D'Urville Island differently. She lived there with her husband and family from 1937 until 1958, at Port Hardy. The mail boat called once a week with provisions, and a generator provided power for the house, though Ariana cooked on a wood stove.

'I loved living on D'Urville,' she told me. 'It was the most beautiful place on earth. The atmosphere was wonderful, the air was so pure and everything seemed to grow. It was like living on a semi-tropical island. We grew all kinds of fruit: grapefruit, plums, apples, boysenberries, raspberries, passionfruit and things I had never seen before like tree tomatoes [tamarillos] pawpaw and guavas that grew profusely. We had tomatoes, cucumbers, kamo kamo and pumpkin growing all year round. As well as the kaimoana [seafood] and the exotic fruit and vegetables we also sought out traditional food. I remember going to gather titi [muttonbirds], pigeons and eels in particular. You never took more than you needed. What we couldn't eat fresh, we dried or gave to our relations.

'My children used to swim all year round. We had a pet tuatara who lived under the root of a pear tree and every time the children went down to the water to swim, they would leave their clothes and towels on the beach. The tuatara would watch over them and if the tide came in he would pull the clothes up the beach, then he would follow them up to the house and make sure they all went inside.'

'We didn't use Pakeha medicine, we used medicine from the bush. We would boil up kopakopa [plantain] and use it as a poultice. It was very effective for cuts and scratches. Kawakawa was good for tummy infections and I'd roast roots of flax, pound them and make them into a tea and use that for period pain.

How to get there

D'Urville Island is open to visitors.
There is accommodation on the island and water taxis operate out of the town of French Pass.

'My husband was farming on D'Urville Island on ancestral land that he and his brother had inherited and which had been leased out for years. He had sheep and cattle. The wool would be taken down to the shore, loaded onto a barge and taken to Havelock. From there it would be carted to Blenheim for sale. In 1951 we got the top price in the South Island for our wool.'

Ariana never felt isolated on D'Urville. Sometimes they would visit Madsen at the southern end of the island for a funeral and they went back and forth to Porirua for special events.

'We had plenty of visitors. We made friends with quite a lot of well-known people who came over from Wellington every year. Boaties would come and moor in the bay near our house. They'd smell the bread I was baking and I'd give them a couple of loaves. They'd take on water from our tanks.'

Like other island women Ariana talked about the difficulties of going to the mainland to have a baby. 'After you had the baby in Wellington, and took the ferry home, the ferry used to go back via French Pass at one o'clock in the morning. They'd come and wake you up and you'd have to tie your baby to your front and get in the basket and be lowered down on the rope into the boat below. You'd look down and see this little wee boat way down and know that you had to be lowered all the way in.'

Schooling, as always, posed problems. 'There was no school where we lived on D'Urville, so my kids had to go back to Porirua. Other kids learnt by correspondence and there was a school at Madsen, but it was too awkward for my kids to get there. The children could never get home fast enough in the holidays.'

When their land on D'Urville was finally sold, and Ariana moved back to the mainland, she was very distressed. 'They were the happiest years of my life,' she said simply.

View north from French Pass settlement, looking into Catherine Cove, D'Urville Island, with Anatakupu Island in the middle distance.

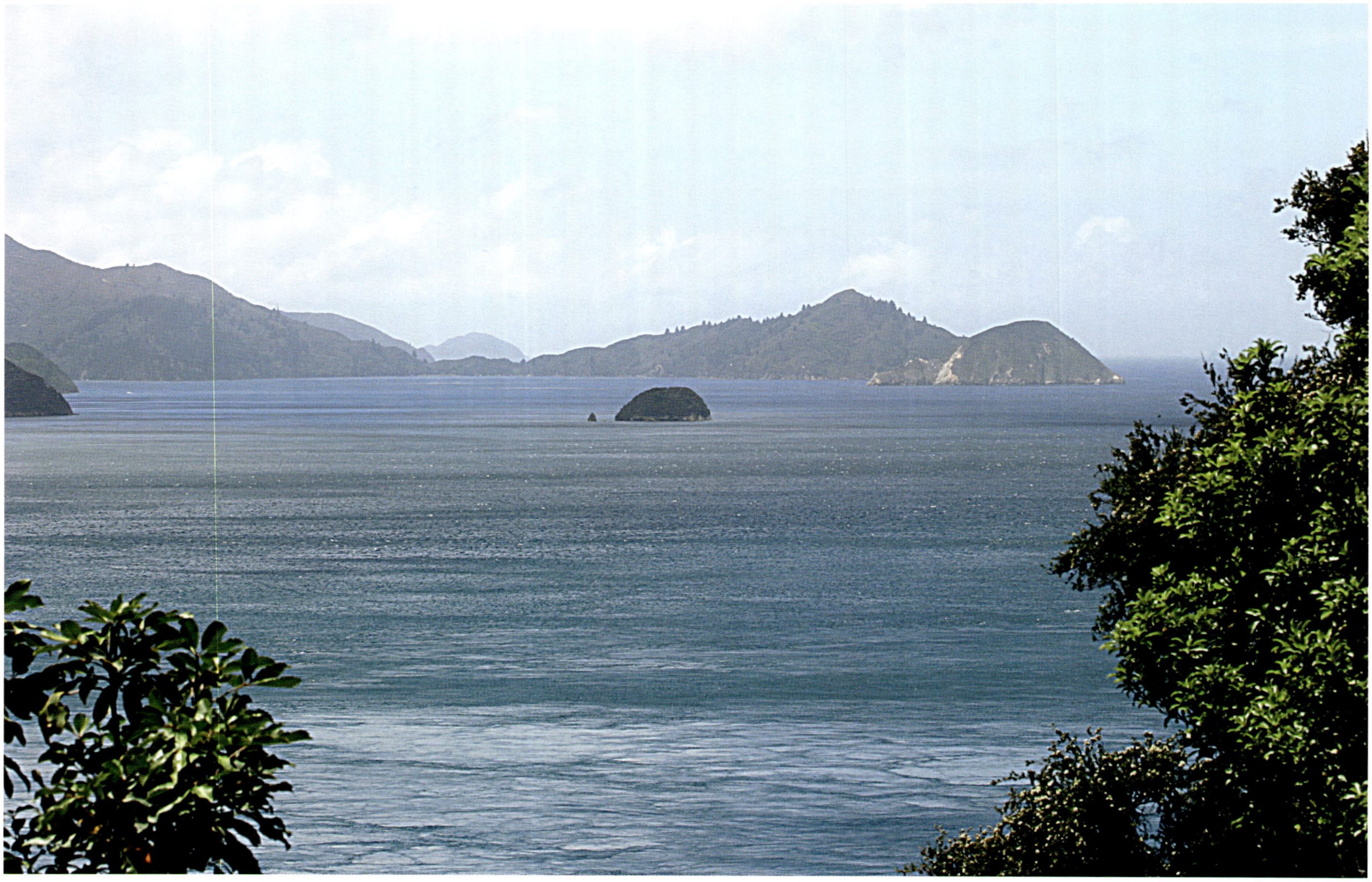

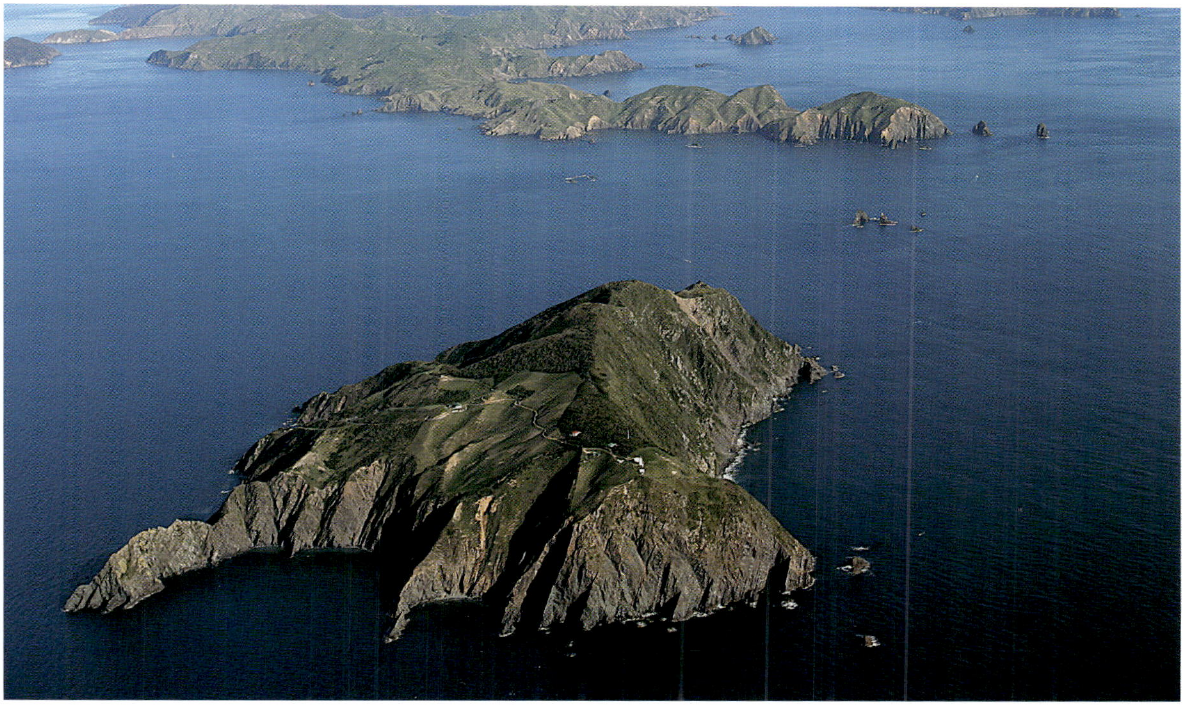

Stephens Island, closed to public access to protect its status as a conservation island, sits far out in Cook Strait with D'Urville Island in the distance.

Stephens Island

North of D'Urville Island is Stephens Island or Takapourewa, a precipitous rocky outcropping set well out in Cook Strait. It is unlikely that Maori lived permanently on the island but archaeological evidence of campsites and middens has been found. Most likely, people from nearby D'Urville went to Takapourewa for the seasonal harvest of shearwater (muttonbird) chicks.

In 1883, members of Ngati Koata were granted clear title to D'Urville Island and Takapourewa but about a decade later the latter was taken under the Public Works Act for the purpose of building a lighthouse – an important navigational landmark for shipping in Cook Strait. Lighthouse keepers and their families lived a lonely existence there in the days before lighthouses were automated.

Closed to casual visitors because of its important conservation value, Stephens Island is considered by many to be 'the home of the tuatara'.

Although tuatara exist on a number of other small islands around the country, estimates suggest there could well be more than 50,000 of these prehistoric animals roaming the island. Scientists visit from all over the world to study the reptiles. We tend to think of tuatara as museum pieces, but here in their natural habitat they are a vital part of the ecosystem.

Early lighthouse keepers observed kingfishers feeding baby tuatara to their fledglings. Australasian harriers feasted on them, causing a bounty to be placed on the birds and incidentally supplying the meagerly paid keepers with another source of income. Black-backed gulls are also a threat to tuatara. A lighthouse keeper in the past reported often seeing gulls grab a reptile, fly out to sea and drop it in the water, hoping to injure it. The gull would then swoop in for the kill and eat it. Occasionally a tuatara was seen to escape and swim back to shore.

Also special to Stephens Island is the Hamilton's frog (*Leiopelma hamiltonii*), discovered in 1915 by a lighthouse keeper's son. These frogs have origins as ancient as those of tuatara. Unusually, it has no tail and no tadpole stage and spends most of its life underground.

Cohabiting with the tuatara and the frogs on Stephens Island are skinks, tree and giant weta, several species of gecko, including a distinctive gecko identified in the 1950s and eventually designated as a new species in 1980, many rare insects and millions of seabirds. These include sooty shearwaters, prions, fluttering shearwaters and diving petrels, which nest on the island and riddle the cliffs and cliff-tops with their burrows.

Islands provide a protected environment for endangered species such as the giant weta. Many of these enormous insects live on Stephens Island.

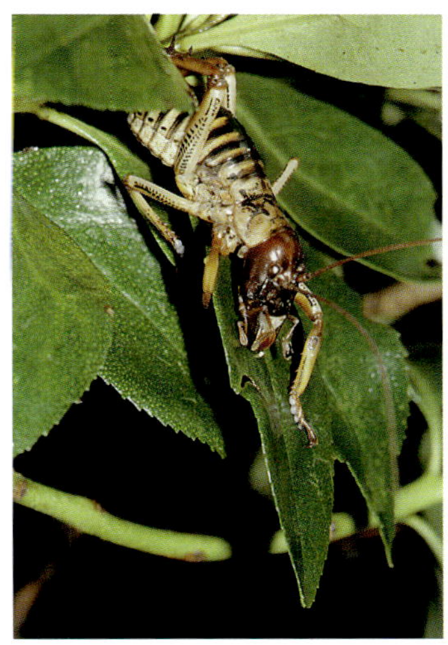

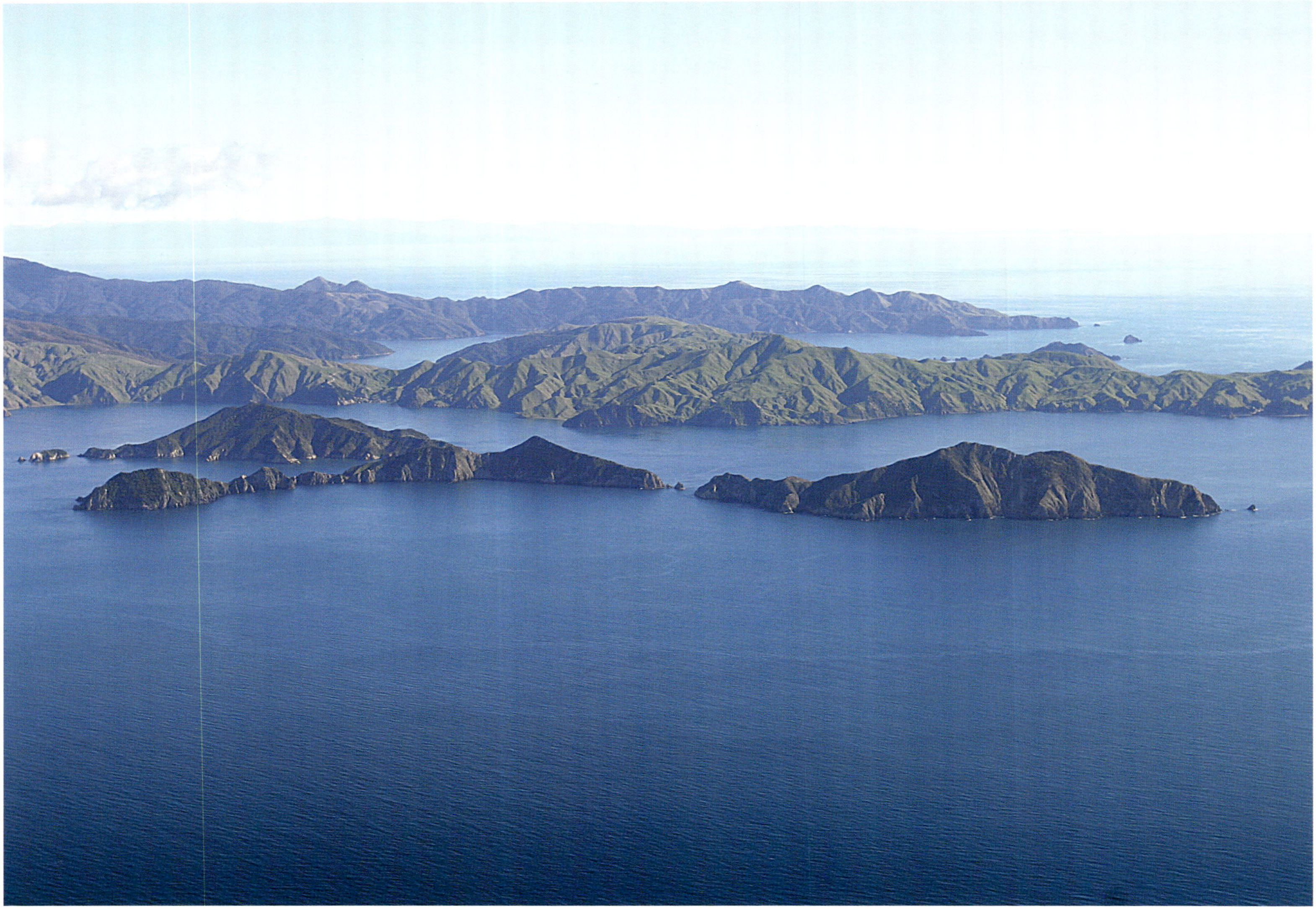

Rangitoto Islands, off D'Urville Island, with Whakaterepapanui at right, Puangiangi front left and Tinui Island just behind.

Rangitoto Islands

The proliferation of wildlife on Stephens Island is so successful that recently a large number of tuatara, giant weta and green geckos were taken to nearby Whakaterepapanui Island, in the Rangitoto Islands group off the north-eastern coast of D'Urville Island. A rat-eradication programme in 1999 cleared the island of predators, the first step in restoring this island to its native state. It is planned to follow up with a future transfer of birds.

Tinui, another island in the group owned by several Maori families and Puangiangi Island, also privately owned and recently sold, were made rat-free at the same time to ensure that no rats could re-invade Whakaterepapanui by swimming from island to island.

Maud Island

Pelorus Sound, accessed from Cook Strait, is home to several islands. Most are now dedicated to conservation, though Maud Island was virtually stripped bare of native forest and farmed for many years. Fortunately, a small vestige of forest remained and it was here that the pakeka frog – a separate species from the frog on Stephens Island – was formally discovered in 1958, though the presence of an unusual frog had been noted as early as 1940.

In 1971 Jack Shand, then owner of the island, gifted the fenced-off portion of bush to the Crown as a reserve for the preservation of flora and fauna and it became a refuge for Fiordland kakapo in 1974. The following year, thanks to the generosity of Shand, the remainder of the island was bought by the Crown with financial help from the Royal Forest and Bird Protection Society and was designated a nature reserve, later being changed to a scientific reserve. Visiting the island is forbidden without a permit.

Since then much conservation work has been carried out on Maud Island with introductions of kakapo, saddlebacks, giant weta, little spotted kiwi and takahe among other species. Although largely free of major predators, stoats have been a recurring nightmare due to their impressive ability to swim long distances and re-invade the island from the nearby mainland. As a result, all kakapo have been removed from Maud and transferred to either Codfish Island (near Stewart Island) or Fiordland.

Chetwode Islands

The Chetwode Islands on the northern fringe of the Sounds consist of two large islands, Nukuwaiata and Te Kakaho, plus three smaller islands. They are maintained as a wildlife sanctuary with landings by permit only. Named in 1836 after Lieutenant Chetwode, acting captain of HM Survey Ship *Pelorus*, the islands are home to kakariki, bush robins, bellbird, tui and New Zealand pigeons and the surrounding waters are visited by king shags, dolphins and seals.

In total contrast Forsyth Island, south of the Chetwodes, is privately owned – at the time of writing by an Iranian-German who specialises in buying and selling islands worldwide. It is advertised as a luxury holiday destination where visitors can hike, fish, swim, snorkel or just enjoy the tranquil trappings of isolated island life.

Looking northwards along the Chetwode Islands.

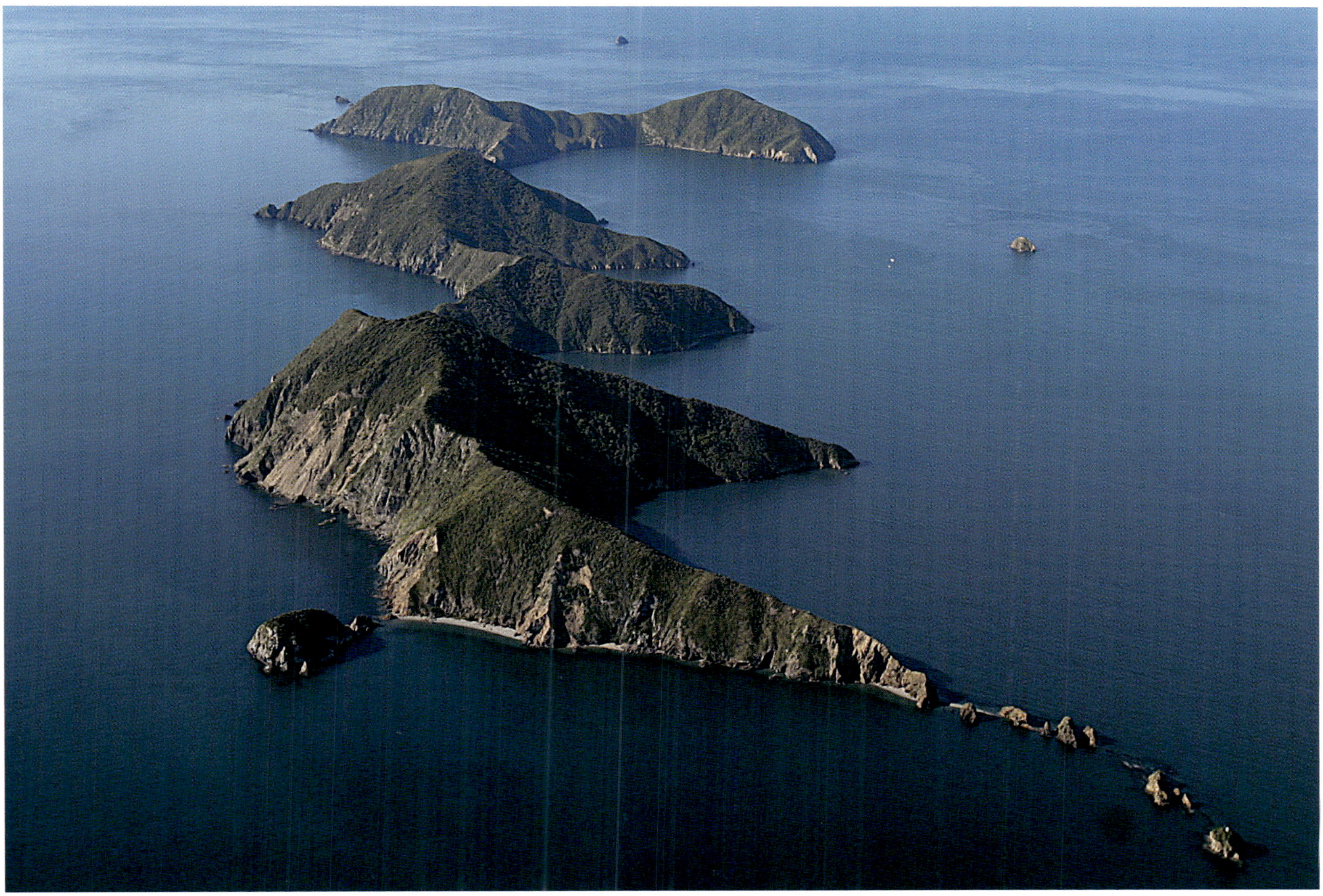

Quarantine Islands

THE YEAR: 1863. THE PLACE: PORT CHALMERS, NEAR DUNEDIN. SOME DISTANCE FROM THE SHORE, A SHIP LAY AT ANCHOR, A LARGE GROUP OF IMMIGRANTS ON BOARD. AT THE MAIN MAST FLEW A TELLTALE YELLOW FLAG THAT STRUCK TERROR INTO THE HEARTS OF ANY OBSERVERS. IT INDICATED THE PRESENCE OF INFECTIOUS DISEASE ON THE SHIP.

That disease was smallpox, as virulent and feared then as SARS is today. Several of the passengers were ill; some had died during the trip. On shore people were frantic. No one wanted passengers or cargo to come ashore. Eventually, all the passengers were off-loaded onto a nearby island, there to wait out the course of the disease, and the ship was kept in quarantine for more than a month – the standard period of 40 days, the term derived from the French number *quarante*.

Quarantine islands were set aside in all four main ports during the 1860s and 1870s. Motuihe, in the Hauraki Gulf, seems to have had few residents and the only graves there are those of crewmen from a merchant ship who died of influenza on the island during the 1918 epidemic. But Quarantine Island, the first to be established, was host to passengers from more than 40 ships.

On a sunny day, the view from the island in the middle of Otago Harbour is idyllic. The sea sparkles, sun bathes the city in the distance and the nearby hills on the mainland are green and inviting. But in the 19th century, few new immigrants who arrived at any quarantine island at the end of a tedious voyage from the other side of the world would feel similarly about their place of incarceration so close to their destination. At best they faced a long delay. At worst, they would

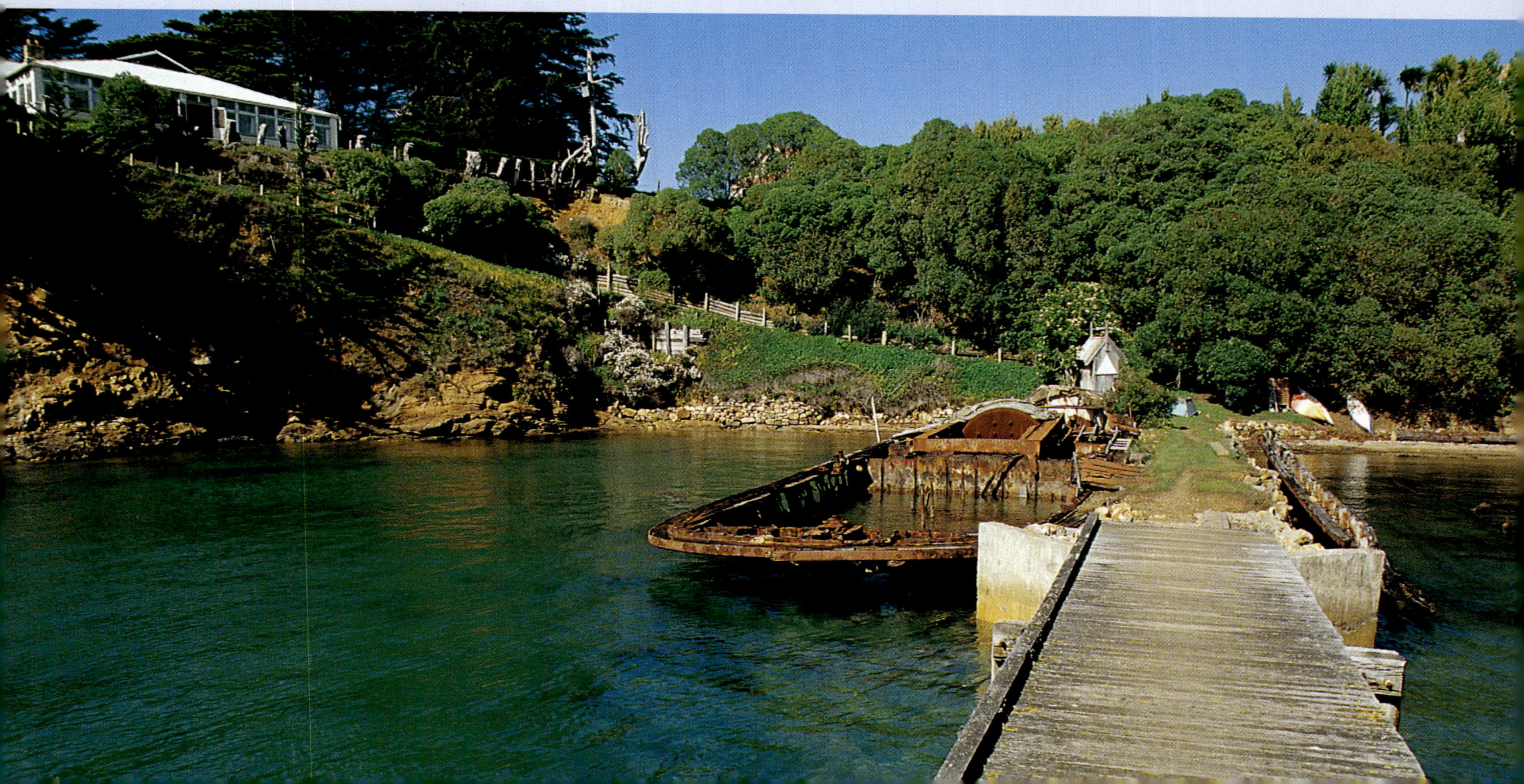

Quarantine Island in Otago Harbour looks appealing on a bright summer's day but to new settlers off-loaded there because of the risk of infectious disease, it must have seemed like a prison.

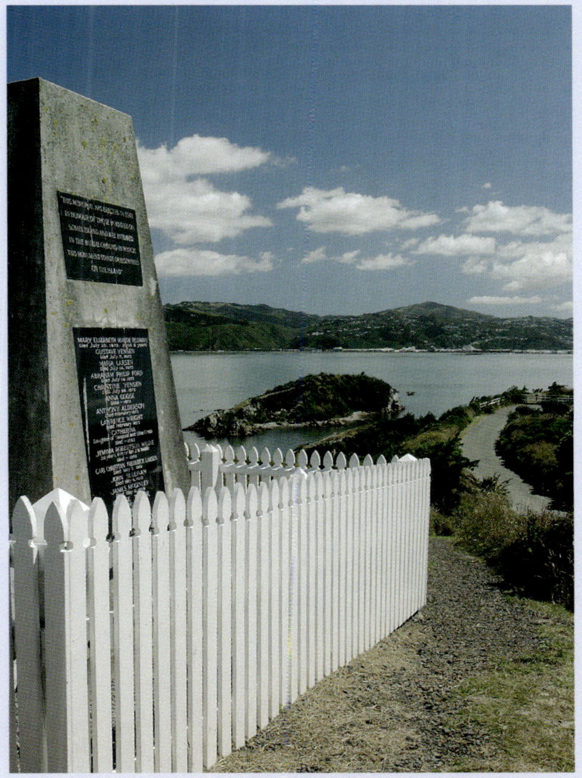

A memorial on Matiu/Somes records the names of those who died there during its period as a quarantine island. The small island behind is where Kim Lee, suspected of leprosy, was confined and eventually died.

be buried on the island. Wandering among the graves, it is only too easy to imagine the fears, the reality.

Victory, the first ship to be detained in Otago, arrived in New Zealand with an outbreak of smallpox on board, although scarlet fever eventually cut the greatest swathe among the island residents.

By the 1890s, the need for quarantine had tapered off. John Dougall, the Keeper of Quarantine since 1863, died at the beginning of the decade and his widow continued farming on the island, helped from time to time by their only surviving son, who was employed on the harbour ferries. Later, he returned to live on the island with his wife and family.

For the next 30 years, the island was farmed by a succession of tenants.

In the 1950s, the purpose of the island changed yet again, and now people visit seeking peace and spiritual refreshment.

Settlers who arrived in disease-affected ships bound for Christchurch were sent either to Quail Island or neighbouring Ripapa in Lyttelton Harbour. A newspaper account of the arrival of the ship *Waitangi* in 1878, with scarlatina on board, states that single women and families were landed at Ripapa, while single men were taken to Quail Island.

A lonely grave on Quail Island is a sad reminder that leprosy was a disease that caused people to be quarantined for much longer than 40 days. By 1924, when the station closed, there were nine lepers on the island, each with his own cottage. The caretakers supplied them with cooked food and a Lyttelton doctor was a frequent visitor. Residents in the town provided the lepers with a variety of gifts including books, a radio and a billiard table but life must have been grim for those unfortunate men.

After 1885, Ripapa became part of the nationwide system of coastal defences, built to defend the country against the perceived 'Russian Scare'. A walled fort was built on the island and exists still. The fort was occupied by the army until the end of the First World War and again during the Second World War.

Matiu/Somes Island has the longest history as a quarantine station, being designated as such in 1869 and finally closing in 1995. The first ship to disgorge passengers arrived in 1872 carrying smallpox. Buildings were hurriedly provided to house the new arrivals and the armed constabulary was issued with orders to shoot anyone who tried to escape. Biosecurity measures, 19th-century style, were rugged.

As on Quarantine Island, the main victims among ships' passengers were the children. Their actual graves have been lost but the names of many who died are recorded on a memorial on the island. While smallpox and scarlet fever were the most common diseases causing confinement, a bubonic plague scare in the early years of the 20th century kept the station in a state of permanent readiness.

From 1893, human and animal quarantine facilities existed side by side and the Department of Agriculture took on a share of the responsibility for the buildings. After the Second World War, the island was used only for animal quarantine and this included the construction of a maximum-security block built in 1968.

Perhaps the saddest case of quarantine concerned a Chinese man named Kim Lee, who was suspected of having leprosy, though his symptoms were never confirmed. He was isolated on Mokopuna Island at the northern end of Matiu/Somes in 1903, where he lived in a cave. The lighthouse keeper on Somes sent him food and water by flying fox. Not surprisingly, he did not survive long.

Looking out into the harbour from Quarantine Island.

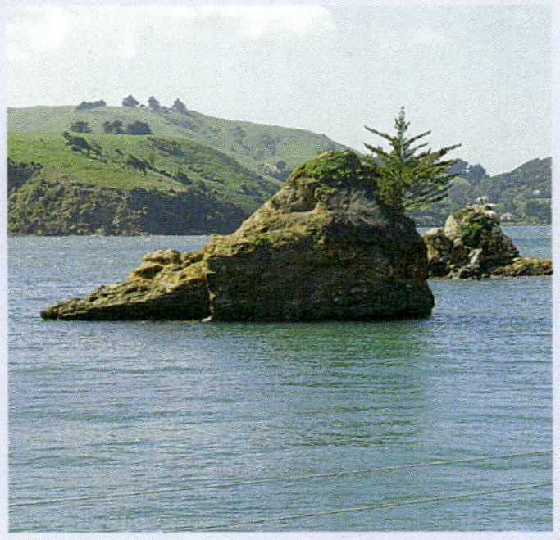

Islands of the South

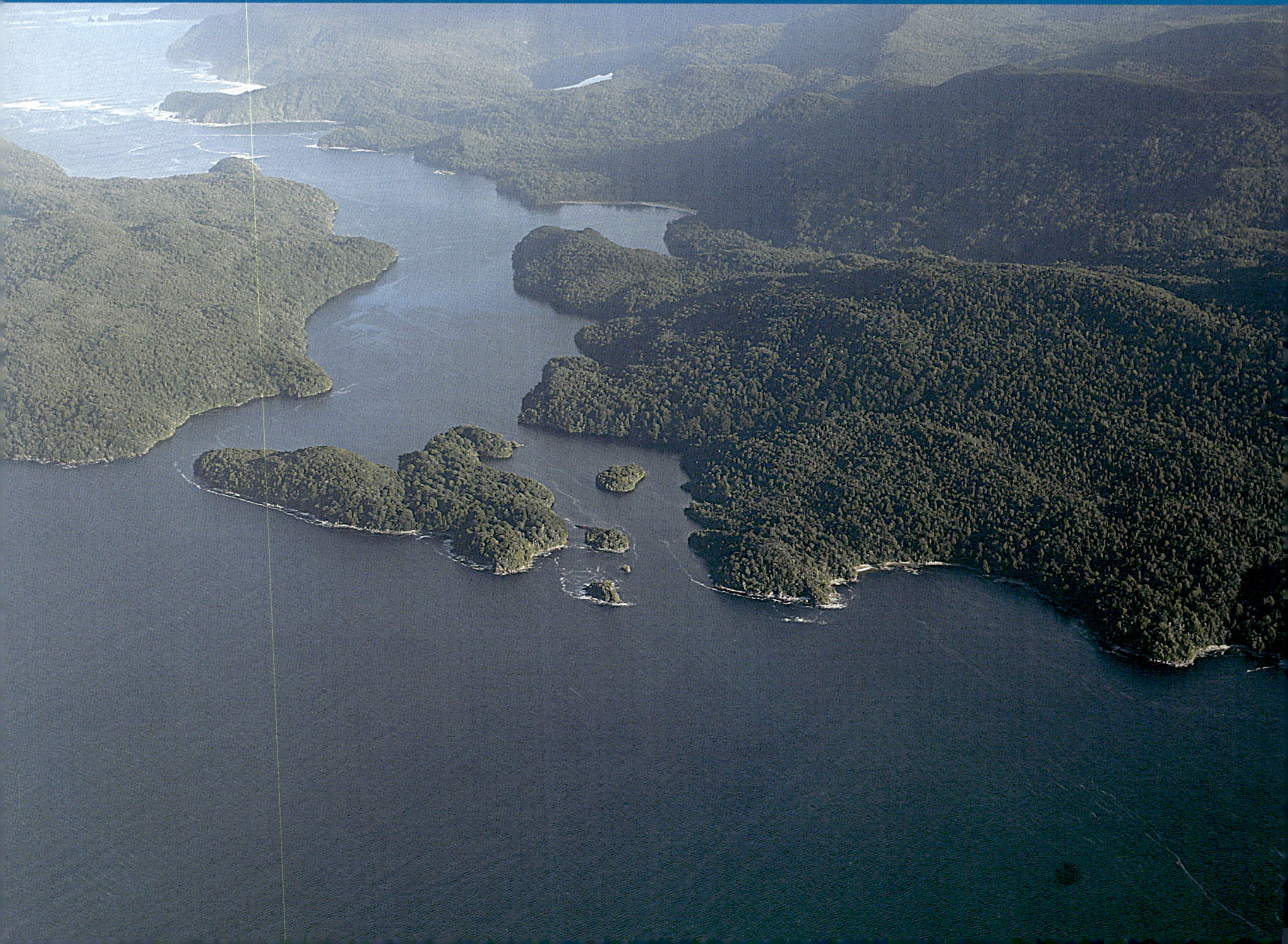

Frequently rugged, wind-ravaged and inhospitable, the islands in Foveaux Strait and Fiordland nevertheless attracted the first European settlers in New Zealand. Seals and whales abounded in the waters around the islands and in the last decade of the 18th century and during the early decades of the 19th, hard-bitten, sea-faring characters lived on these islands for months at a time, often enduring hunger and privation for the sake of the meagre earnings they eventually pocketed. Stewart Island is still home to some of their descendants.

Islands of Fiordland

Dusky Sound is one of the most beautiful places in New Zealand. It is also one of the wettest and loneliest. No road winds into the area; walking there takes four or five days' hard slog and launch trips are few. Yet paradoxically it takes 15 effortless minutes to get there in a small plane from Lake Te Anau.

Left: Little Island, with Great Island behind, sits near the mouth of Chalky Inlet in Fiordland.

The flight itself is awe-inspiring. Mountain peaks, powdered with snow in winter, stretch to infinity. Glittering threads of river curl their way through shadowed chasms. The bush is dense and dark. Mist steals through the valleys carved out by long-gone glaciers, though according to Maori myth it was Tu Te Raki Whanoa who gouged out the fiords and lakes of Fiordland with his giant digging stick as he stood with one foot on Resolution Island and the other on Secretary Island.

Everything in sight is measured by height and depth until the plane reaches open sea and the view flattens out. As it turns lazily inland again, my mind struggles to map the web of waterways and jumbled land mass that is Dusky Sound. Innumerable islands are splattered across the water – droppings of a giant flying creature maybe, who lost his bearings as he passed overhead. Resolution Island at the mouth of the sound bulks large, bigger than Waiheke, smaller than Great Barrier. To the south are Chalky Inlet and Preservation Inlet, each with their own raft of islands. To the north is Doubtful Sound.

People come and go in this area of New Zealand but no one stays for long and the imprint of humans is negligible.

If this seems like another world to those of us who have come to know Fiordland from occasional visits and photographs, it is hard to imagine the reaction of Captain Cook and his men when they first sailed past Dusky Sound in 1770. They must have liked what they saw for Cook returned three years later after months spent exploring the Southern Ocean. He beached his ship, *Resolution*, for five weeks in Pickersgill Harbour on the mainland opposite Crayfish Island and not far from Resolution Island to the west. There his crew enjoyed some hard-earned rest and recreation while they carried out running repairs. And it was there that New Zealand's first ever brewery was established. Cook's men concocted a brew from rimu and manuka, and while the taste may leave something to be

How to get there

Scheduled cruises to Dusky Sound and Preservation Inlet are available several times a year. Chartered cruises, including kayaking, are also available. Most leave from Doubtful Sound and transport is provided from Manapouri or Te Anau.
It is also possible to tramp into Dusky Sound.
Chartered flights over Fiordland leave from Te Anau or Invercargill airports.
For further information contact the Visitor Centre at Te Anau.

Dusky Sound is a mass of complicated waterways, dotted with numerous islands. Looking westwards towards the sea, Cooper Island sits in the centre of the picture with Long Island in the background.

Right: A sailing ship is dwarfed by the steep, forest-clad mountains that surround Doubtful Sound.

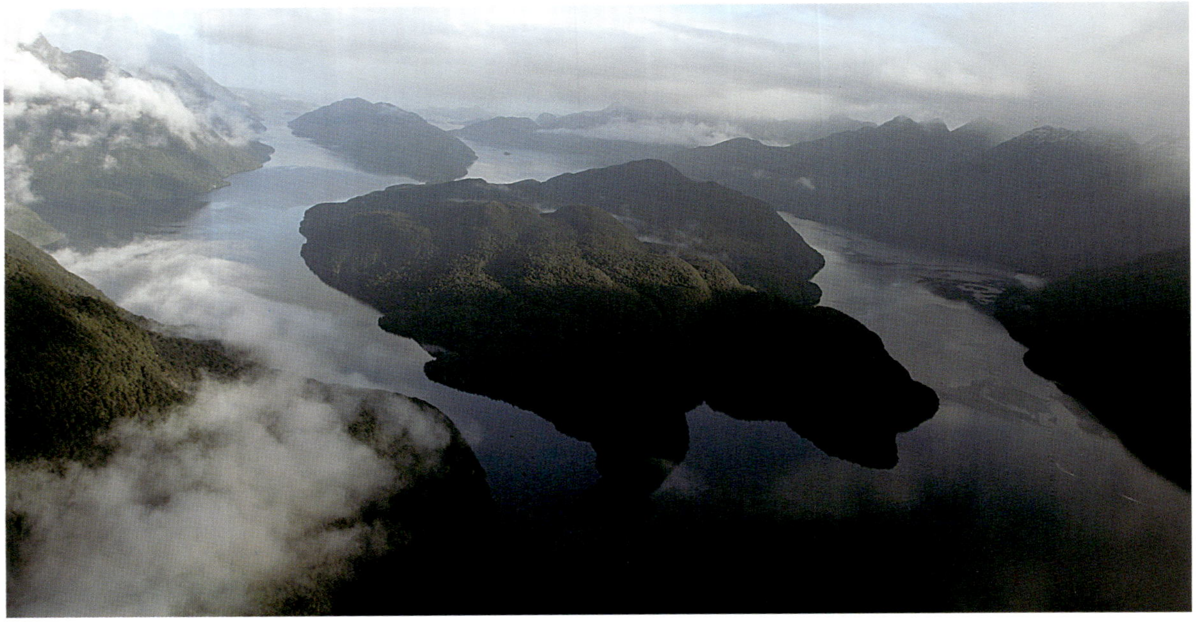

desired by today's standards, the fresh vitamins it contained helped keep scurvy at bay.

In small boats they explored the area, naming many of the islands and charting the waterways with Cook's usual accuracy while the botanists on board made paintings of the plants and birds they found, as well as writing descriptions of them for posterity.

They even made contact with a small, probably family, group of Maori living in the area. Initially wary, they were won over by Cook's overtures and the leader even ventured aboard the ship. Presents were exchanged but the group suddenly disappeared before Cook departed and no trace was ever found of them again, although human bones discovered in a cave by later arrivals in Dusky may have been those of the 'family'.

Several other European explorers arrived shortly after Cook. Bauza Island in Doubtful Sound, where a marine reserve has been established between it and neighbouring Secretary Island, gets its name from Spanish explorers who arrived in Fiordland in 1793.

But as in other parts of New Zealand the first Europeans to actually live in this remote corner of the world were sealing gangs who arrived 20 years after Cook. New Zealand's first sealing gang set up camp on Anchor Island at the entrance to Dusky Sound in 1792, an island now targeted as a future home for a colony of threatened kakapo. The first European house in New Zealand was built there as was the first sailing vessel. By the 1820s the seals had been nearly annihilated and the whalers arrived, setting up a whaling station in Preservation Inlet where indiscriminate slaughter of the beasts once again led to a short-lived industry.

Visitors to Fiordland for the remainder of the 19th century were few. Explorers established several routes through to the West Coast and one or two hardy prospectors sought precious minerals in the area without conspicuous success. From Cook's earliest reports it was obvious that Fiordland was home to many of New Zealand's birds, including the rare takahe and increasingly rare kakapo. Andreas Reischek, the infamous Austrian bird collector, spent months exploring the wilderness and adding to his tally. However, with the gradual awareness of the consequences of large-scale clearing of native habitats throughout the country and the need to create bird and bush sanctuaries, early conservationists turned their

Breaksea Island, with its distinctive cap of cloud, is reminiscent of Little Barrier at the other end of the country.

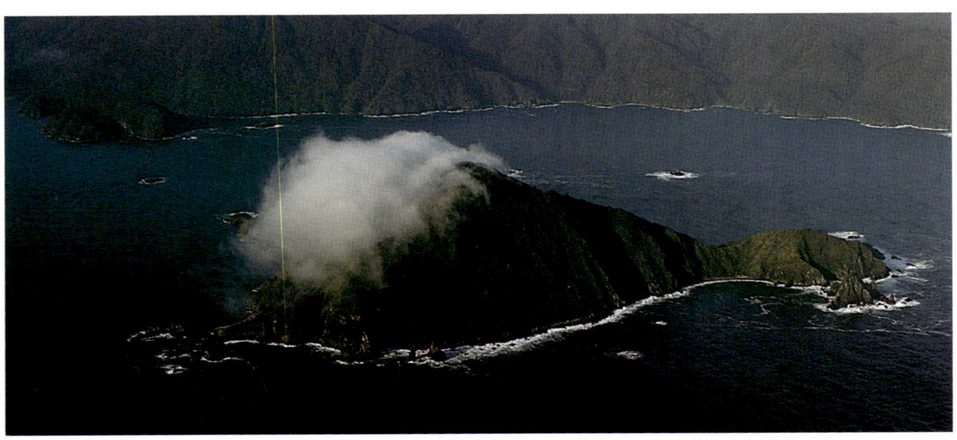

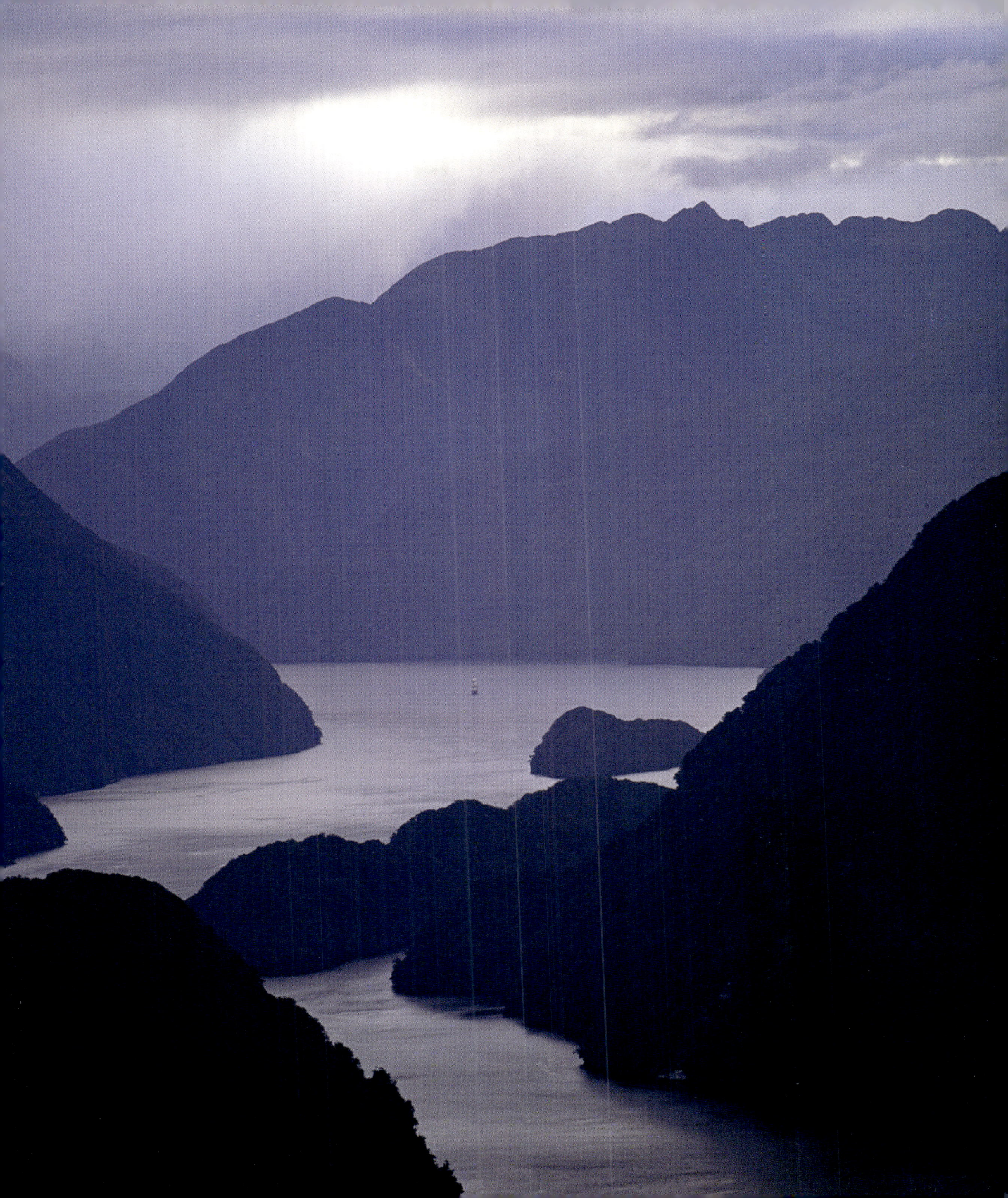

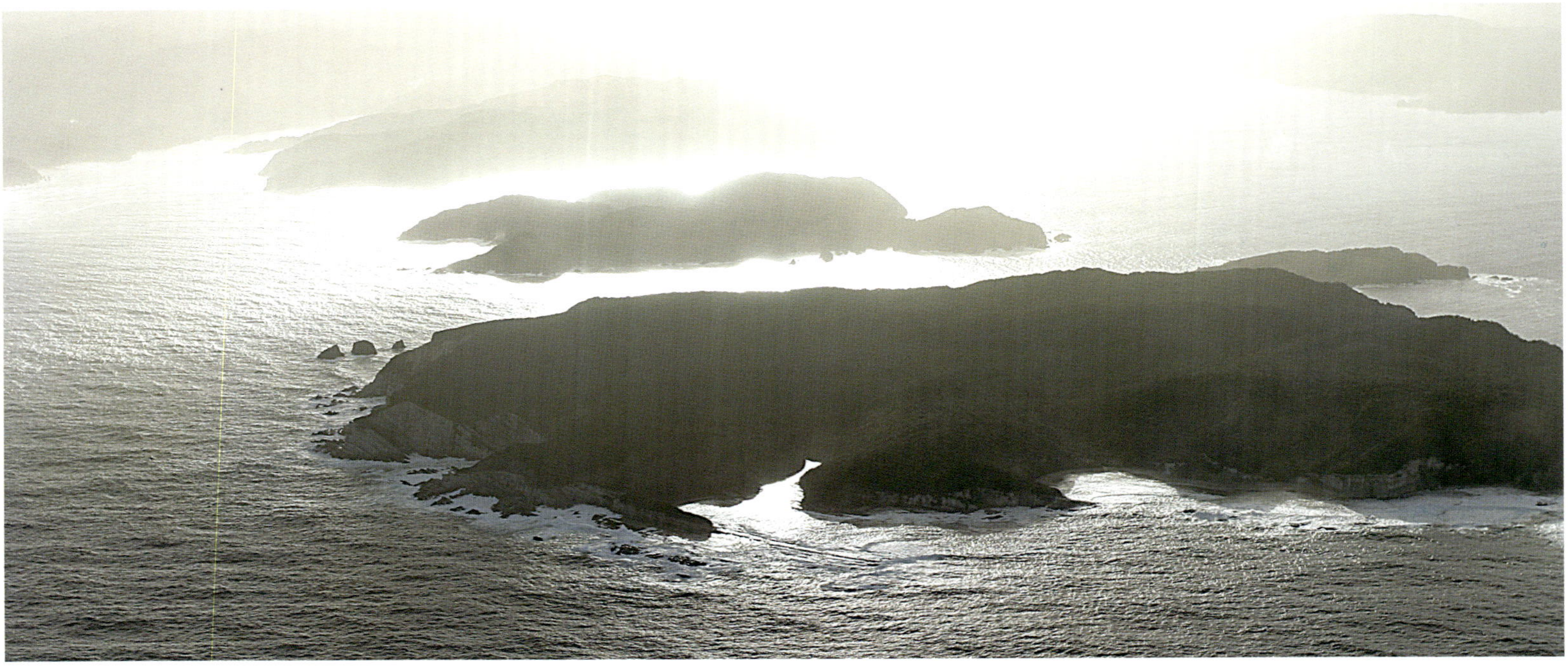

A view into the sunlight with Chalky Island in the foreground, Passage Island and Great Island beyond.

The entrance to Dusky Sound is liberally sprinkled with islands. Anchor Island is in the foreground with the Seal Islands in mid-picture and Indian Island in the background.

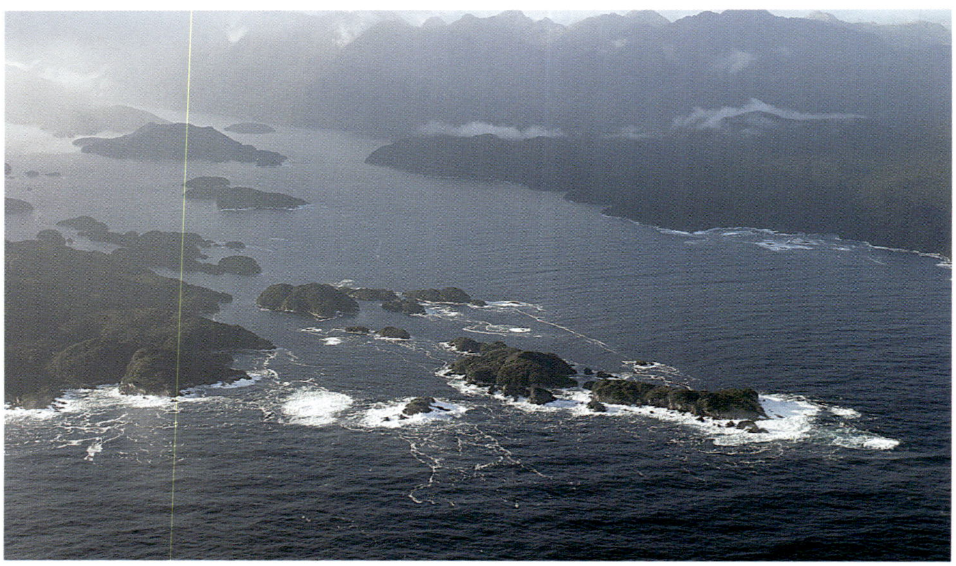

view towards Fiordland. In 1891, Resolution Island, not far from Cook's anchorage in Dusky Sound, was declared the colony's first flora and fauna reserve. Secretary Island, where the second takahe capture was recorded in 1851, became a reserve two years later.

Richard Henry, an enthusiastic naturalist, was resident in Te Anau at the time. Already fascinated by the abundant and varied birds in the area but worried about their ability to survive in the face of increasing predators, he was a natural choice as the first caretaker on Resolution Island. In 1894 he built a hut on nearby Pigeon Island and, often working alone during the next 15 years, he transferred more than 400 kakapo and numerous kiwi to Resolution, studying and writing about their habits when not actually involved in trapping the birds. It is difficult to imagine his despair when he finally discovered that stoats could swim to the island from the mainland and realised that all his work had been in vain. He left Resolution in 1909 and became custodian at Kapiti Island for a further two years before retiring.

Were Richard Henry alive today, he would be delighted to know that both Resolution Island and Secretary Island are the latest island sanctuaries to be targeted by the Department of Conservation for restoration, involving a pest-eradication programme lasting several years. Already the islands are home to species such as the Fiordland crested penguin, the New Zealand falcon, weka, kaka and kea. It is hoped that kakapo, yellowhead (mohua), rock wrens and robins may also eventually live on the islands and that Resolution Island, with its large river systems, may be able to support a population of the endangered blue duck or whio.

Chalky Island in Chalky Inlet has recently become a sanctuary for kakapo. Hopefully its beech forest will supply plentiful food and so encourage the birds to mate and produce young.

No longer just a destination for overseas tourists and hardy trampers, Fiordland is taking its place in the conservation network of the country.

Islands across Foveaux Strait

Stewart Island

Early morning and I'm awakened by a manic screeching in the distance. A moment's return to peace, then it comes again, a tortured shriek this time, right outside our window. I tiptoe across the floor, quietly pull the curtain aside, and there, sitting on the balcony, is a kaka – cruel hooked beak, sardonic eye and feathers the colour of moss – a bird I've only read about up till now.

We're on Stewart Island, in Halfmoon Bay, which sits at the entrance to Paterson Inlet, a large protected waterway that cuts deeply into the island on its eastern side. Oban, as the town is called, is the only permanent community and the rest of the 1700-square-kilometre land mass is virtually unchanged from its original state.

Fly over the island and it looks like a sprawling sea creature, covered in dense fur. Brilliant blue bays are cradled within rugged limbs of land that extend out from the body of the creature, and its more vulnerable parts are fringed by gleaming pristine beaches. Spending time here is probably the closest anyone can get to experiencing New Zealand before the arrival of humans.

Had Captain Cook been able to charter a flight when he first arrived here in 1770, the land

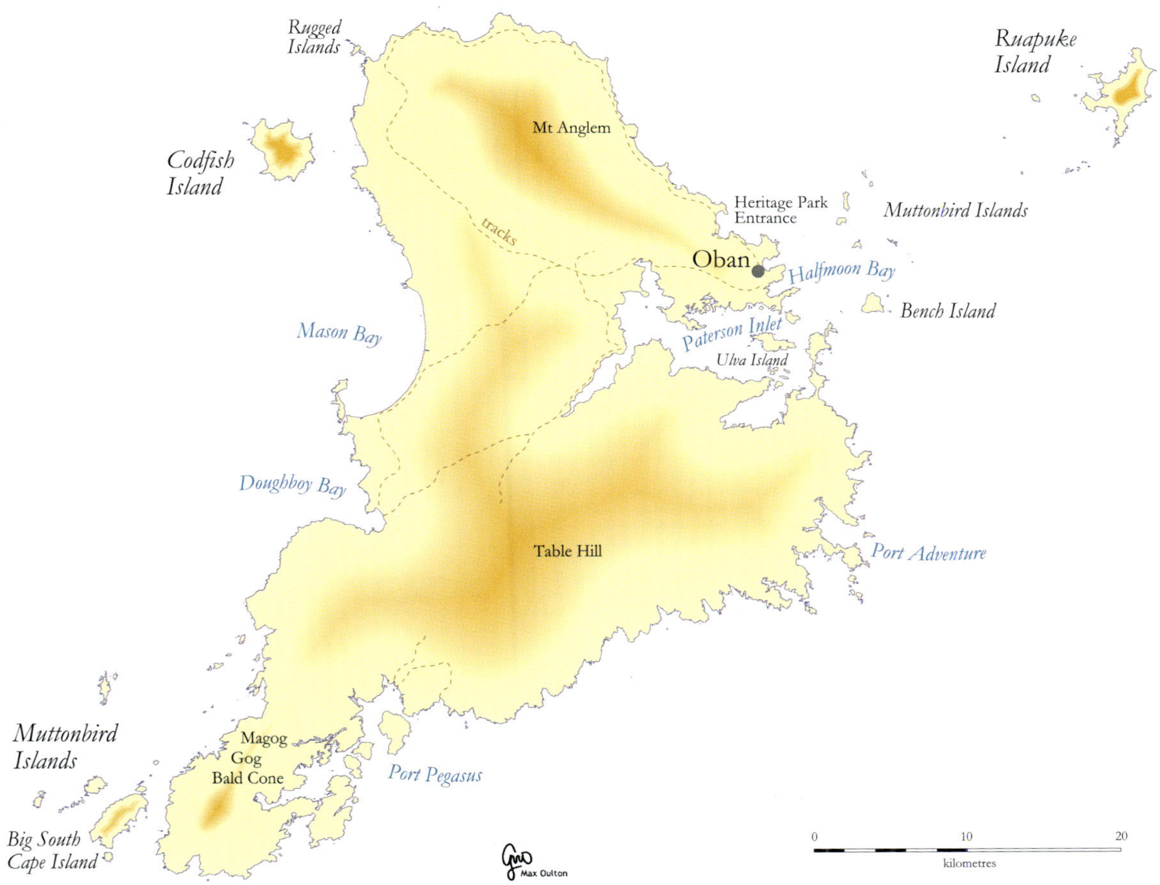

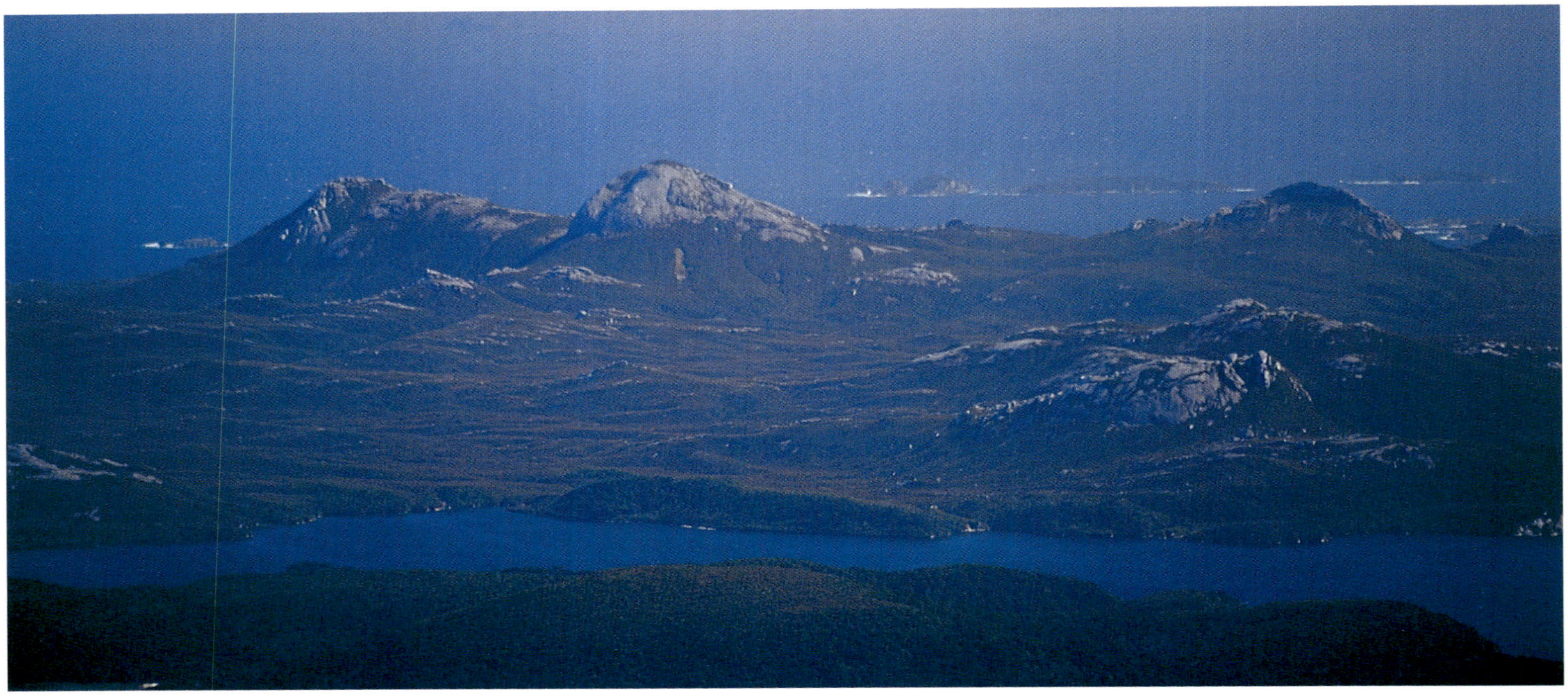

The granite domes of Magog at left, Gog and Bald Cone, right, rise above the dense bush-covered hills at the southern end of Stewart Island.

then would have looked much the same as it looks now. He would have clearly seen Foveaux Strait too and known that he was looking at an island. Unsure of the existence of the often turbulent strip of water, Cook linked Stewart Island to the mainland by a dotted line on his chart. It was not until 1804 that a sealer, exploring the coast of Stewart Island in a whaleboat, definitively discovered the strait, although Maori had known of its existence for generations. Foveaux was the Governor of New South Wales in 1809 and for reasons that are unclear his name has lived on.

The earliest Europeans in the Stewart Island area were whalers and sealers prowling the Southern Ocean in search of prey. The men brought with them a heritage of independence from their own island homes of the Shetland and Orkney Islands. This trait, along with qualities of resilience and self-sufficiency that the Maori inhabitants possessed in large measure, has endured in their joint descendants and is shared by others who have made the island their home.

In 1826, a team of boat builders from the Bay of Islands, under the leadership of William Cook, was enticed to Port Pegasus, by the offer of ship-building work. It is a spectacular, peaceful, land-locked harbour at the south end of the island where the present-day inhabitants of Oban go for rest and recreation and an escape from the summer influx of tourists. Sheltered anchorages make it an ideal haven for cruising boats and it is renowned as a place to view the magnificent displays of the Aurora australis. Weather permitting, trippers can go there and back in a day from Halfmoon Bay.

But the weather makes it an inhospitable place for permanent living and the originator of the boat-building scheme so long ago was a scoundrel. The colonists were virtually abandoned. They survived there for seven long years, sometimes close to starvation, before the opportunity arose to return north. (Their sponsor, William Stewart, traded on the reputation of a contemporary of the same name who had earlier charted Port Pegasus and whose name the island carries.)

Boats form an integral part of island life and Stewart Island had wood aplenty for building them. Following on from the lonely boat builders of Port Pegasus, other settlers turned their hands to the craft. Early photographs show boats in varying stages of construction, on beaches in the vicinity of Halfmoon Bay.

Charles Traill, whose descendants still live on the island, commissioned the building of a cutter that became Stewart Island's first official mail and passenger carrier. The keel of the *Ulva* was laid on Ulva Island in 1875 and was launched at the end of the following year.

Sawmilling was one of the industries that employed men on the island for a period of about 70 years from 1861. Small colonies of sawmillers' cottages sprang up around the mills. Tramlines to

Freight ferry alongside the wharf at Oban on Stewart Island.

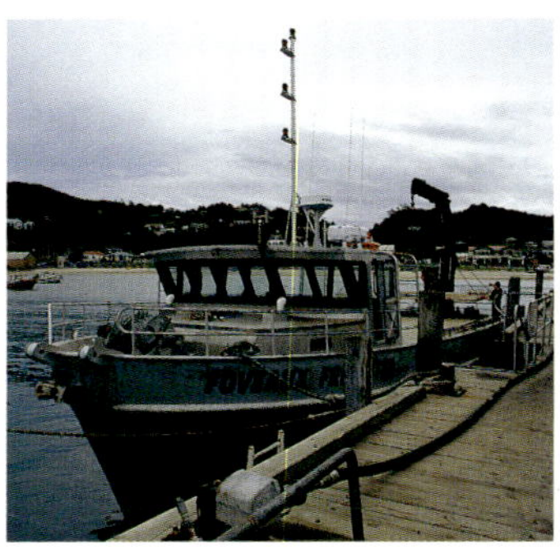

carry the logs from the bush to the mill – usually at the mouth of a stream – and a long wharf to ship the timber out became features of the landscape in various bays around Paterson Inlet.

Perhaps the biggest impact from the past on the island we know today is the lack of any large-scale successful farming. Elsewhere in New Zealand huge tracts of forest were cleared to make way for sheep and cattle. Although farming took place at various locations on Stewart Island, little native bush was cleared to make way for it, with positive results for later moves towards conservation.

I first went to Stewart Island in the 1950s. My grandfather spent his annual holidays on the island and he introduced my brother and me to this, his favourite place. I suffered my one and only bout of seasickness on the ferry *Wairua* as it buffeted its way across Foveaux Strait for three long hours.

In those days there was no electric power laid on – people used candles, kerosene lamps or had their own generators. The thump of motors at twilight was a common sound.

We went out on a boat and pulled up endless fish. I hooked a groper so large that my grandfather had to haul it aboard. A friend gave us a sack of live crayfish and we boiled them up in a kerosene tin on a primus stove. We went on bush walks and discovered a house built like a boat.

Times have changed. Now visitors to the island travel aboard a modern catamaran that takes just one hour to make the crossing. Fishing-boat skippers turned tourist operators run fishing trips and the catch is limited. Crayfish are exported and rarely land on the tables of casual arrivals.

But bush still cloaks the island, the 'boat' house is alive and well and the birds have increased. Tui, kaka and bellbird flood the town with their shrieking, trilling, chattering, clacking and warbling. Fantails dart through the trees and New Zealand pigeons crash from branch to branch. Life here moves to an island rhythm. The magic is intact.

Oban is a small village tucked into steep hills. Scattered cottages punctuate the bush and the one hotel stands solidly on the foreshore as it has for decades. Seabirds soar overhead. Pleasure craft and the occasional fishing boat moor in the bay. Once it would have been the other way around – fishing boats and the occasional pleasure craft. But the advent of expensive fishing quotas has changed the economy of the island and tourism now is an important source of income for increasing numbers of islanders.

As in any island community the wharf is the hub of the town. Upstairs in the ferry building there's a café with a fine view. Boats come and go. When the twice-daily passenger boat from Bluff arrives, the town springs to life. Two or three mini-buses arrive to ferry visitors to their destination, returning islanders are greeted with warm hugs and young backpackers straggle off to find accommodation. A forklift darts back and forward dealing with the luggage.

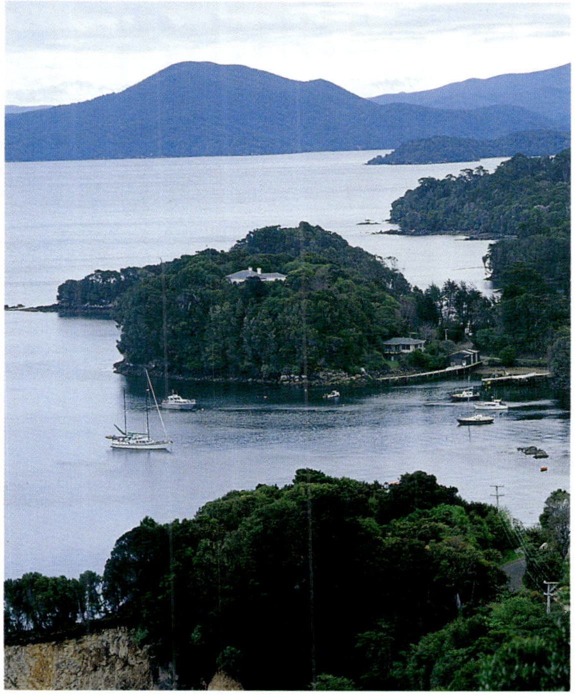

A quiet anchorage at Thule in Paterson Inlet, Stewart Island.

How to get there

Daily ferries run between Bluff and Stewart Island.
Chartered cruises to Stewart Island and around its coastline also depart from Bluff.
Chartered flights depart from Bluff Airport.

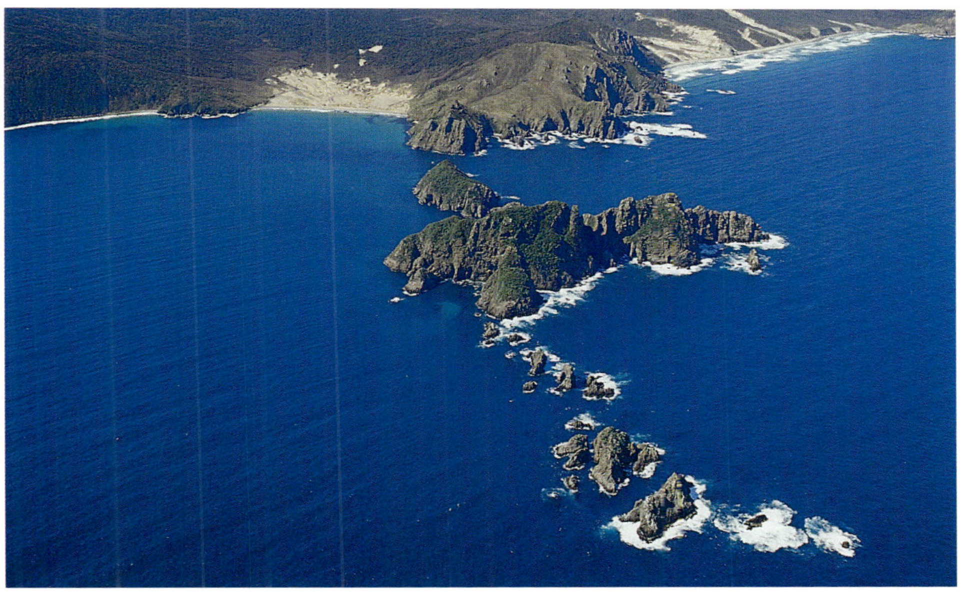

Rugged Islands off the north-west corner of Stewart Island.

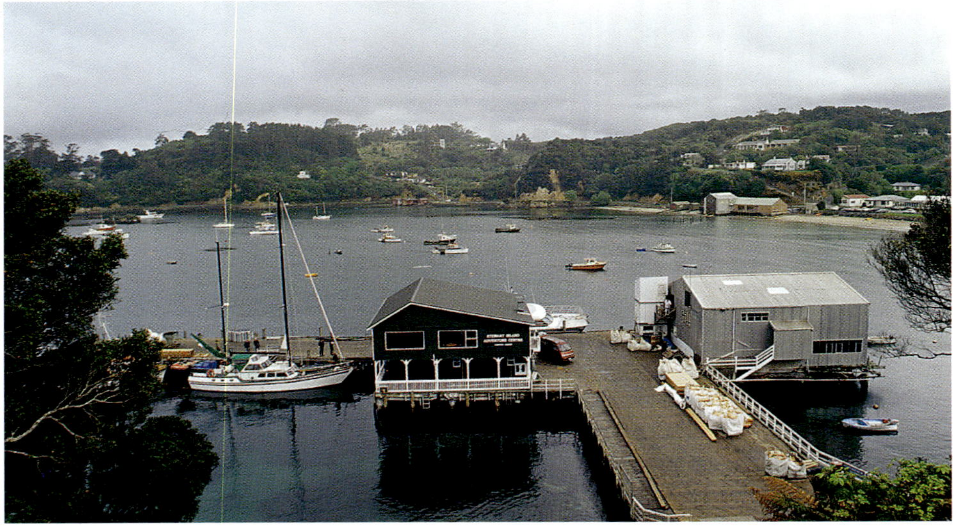

Oban in Halfmoon Bay is the only town on Stewart Island and there is a flurry of activity at the wharf every day when the ferry arrives from the mainland.

Later a freight ferry hoves into sight and a couple of hi-abs swing into action ready to handle the day's cargo – an incoming load of timber, a new vehicle for the island, maybe drums of diesel for the town's relatively recent electric power plant or bags of cement for a building project. Crates of chilled blue cod and paua are loaded straight from the Fisherman's Co-op processing plant on the wharf for the return journey to the mainland.

At the end of the season fishing boats pull in, their quotas filled, ready to off-load their cray pots or cod pots, a smelly cargo that has to be stored on shore to await the next season. A charter boat ties up and its skipper passes the time of day with a friend before catching up with routine maintenance. It's a place where people have time to talk.

With a total of 400 inhabitants, the only strangers are arriving visitors – and they are drawn into the relaxed, friendly atmosphere of the local population. There was a time when you could live here 25 years and still not qualify as a local. Now they say, if you get off the boat in the morning and you're still here in the afternoon, you're a local.

Leonard Cockayne, a visionary botanist and an ecologist long before the term became current, foresaw the importance of the island for recreational uses as early as 1909 when he visited Stewart Island and carried out a botanical survey.

'The capabilities of Stewart Island as a pleasure resort can hardly be overestimated. That it will eventually be celebrated, not only in New Zealand, but throughout Australasia, is certain,' he wrote in his report. He went on to talk about the hundreds who visited the island even then, to enjoy boating, excellent fishing, bathing, mountain climbing or just picnicking in the midst of beautiful scenery.

Some areas on the main island were first put under protection as early as 1903 and, until the Department of Conservation was established in 1987, administered by the New Zealand Forest Service. But it wasn't until 2002 that Stewart

Bragg's Bay. Stewart Island and its outliers are not all rugged coastline; there are also many sheltered inlets and sandy beaches.

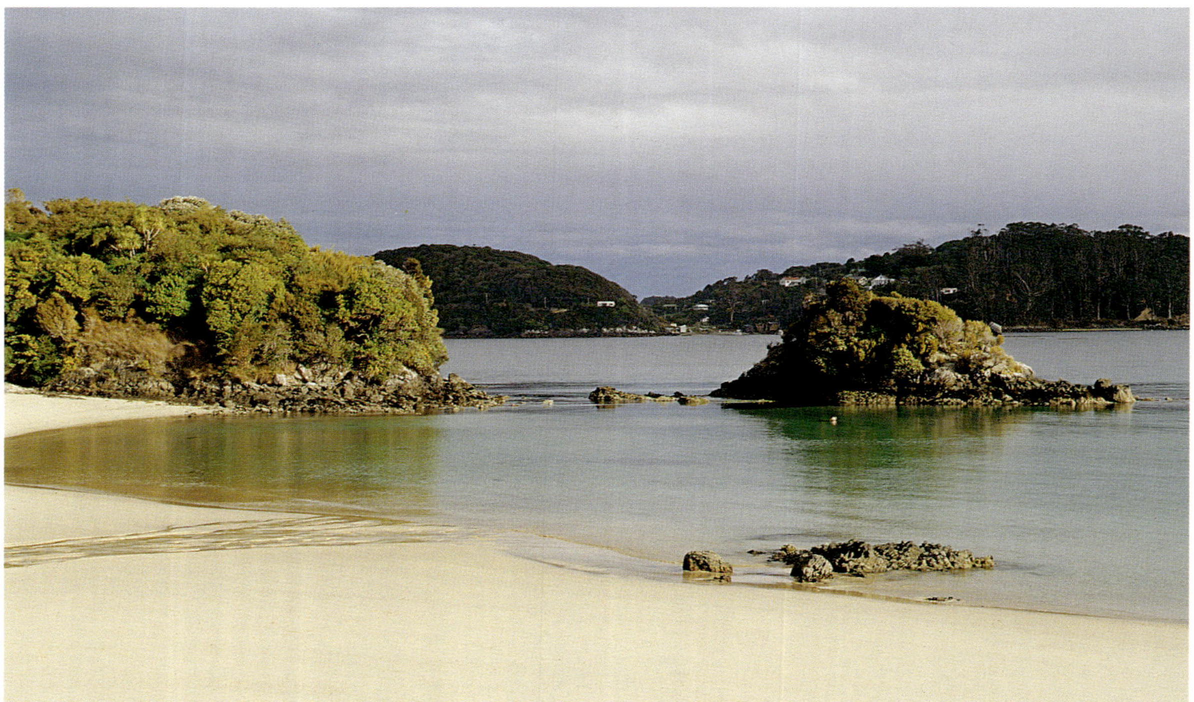

Island or Rakiura – 'Land of the Glowing Skies' – was designated a national park with nearly 90 percent of the land, including 170 islands and rock stacks in Foveaux Strait, administered as conservation estate under the aegis of DOC. Perhaps that time, which Cockayne foresaw nearly a century ago, has arrived.

People visit the island today to escape the rigours of modern life and to enjoy the undisturbed natural beauty. The vegetation is lush. It is possibly the southernmost rainforest in the world and although there are few plants endemic to the islands, many native species thrive. Orchids drape themselves from tree branches and several varieties of ferns enjoy life beneath the tree canopy. Podocarps and hardwood – rimu, kamahi, totara, kahikatea and southern rata – dominate the forest areas in the northern half of the island while the coastal forest includes windswept manuka, hebes, dracophyllums and several different 'daisy bushes' often seen clutching onto cliff faces, flourishing in the face of gales and salt-laden spray. Muttonbird scrub is used variously to describe at least two of these species. In days gone by, a novel use was made of their large leaves. People treated them as postcards, writing on the downy, white surface and sticking a stamp to it. Unfortunately the post office no longer accepts these as legal mail.

The remaining areas are covered in grasses, swampy wetlands and, as the altitude increases, herb-fields take over and interesting alpine species grow. In 1881, during the first recorded ascent of Mount Rakeahua on the western side of the island inland from Mason Bay, two botanists found previously unknown species of both *Aciphylla* and the cushiony clumping *Raoulia*, often called vegetable sheep.

Mason Bay, a magnificent deserted beach on the western side of Stewart Island, curls into the Ernest Islands at its southern tip.

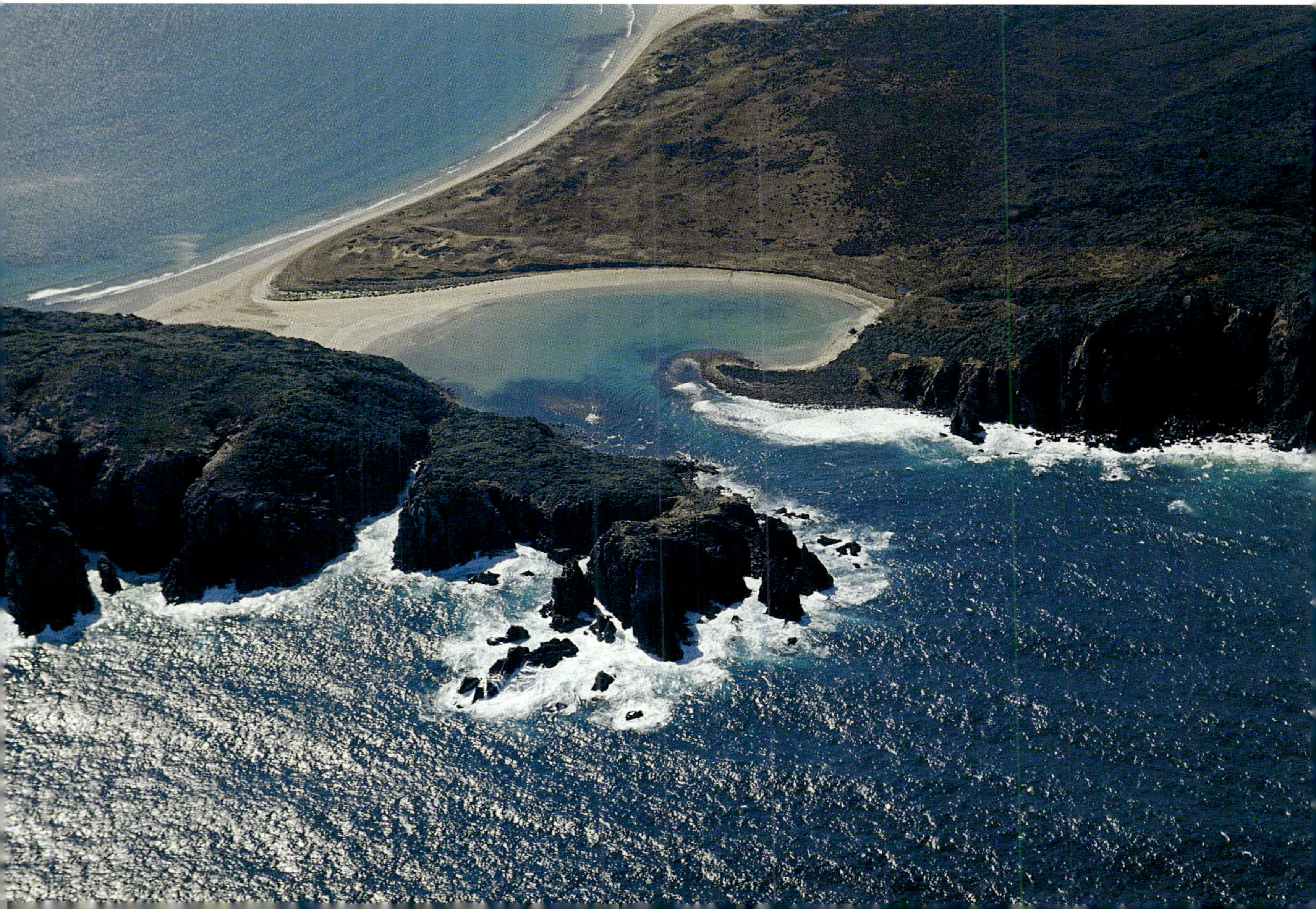

Many visitors go there to see real live kiwi in the wild. It's said that more kiwi than people inhabit the island. The brown kiwi or southern tokoeka is a sub-species endemic to Stewart Island with its own particular habits – they tend to maintain family units and it's not uncommon to see these usually nocturnal creatures foraging in daylight hours, particularly in the Mason Bay area.

Other visitors come to hunt, dive, fish, or go kayaking and exploring. With only 25 kilometres of road and more than 250 kilometres of tracks, Stewart Island has long been known as a tramper's paradise. Visitors come from all over the world to roam through the forest, enjoying the fascinating variety of birds and marvelling over the spectacular deserted beaches. Few depart without also mentioning the mud that clogs their footsteps although many of the tracks include long spans of boardwalk, built by DOC to protect the vegetation as much as trampers' boots.

Day-trippers can explore the bush or wander on lonely beaches of golden sand within a stone's throw of Halfmoon Bay. Serious trampers can choose to follow the North West Circuit that takes between eight and 12 days to complete. It circumnavigates the northern part of the island following the coast around the top half of the island and continuing to Mason Bay. From there it's an overland trek to North Arm in Paterson Inlet and so back to Halfmoon Bay. A short side trip leads to the summit of Mt Anglem, the highest point of the island at 980 metres.

In the south Gog and Magog rise behind Port Pegasus, distinctive on a clear day with their bald granite domes standing above the dense covering of bush at lower altitudes. Joseph Banks, who accompanied Cook on his first voyage to New Zealand, commented that they appeared to be paved with glass.

The large tracts of forest wilderness have always been home to many birds. While rats and cats became a threat after the arrival of Europeans, fortunately no ferrets, weasels or similar predators have reached the island and it has become a haven for threatened species from the mainland.

Stewart Island is in the midst of radical change. Long-term residents express different views about the island's future. Some fear its popularity as a tourist destination will spoil the character of the island as they have known it. Jo Leask is a member of one of the oldest European families on Stewart Island. The name is legendary: they claim seven generations in a direct line from Orkney Islander Tom Leask who gave his name to Leask's Bay about 1866, and it was members of her husband's family who farmed at Mason Bay for many years. 'We all feel very protective about our island but some of us realise we rely on visitors for our livelihood. Traditional islanders are reluctant to admit it has to change and that tourism is the way of the future.'

But Jo herself has embraced the new way with enthusiasm. 'I love change,' she admits. She has recently opened a new business called Empress Pearls, which sells jewellery made from paua pearls.

Dil Belworthy does not claim the island heritage of the Leasks. About 10 years ago he was living in Auckland, he was at a loose end and his brother-in-law, a Stewart Island fisherman, persuaded him to come south and work on the island. Now he and his wife run a business, which employs other islanders, screen printing and selling high-quality T shirts they design themselves. In addition to their shop in Oban they supply several other gift shops throughout the country, including Auckland Airport and Te Papa Museum. 'Some people see DOC and tourism as the enemy,' he comments. 'In reality, they are the saviour.'

Muttonbird Islands

Stewart Island, the third biggest in New Zealand after the North and South Islands, is itself surrounded by numerous small islands. Maori came here long before Europeans arrived on the scene. Perhaps they initially fled the mainland to escape enemies but they also hunted the fur seal and collected shellfish for food. They also gathered muttonbird chicks as a special delicacy – a tradition that lives on today on the Muttonbird or Titi Islands at the southern tip of the main island and those dotting the sea at the approach to Halfmoon Bay.

Several of these islands, owned by Ngai Tahu, are of conservation value because of their diverse range of habitats and mix of species. Close co-operation between Maori and DOC has seen the recent eradication of rats from Putauhina Island, followed by the re-introduction of fernbirds and robins.

Conservation, however, has not stopped the age-old harvest of muttonbird chicks that still

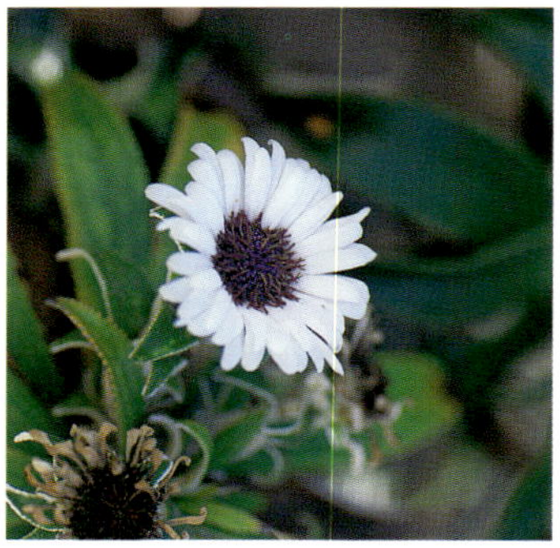

Muttonbird scrub is a common name used to describe at least two of the coastal 'daisy bushes' that hug the cliffs around Stewart Island.

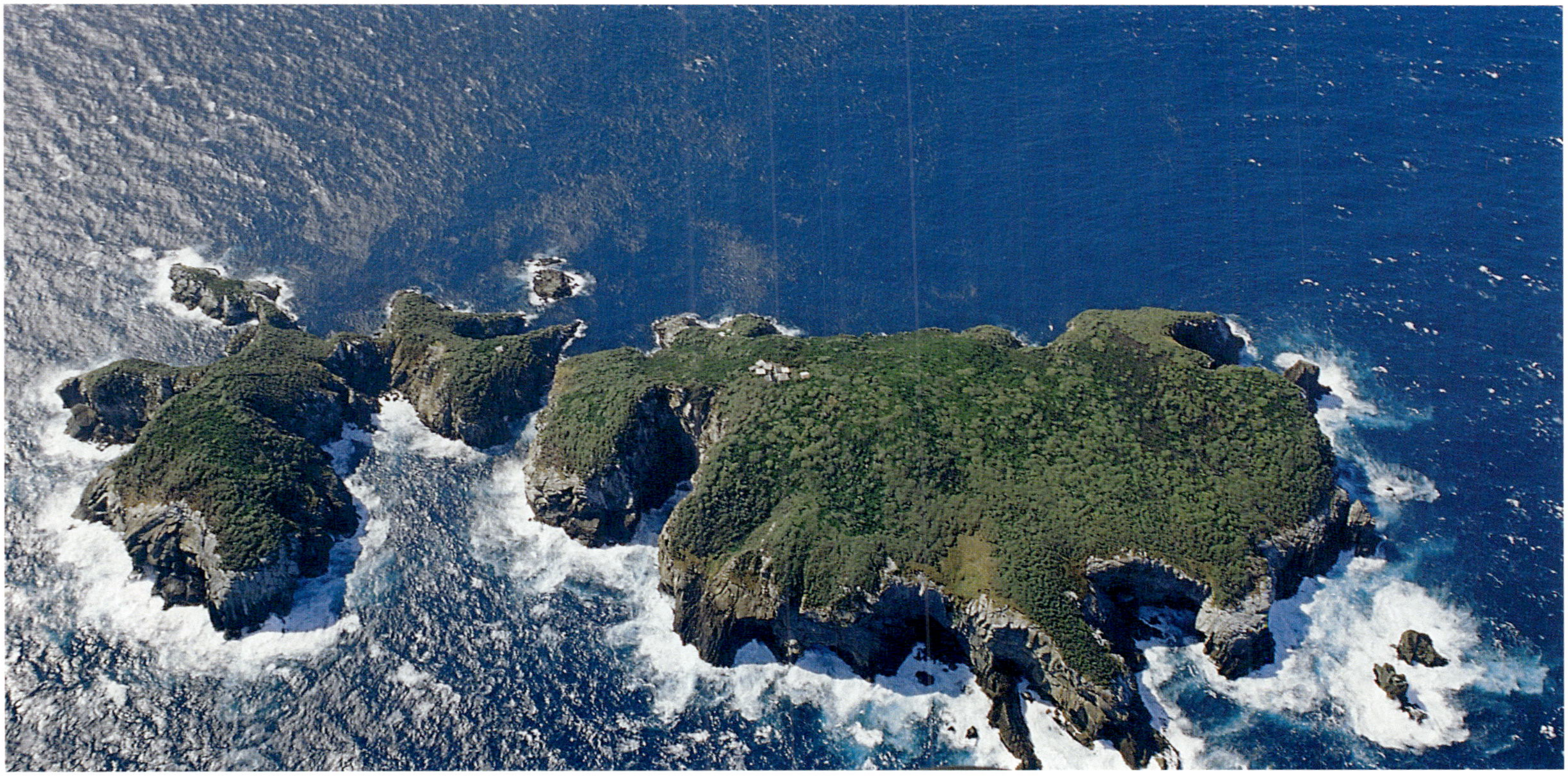

Pohowaitai Island is one of the Southern Muttonbird Islands. Maori families come here in April to harvest the muttonbird chicks – the houses they stay in are visible at the top of the island.

takes place on the Titi Islands in autumn. Rights to the delicacy are strictly traditional and jealously guarded, with individual hapu or Maori family groups claiming the harvest on different islands.

Codfish Island/ Whenua Hou

Yet other islands, some little more than a steep chunk of rock covered in bush and pounded by the sea, are scattered along the western coast of Stewart Island. The largest of these is Codfish or Whenua Hou, now an important conservation island.

In the late 1970s an exciting discovery was made in the mountainous Tin Range in the south of Stewart Island. A colony of extremely rare kakapo was discovered and these ground-dwelling parrots remained living in the wild until 1997 when they were removed to the sanctuary of Little Barrier Island in the Hauraki Gulf to ensure their survival. Birds bred in captivity there have since been brought back to Codfish, an island that is rigorously monitored and predator-free. Even the resident competitive weka were removed to Stewart Island to give the kakapo optimum chances of survival.

Codfish Island, home to a colony of threatened kakapo, with Stewart Island in the background.

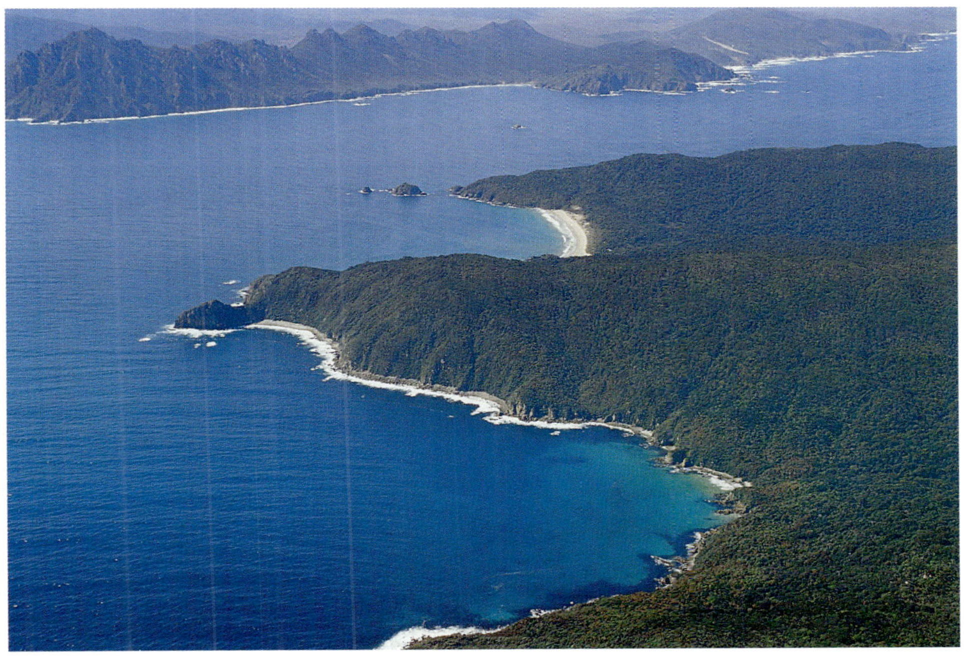

Ulva Island

I spent a morning fishing in Paterson Inlet a few years ago. The blue cod were biting and our party stopped at a patch of rocks near the shore to gather mussels before the boat pulled into the jetty at Ulva Island for a lunch of fresh fish and home-made bread. Bellbirds flooded the air with song. Mere talking seemed crass.

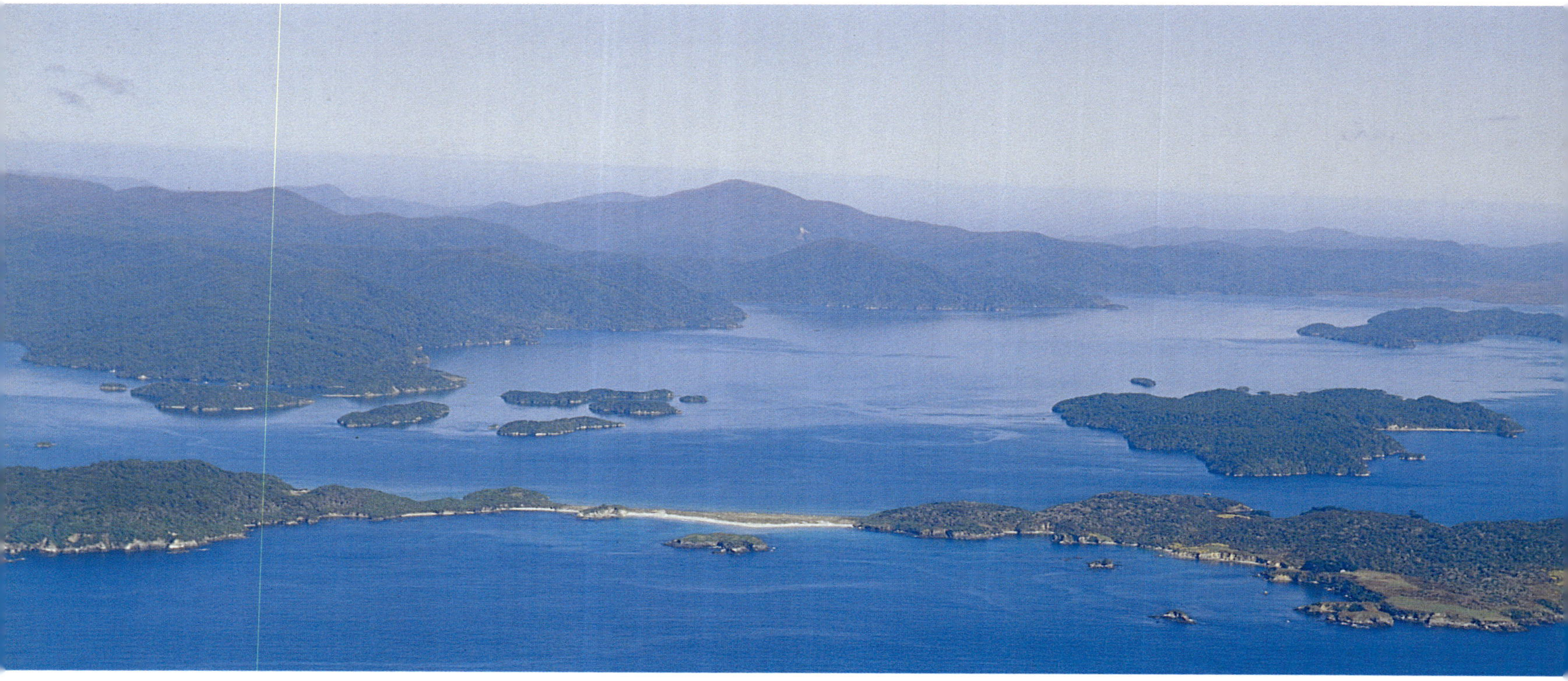

A view across Paterson Inlet showing The Neck, centre front, and the Bravo Islands on the left. Ulva Island is at the right.

Ulva Island is a haven for native birds. Like Tiritiri Matangi it is run as an open sanctuary – a place where birds and plants can live a safe, protected life but members of the public have free access to enjoy the wildlife at close quarters. It also serves as a scientific reserve, with students going there in the holidays to do research on bird populations. Ensuring that all arriving boats, kayaks, picnic baskets and even trampers' packs are rat-free before people step on the island is an important task. The lives of the birds depend on their habitat remaining predator-free.

People have been coming here for generations. Ngai Tahu Maori visited periodically in the past to strip bark from totara trees which they used in the storage of muttonbirds. Charles Traill, he who established the first Stewart Island post office that is still there, created an extensive garden on Ulva and some of the exotic trees he planted exist today. As early as the 1880s, the Tourist Department provided funding for tracks. Now visitors arrive by water-taxi and wander all over the island.

Ulva Amos, who delights in being described as a radical greenie, is a direct descendant on her mother's side from the first people on Rakiura, the Waitaha. On her father's side she has six generations of European island ancestry. Ulva was named after her mother's favourite island. 'It meant that my grandmother allowed me all sorts of privileges my brothers didn't have,' she chuckles, remembering her childhood. 'I went away for 14 years, while my children were being educated. Now I'm back on Stewart Island and working as a Maori nature tourism operator.'

Ulva takes visitors on guided tours of her namesake island and is involved in the bird-retrieval programme. She goes to Freshwater to get robins and to Codfish to collect riflemen, all of them released on Ulva. 'The riflemen are as big as your thumb,' she says. 'The babies are like bumblebees. They make slits in the trees for nests. And from October to December you see little bundles of fluff that turn out to be weka chicks.'

There's no doubt where Ulva's (the person's) heart lies. 'I want everyone to know about our wonderful place and I want to show it off,' she says, and then hesitates. 'Oban used to be a sleepy fishing village. That's changed. People are realising Stewart Island is a treasure. Building permits are increasing and we're getting some quality accommodation now,' she continues. 'But I just hope it doesn't happen too quickly, that we're not inundated with visitors. Now in January and February, Ulva – the island, not me – is almost getting too busy.'

Ruapuke

Ruapuke Island sits in Foveaux Strait to the east of Stewart Island and is in fact closer to Bluff. In contrast to its neighbour, it sits low in the water with spreading tentacles of land that creep out into the sea and create sweeping bays of calm blue water on a fine day. The island is privately owned, and apart from some areas of bush in the centre, the land is mainly cleared and used for farming.

Visitors nowadays come to Ruapuke for the scuba diving and snorkelling in the clear waters around the island. There are interesting reef areas, the fish life is prolific and the wreck of the *Awarua* is an added underwater attraction.

In the early 1800s the picture was very different. A sub-tribe of Ngai Tahu lived here and the island supported one of the largest Maori populations in the south. Hone Tuhawaiki, known as 'Bloody Jack' for his continuing use of the colourful adjective when he spoke English, was their chief. He was a skilled warrior, managing to repulse Te Rauparaha in battle, and an intelligent, able negotiator with the Europeans.

Recognising that the Europeans were a threat to the traditional Maori way of life, but realising they were in New Zealand to stay, he requested a European missionary at Ruapuke to teach his people more of the foreign ways. In May 1844 Reverend Wohlers arrived, sent by the German Lutheran Mission, and he performed the first marriages in the area. The Reverend Wohlers later moved to Stewart Island and became a well-respected member of the island hierarchy. Tuhawaiki, however, did not live to see the results of this forward-looking strategy. He was drowned later in the same year and little more is recorded of his island stronghold, though his name lives on in a tiny island off the Catlins Coast.

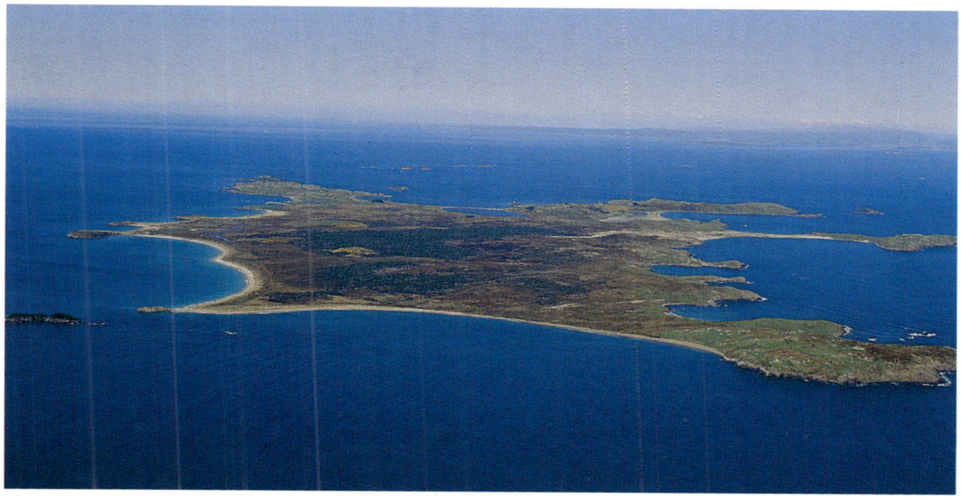

Low-lying Ruapuke Island sits in Foveaux Strait to the east of Stewart Island and was the home of Hone Tuhawaiki, also known as 'Bloody Jack', a skilled Maori warrior of the early 19th century.

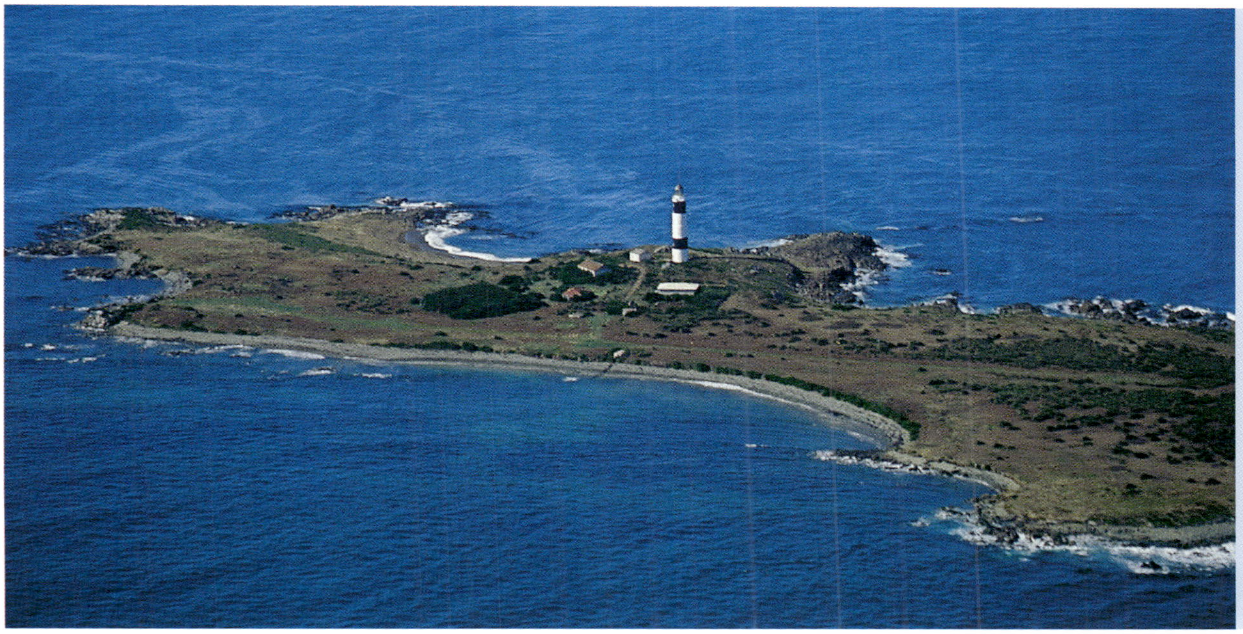

Dog Island, unlike so many islands around the coast of New Zealand, is flat, open land and sits only two metres above sea level. It sits so low in the water that the lighthouse built there in 1865 from rock quarried on the island, had to be abnormally tall – 36 metres – in order to be visible to mariners approaching Bluff Harbour. Although only five kilometres from shore, life on the island was lonely for the lighthouse keepers and it was one of the last lights to be automated in 1989.

Islands in Lakes

THERE'S AN ISLAND, AND IN THE MIDDLE OF THE ISLAND THERE'S A LAKE, IN THE MIDDLE OF THE LAKE THERE'S AN ISLAND AND IN THE MIDDLE OF THE ISLAND THERE'S A LAKE. CONFUSED? YOU'RE MEANT TO BE. THIS WAS A RIDDLE WE RECITED AS CHILDREN TO CONFUSE THE UNINITIATED. THE ISLAND IN THE LAKE IS PIGEON ISLAND AND THE LAKE AT THE VERY TOP OF PIGEON ISLAND IS PARADISE LAKE, SOMETIMES CALLED THE ARETHUSA POOL. PIGEON ISLAND SITS IN THE MIDDLE OF LAKE WANAKA, WHICH SITS IN THE MIDDLE OF THE SOUTH ISLAND. IT'S A FAVOURITE PICNIC SPOT FOR LAKE BOATIES.

We usually think of islands as the home of screeching seagulls, with tidal beaches and rocky shores pounded by the sea. But New Zealand is also a land of lakes, and the islands in them offer the same mystique, the same attraction and the same sense of security as an island in the sea.

Alfred Pinn must have felt this way about Pigeon Island, hemmed in by mountains and out of sight of the tiny settlements that existed at each end of Lake Wanaka. He built a cottage there in 1863 and established a boat-building business. He also found time to run a couple of hundred merino sheep.

Autumn touches the trees on Ruby Island in Lake Wanaka where, in days gone by, New Year's Eve revellers enjoyed the sprung dance floor that existed at the top of the island.

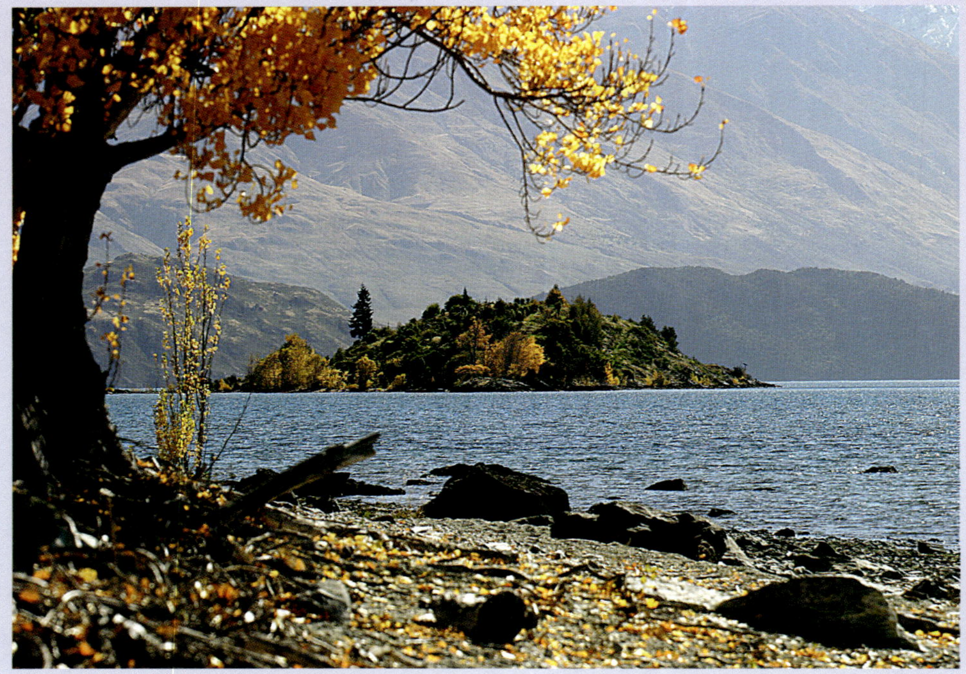

No one followed in Pinn's footsteps but the garden he planted lingered on and years later vestiges of it could still be seen by excursionists who travelled up the lake to picnic on the island, now re-christened Mou Waho.

Ruby Island is visible from the town of Wanaka. It was designated a reserve in 1886 and became a popular picnic spot with early settlers. In the 1920s it became known as the 'good-time island'. An outdoor cabaret existed on its summit and as many as 300 revellers would spend New Year's Eve dancing on the wooden floor, especially sprung for the purpose on hundreds of old car tyres. Ruby Island was never inhabited, except by a plague of rabbits, and they were poisoned off by 1985. Revegetation began on the three hectares of glacial rock remnant but in 1992 a fire broke out and burned everything from end to end. Since then volunteers have made regular trips to the island to replant it. As well as a jetty, there is a toilet and barbecue facility for picnickers.

Another island in Lake Wanaka has been designated a bird sanctuary. In 2002 15 pairs of buff weka were transferred from the Chatham Islands to Stevensons Island (Te Peka Karara), about 10 kilometres from the town of Wanaka. As the birds breed, it is hoped to release more weka on Silver Island in Lake Hawea.

The threat of fire is a constant worry on islands that have been reclaimed for conservation. For the last 10 years or more, botanist Neil Simpson, has been involved in the reforestation of another Pigeon Island, this one near the head of Lake Wakatipu. Twice in that time he has seen fire wipe out the results of much of the hard work that he and other volunteers have put in. Maybe the island should be closed to casual visitors? 'No, I don't think so,' was Neil's answer. 'People should be able to enjoy these places. I just wish they would be more careful!' The fires are usually a result of live embers left after picnickers have built a fire on the beach.

Pigeon Island in Lake Wakatipu was farmed in years gone by, as was nearby Pig Island. The Department of Lands and Survey took over administration in the 1980s. Any goats remaining on the island were destroyed as were possums at a later date. After the pests were eradicated, intermittent planting started and yellowheads or mohua were transferred there. Later it is hoped to introduce other bird species, although the risk of stoats arriving on the island is a concern as it lies within swimming range of the mainland.

Lake Taupo may be the biggest lake in New Zealand but tiny Motutaiko is its only island of mentionable size. Little is known about its history but the remains of a fortified pa exist there. The island is claimed by Ngati Tuwharetoa, and an ancient burial ground is evident there along with a burial cave which is registered as a historic site and regarded as tapu.

Much more is known about Mokoia Island. Tales of tragedy and romance are part of the history of this bigger fragment of land that sits and broods in Lake Rotorua. I step ashore and arrange with the skipper of the boat to pick our party up again in a couple of hours. I came here once as a child and bathed in Hinemoa's pool. In its warm waters, the legendary noble Maori maiden lay to warm herself after swimming from the mainland at night, to meet her forbidden lover. And it was here Tutanekai found her. He was handsome and athletic, but of humble birth and when canoes were sighted the next morning heading for Mokoia, the islanders feared retribution. But common sense prevailed and rejoicing replaced warfare.

Invasion in the early 1820s ended very differently. This time Ngapuhi canoes headed towards Mokoia in the pre-dawn hours. Te Arawa had lived on the island for centuries and cultivated its fertile soil. There was nowhere to flee when the fierce warriors descended on them. But thanks to the raucous warning from a flock of tarapunga, disturbed at their feeding on the beach, some members of the island tribe had time to hide and so escape death. The red-billed gulls of Mokoia Island have been revered by Maori ever since and protected by tapu. This episode was only one in a long history of conflict on the island, mostly among local hapu. In the 20th century the conflict moved to the courts, with a series of court hearings lasting almost half a century, until 1955. Mokoia is now administered by a trust board made up of representatives of four hapu.

Goats, rats and a few remaining sheep from earlier days were finally eradicated by 1990 and the understorey has increased rapidly. In the larger gullies mature forest creates a canopy over more open ground. A likely place to watch for birds, we sit down and wait. Soon fantails arrive. A curious North Island robin appears. There's a rush and a crash through the canopy. It's a couple of tui quarrelling and scrapping over territory. A rustle in the undergrowth and I turn to catch the merest flash of rusty orange. A saddleback or tieke arrives and starts fossicking in the undergrowth.

The more endangered of these birds have been transferred here in recent years. Conservation groups, schools, university groups and volunteer bird groups, under the direction of the Department of Conservation, have all contributed to the success that is Mokoia. Only a 15-minute boat trip from Rotorua, here is an island that provides easy public access to an intriguing environment that would have been unimaginable less than two decades ago.

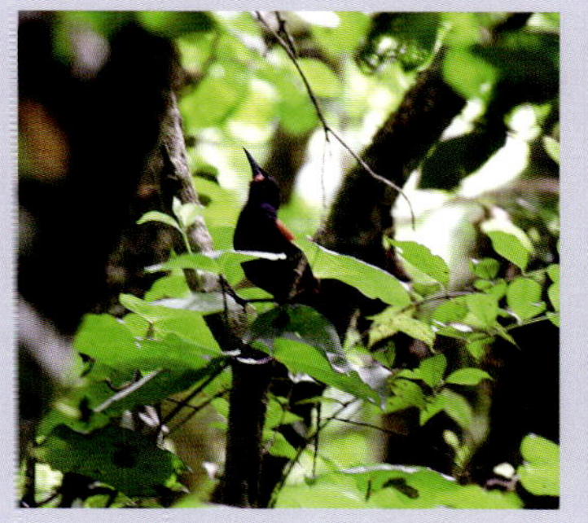

A rare saddleback on Mokoia Island in Lake Rotorua.

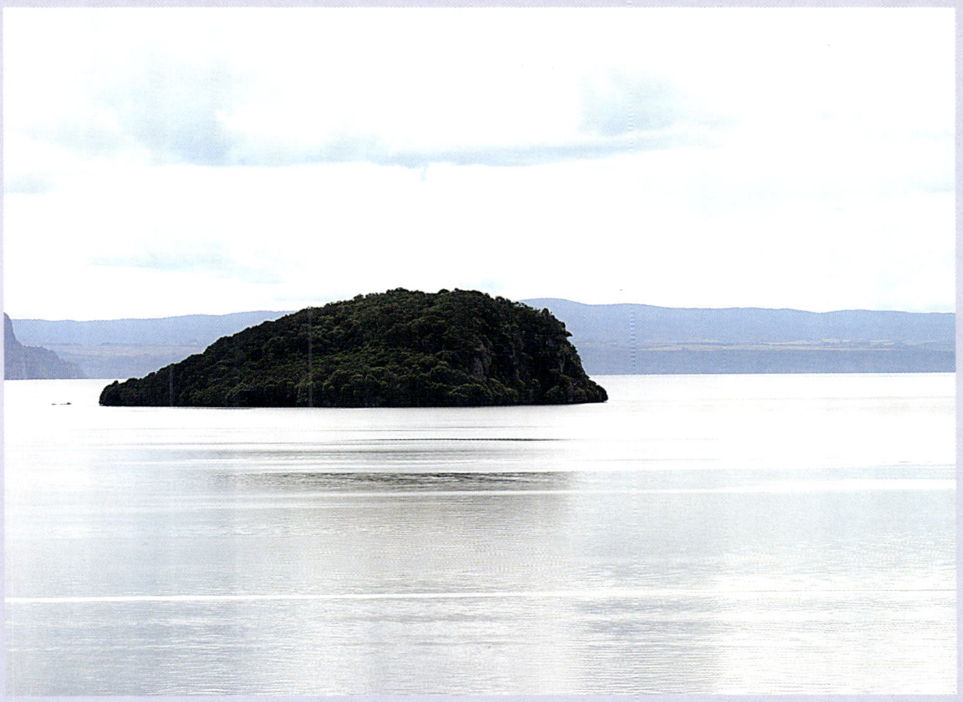

Motutaiko Island in Lake Taupo was once inhabited by Maori and today is still protected by tapu.

Chatham Islands

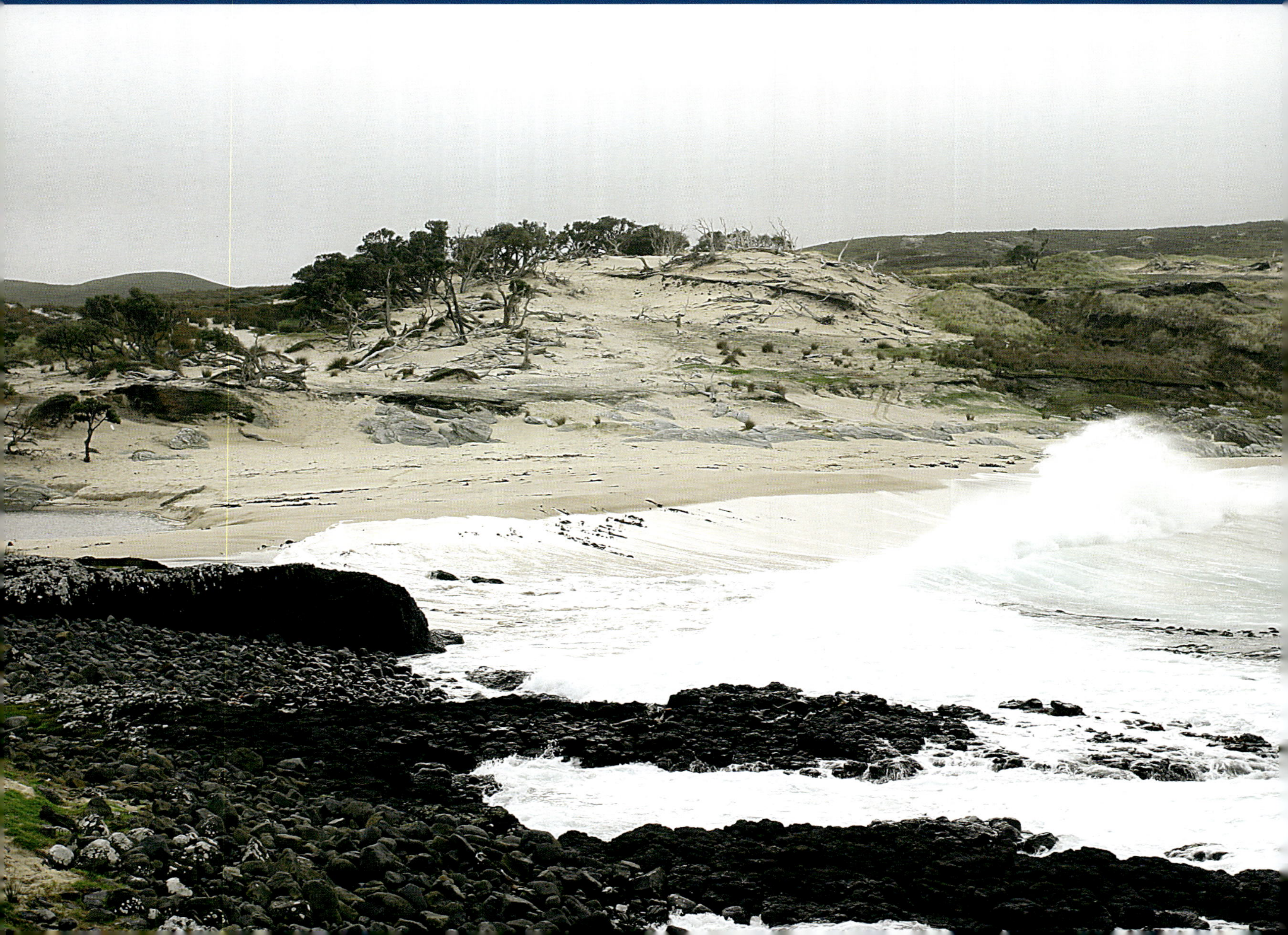

IT TAKES ABOUT ONE AND A HALF HOURS TO FLY TO THE CHATHAM ISLANDS AND IT'S NOT A CHATTY TRIP. I HAD FORGOTTEN HOW MUCH NOISE A PROPELLER-DRIVEN PLANE MAKES. AS WE APPROACHED THE MAIN ISLAND, I MOVED MY WATCH FORWARD 40 MINUTES, SCRAPED ICE OFF THE INSIDE OF THE WINDOW AND PEERED DOWN, HOPING TO SEE THE WHOLE ISLAND BENEATH ME. ALAS, IT WAS FOGGY AND I HAD ONLY THE MEREST GLIMPSE OF TE WHANGA LAGOON BEFORE WE LANDED.

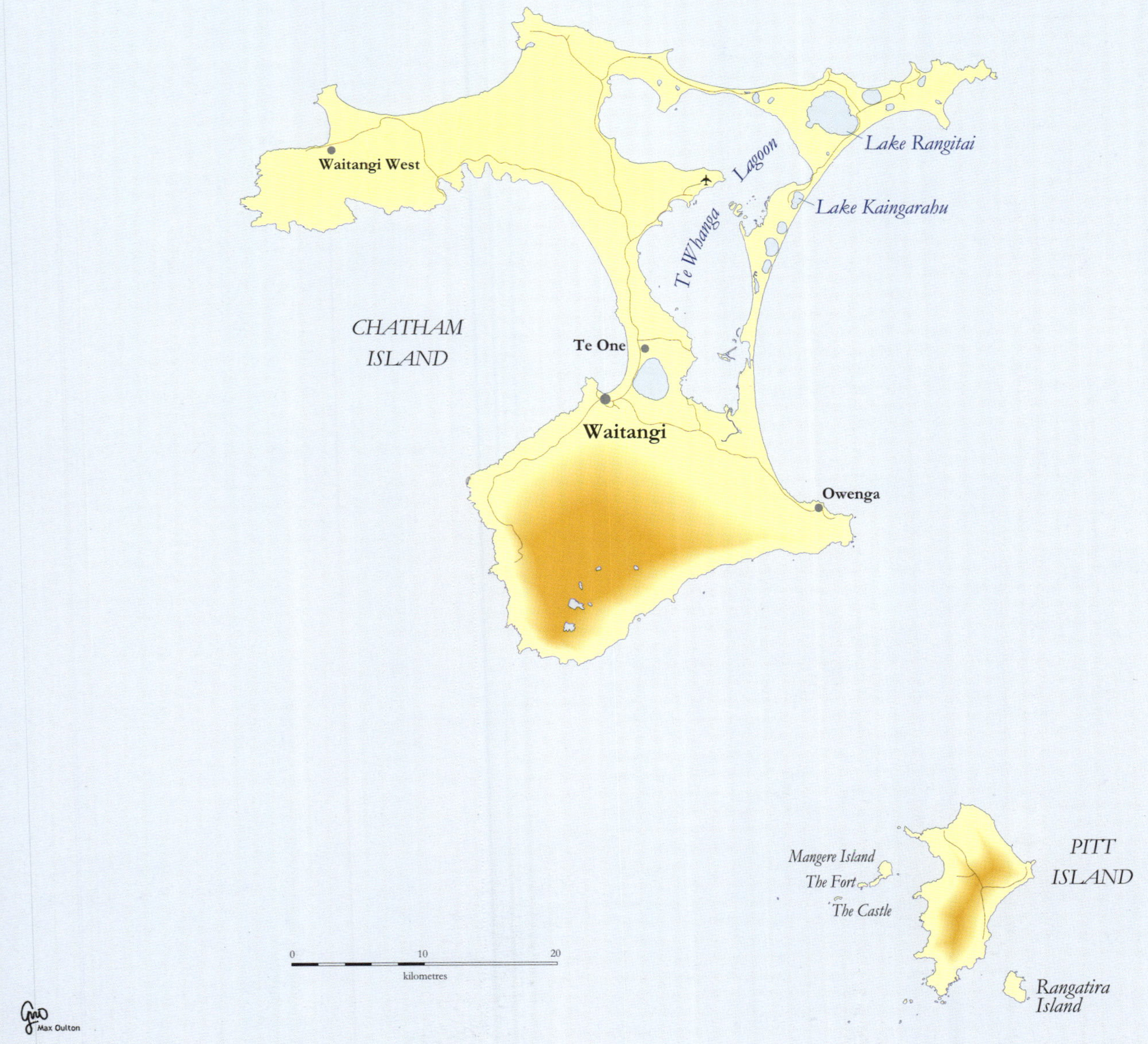

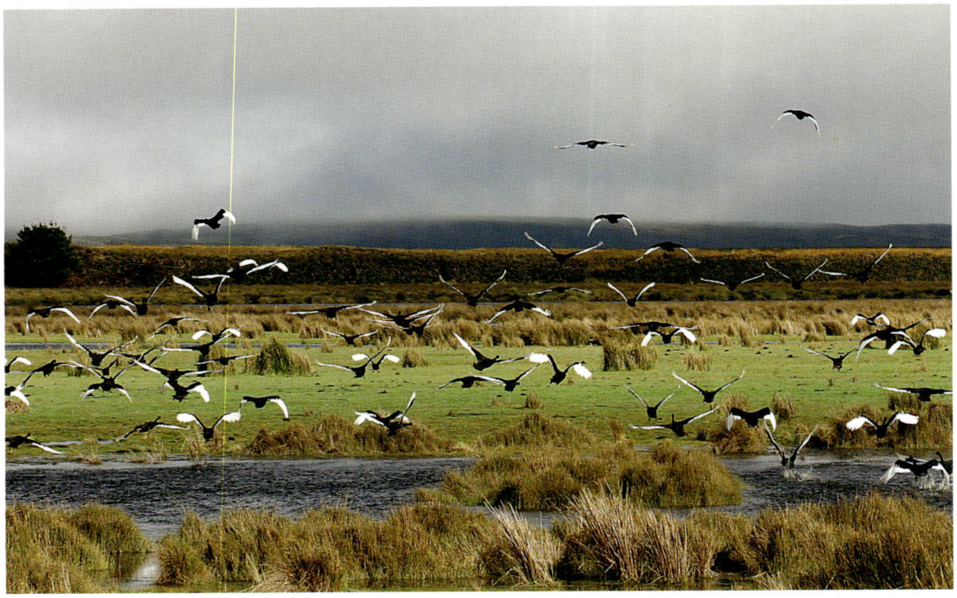

Black swans near Lake Wharemanu at the northern end of Te Whanga Lagoon.

Previous Page: The desolate beach at Paritu is typical of beaches on the western side of Chatham Island.

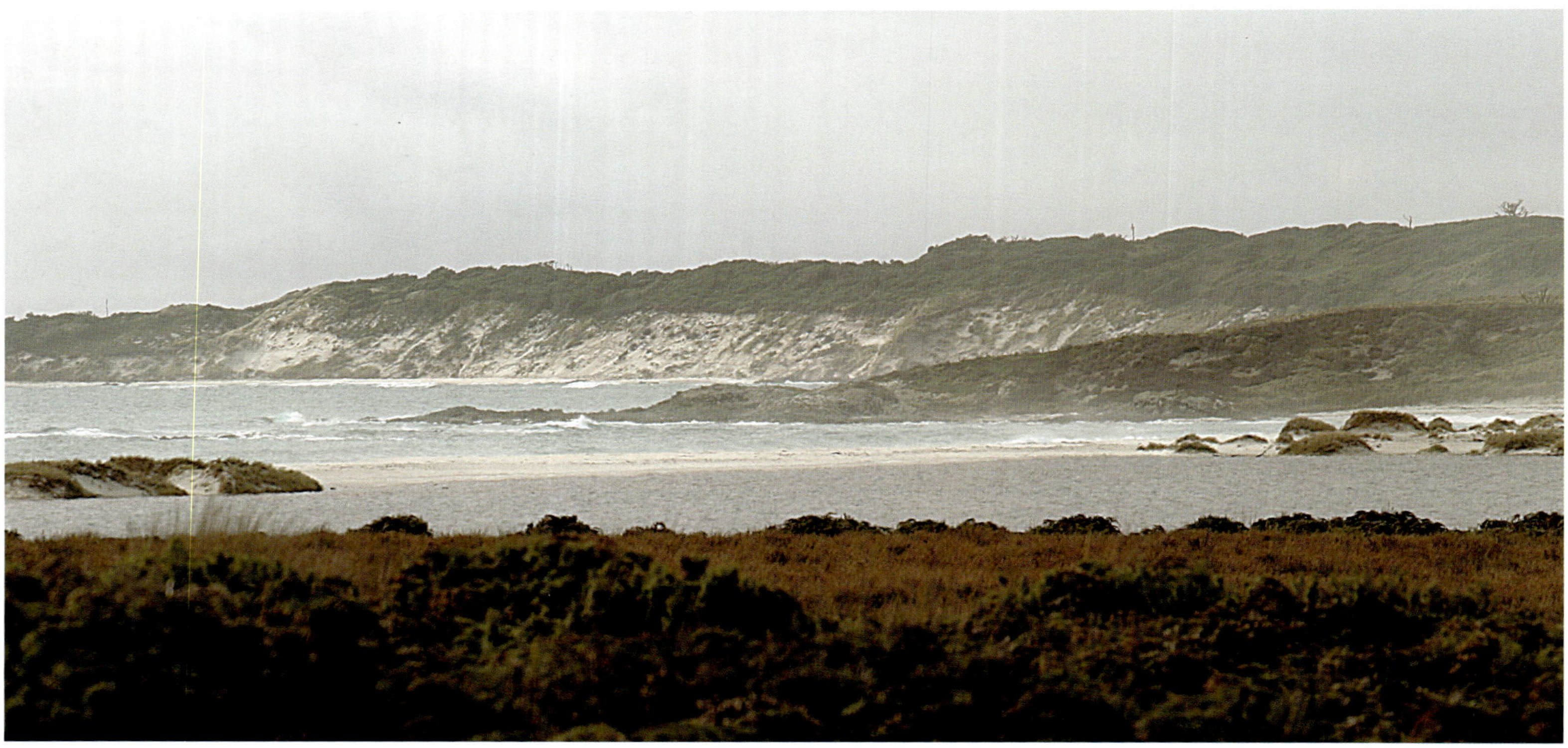

The wild northern coastline.

The Chatham Islands sit at a far-flung point in the ocean, more than 800 kilometres east of Christchurch, close to where the warm streams from the north collide with the Antarctic currents swirling up from the south, resulting in the frequent fogs that drape the islands. They consist of Chatham Island (the main island), Pitt Island and its offspring of South East Island and Mangere Island, plus a proliferation of smaller islets and rock stacks. They all became officially part of New Zealand in 1842 although, interestingly, effective British law did not operate until 1863 when the resident magistrate secured the release of Moriori held as slaves by Maori.

The scenery is unlike that of any other islands I have visited around the country. The landscape of Chatham Island itself is low-lying, broken here and there by the occasional volcanic excrescence and fringed in places with long, sweeping beaches of golden sand. The lagoon is vast, taking up nearly 20 percent of the total area. Along with a string of nearby lakes, it is inhabited by flocks of black swans that scribe graceful hieroglyphics across the surface of the water and then occasionally electrocute themselves as they take off and fail to clear the power lines in their path.

Cut off from the sea four years out of five, the lagoon is nevertheless a saltwater habitat that provides a home also for flounder, whitebait, eels, ducks and migratory wading birds. At its edges, rushes grow thickly, covering huge expanses of boggy peat, formed by the accumulation of dead plant matter over hundreds of years. The rest of the island is divided between pastureland and large areas of bracken. Buff weka dart in and out of the undergrowth and sabotage islanders' gardens.

Many of New Zealand's offshore islands came into being when river valleys were drowned at the end of the ice age, leaving only the peaks above water. In contrast it is less likely that the Chathams were ever connected to the main islands. There is no evidence that tall podocarps, beeches or rata ever grew there. In their place some species that we normally see as shrubs have become bigger to form the canopy cover. Chatham Island now boasts much larger versions of plants that we're familiar with on the 'mainland', such as karaka (corynocarpus), originally brought to the islands by the Moriori, corokia and dracophyllum.

Over time some of the plant species that managed to survive have adapted to the conditions and eventually evolved into new species. They include trees as well as several giant or mega herbs such as the well-known Chatham Island forget-me-not. Even the flax differs from mainland flax. The leaf blades are wider and the whole plant seems more substantial – probably to help it withstand the harsh climate.

Surprisingly, it is not bitterly cold most of the time but the terrain offers little protection from ferocious south-westerlies that rake the island and batter the trees. *Olearia traversii*, called akeake on the islands, has become an icon of the Chathams. It survives in very exposed sites and many are tortured into weirdly arthritic shapes – some even growing parallel to the ground! Gardeners on the island need to establish wind shelter before they can even think of planting ornamentals.

Before the arrival of humans, the islands were covered with a thickly woven canopy of tough trees that sheltered the less robust varieties. Now only a few stands of bush remain and a nikau palm variety, which is endemic to the Chathams, has almost disappeared. A few concerned landowners have fenced off reserve land and, with stock excluded, bush is regenerating. Near the coastline occasional 'sand blows' are glaringly visible, silvery-white in the sunlight and punctuated by the skeletons of trees that the sand dunes have overcome in their relentless advance. They are stark reminders that the landscape has been vastly altered.

The islands the Moriori found when they arrived several hundred years ago must have offered a much more sheltered environment than that which today's residents enjoy. Little is known about these original settlers. Why they finished up at the Chathams is an enigma but current thinking supposes they arrived some time before the 15th century and that they, like the Maori, are descendants of the first Polynesians who came to New Zealand. Over generations in such an isolated community they developed their own version of the originally common Polynesian language and, more remarkably, developed a way of life that eschewed violence. Warfare was unknown in their culture and differences of opinion were solved by hand-to-hand combat. When the first blood was drawn, the issue was at an end.

So they were ill-equipped to cope with a group of Maori who came marauding in the mid-1830s, having hijacked a ship at Somes Island in Wellington Harbour, along with its captain. It is estimated that about 300 Moriori were killed outright and many among them eaten. The rest were enslaved. When the Moriori finally won their freedom in the 1860s, their numbers had shrunk from an estimated 1600 in 1835 to somewhere between 100 and 150. When they attempted to reclaim ownership of their land in the late 1860s a Native Land Court ruling awarded most of the Chathams to Maori, by right of conquest.

Maori in fact were the third in line among visitors to the Chathams. In 1791, the Royal Navy brig *Chatham*, under the command of Lieutenant William Broughton, 'discovered' the islands and gave them their English name. Moriori had christened the main island Rekohu, meaning 'misty skies'.

Little more than a decade after Broughton paid his fleeting visit, European sealers came to the Chathams. Frequently they were deserting sailors or former convicts – from all accounts, a tough bunch of entrepreneurs. By the early 1840s when they had almost annihilated the fur seal colonies, many moved on – a recurring theme on the Chathams – with a few remaining to become long-term settlers. For a few short years, shore whaling replaced sealing and when that too failed, the remaining Europeans joined local Maori in agricultural pursuits.

There is a tradition of boom/bust enterprise in the Chathams. Teresa McDonald, a fifth-generation Chatham Islander, who traces her descent from a Portuguese sailor and from an early Maori arrival on the island, spoke of her grandfather who farmed near Kaingaroa on the north-eastern tip of the island. He grew potatoes, and for several years his business flourished. A jack of all trades, he salvaged timber from two of the frequent shipwrecks on the coast and built

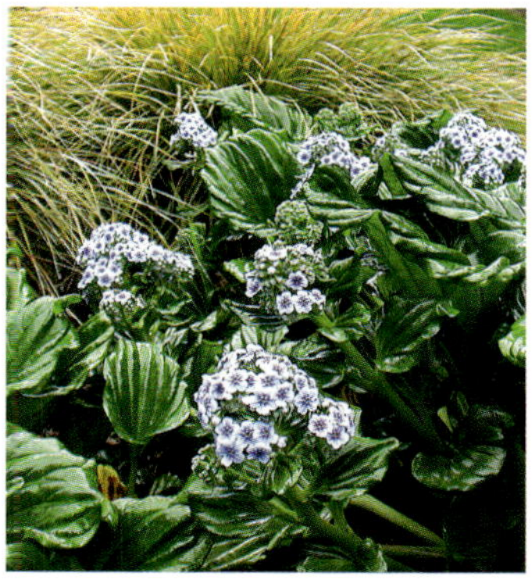

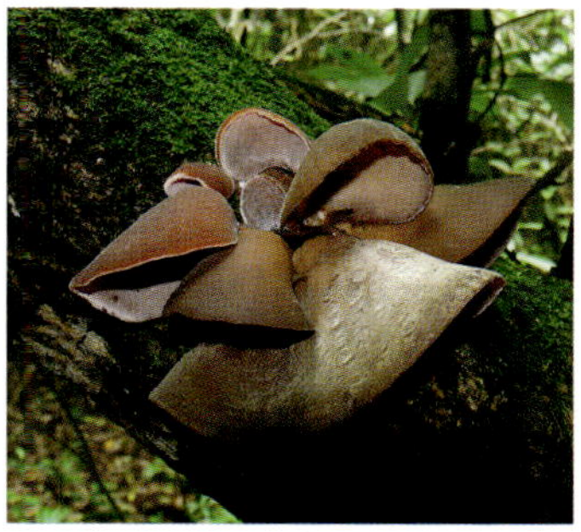

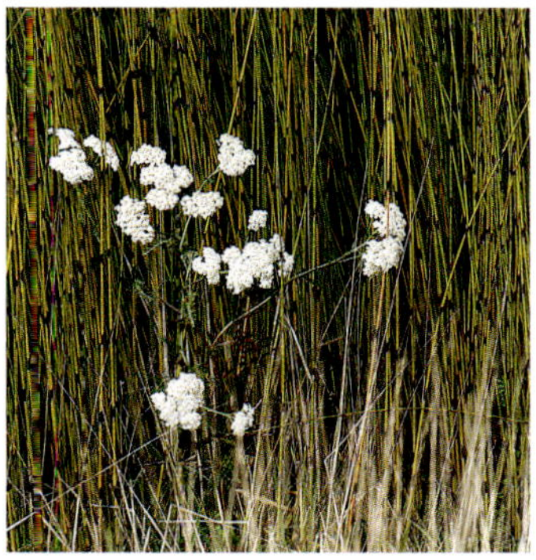

Some of the flora of Chatham Island. From top: the Chatham Island forget-me-not; a spectacular forest fungus; wildflowers growing among the reeds around the edges of Te Whanga Lagoon.

How to get there

Air Chathams operates regularly scheduled flights between Chatham Island, Auckland, Wellington and Christchurch.

his family an elegant 23-roomed house, lit by gaslights from a sailing ship. When his potato crop failed from blight a few years later, he dismantled the mansion and took it to Rapanui where he started from scratch again, buying cows and selling their milk to a cheese factory.

Growing potatoes was big business on the Chathams about the time of the infamous potato famine in Ireland. David Holmes, a long-term resident of Chatham Island who at one stage grew potatoes to supply the cookhouse for a fish-processing plant, once told me that the Chathams were regarded as the Potato Garden of the Pacific in the 1850s and 60s. Grown by Maori, 'hundreds of tons' were exported to Australia and bartered initially for blankets, later for horses. It was rumoured that 30 tons of potatoes would procure one horse. Chatham Island potatoes also fed the gold miners of Central Otago. Nearly a century later, at the beginning of the Second World War, when potatoes were once again in short supply in New Zealand, Chatham Island potatoes came to the rescue.

I stayed at Chatham Lodge, owned and managed by John and Denise Sutherland. They came to Chatham Island nearly 30 years ago because they saw it as an opportunity to get into farming. Cattle graze by the roadside leading up to their lodge and more sleek-looking beasts dot the lush paddocks viewed from the lodge windows.

A view across the township of Waitangi.

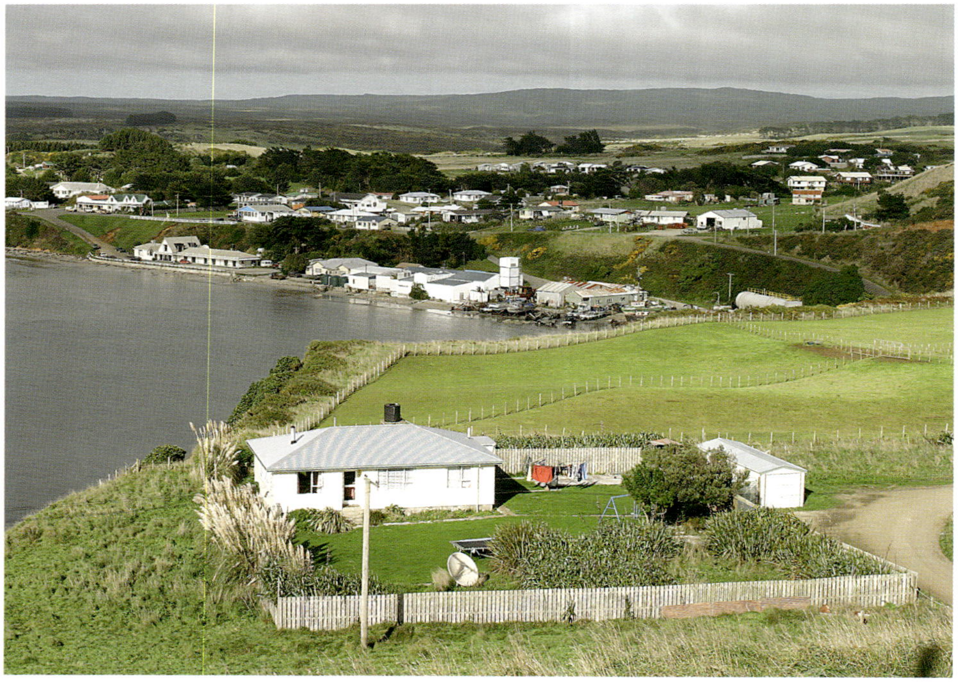

John and Denise branched out into other enterprises and some years ago also dedicated 500 acres of their farmland as a reserve. They fenced it off to exclude stock and now the native bush is regenerating strongly. On the evening I arrived I followed a track they maintain and emerged from the bush to wander over endless sand dunes and watch the sun set in a flare of drama over the sea. Sunsets and sunrises are frequently spectacular; Pitt Island sees the world's first sunrise each morning.

Gravel roads traverse the island and trace a thin line around most of the coast, linking the far-flung homesteads and connecting the settlements of Port Hutt, Te One, Waitangi, Kaingaroa and Owenga. There are few cars on the road – though numerous dead cars litter fields and grass verges – and moving around the island is a friendly business for every driver waves to every other driver on the road. Lots of horses graze in the paddocks. They're still used for mustering, and horse racing has been a popular pastime since horses were imported from Australia in the 19th century.

St Augustine's Anglican Church, the oldest church on the Chathams, is at Te One, as is the largest primary school. Waitangi boasts a couple of shops, a museum, a fish-processing plant, a café that's open most days and serves superb fish and chips, and a hotel on the foreshore that is the life and centre of the community – though less so now that Sky television has come to the island.

In the early days of European settlement there were stresses and strains among the various races on the island. Over time, with intermarriage, many of the differences were smoothed out. "When we arrived in the 1970s," John Sutherland remembered, "everyone considered themselves just islanders. But it's changed again. Many people now claim their Maori or Moriori roots, or both."

Tommy Solomon, reputed to be the last full-blooded Moriori, died in 1933 and a statue was erected in his memory in 1986. He stands tall, overlooking the sea, at Manukau, the last Moriori reserve on the Chatham Islands.

Teresa McDonald affirmed however that it is still a close-knit community. "If something bad happens, everyone pulls together. There is a strong sense of belonging," she said.

As it has through most of its history, the livelihood of the island depends in large measure on its coastal resources. For the last century, fishing has been an important income earner. A

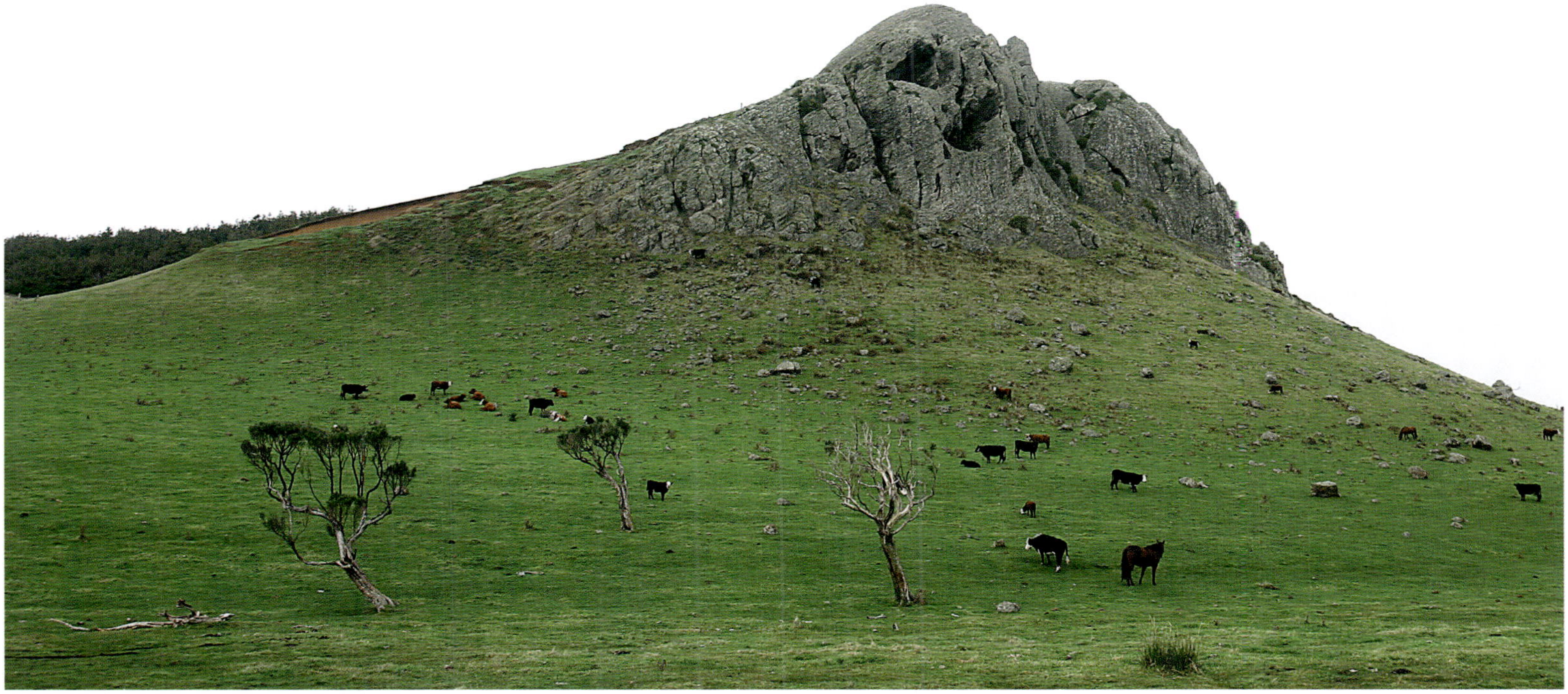

Volcanic outcropping near Waitangi West.

boom time hit for a few short years between 1966 and 1970 when crayfishing was at its height. Eight new factories opened on the island and helicopters were kept busy lifting baskets of crays off the boats – as many as 147 at the height of the frenzy – and ferrying them to the processing plants. The population of the islands grew, there was more work than ever before and more money to spend.

However, as with sealing and whaling, the resource was exploited way beyond a sustainable level and by 1976 all the helicopters had gone, most of the profits likewise and the population had returned almost to its former number. Nowadays it hovers at about 700 residents.

A radio-telephone link was established with New Zealand in 1953 and following the building of a flying boat base during the Second World War – constructed by community effort in two short weeks – the island has been linked to the mainland, more or less continuously, by an air service.

There's a haunting loneliness about Chatham Island that reminds me of the Isle of Iona off the coast of Scotland. It's a loneliness that both attracts and repels. It's easy to imagine it becoming obsessive.

If Chatham Island is lonely, then Pitt Island ought to be deserted. Only a few hectares in area, it lies south of Chatham Island, situated squarely in the path of the Roaring Forties, a belt of continuous gales that rage around the globe below the 44th parallel. The only regular access to the island, across 20 kilometres of sea, is by plane. And if the mist creeps in, the plane does not fly. Any plans I had to go there were frustrated by the weather – I had no desire to be marooned there for three weeks as my hosts at Chatham Lodge had once been.

The island is hilly, open country, grazed by farm animals, with a few patches of regenerating forest. The coastline is more rugged than that of Chatham Island and there is no harbour. Incredibly, people have lived there continuously for the last 160 years, engaged in fishing and farming.

In 1843 Frederick Hunt, not long arrived in New Zealand from England, went to the island with his wife and young family. He set about building a primitive hut for them, before he departed for the mainland to organise supplies for their future life on the island, leaving his wife to cope alone, until he was able to return. They went on to have six children, establish a thriving farm and vegetable garden from which they supplied visiting whalers, and his wife never left the island again. Eventually he built a substantial house from materials salvaged from a wrecked whaler.

In the 1860s, Hunt encouraged two young farmers, James Langdale and Matthew Gregory, to take up land on Pitt Island and help him raise sheep. They married two of Hunt's daughters and their sons inherited the estate when Frederick Hunt died in 1891 – on the proviso that they added Hunt to their names.

In the 1970s there was a large flock of feral sheep on the island, most probably descendants of the Saxony merino sheep that Frederick Hunt imported to the island in the 1840s. A reserve for 300 of the animals was created on Pitt Island in 1981 and rare breed owners around the country have also established small flocks of the sheep.

Today the island is still farmed by descendants of the Gregory-Hunts and Langdale-Hunts and it is their children who keep the Pitt Island school alive.

Chatham Island, at a little more than 90,000 hectares, is by far the largest of all the islands that sit around the 44th parallel. Pitt Island is smaller in area than the Te Whanga Lagoon on Chatham. The Sisters are two tiny rocky islets, so small that it's almost impossible to see them on a map, situated north-west of Chatham Island. Wandering albatrosses and royal albatrosses nest on these islands and on The Forty Fours to the east, although the birds' habitat on The Forty Fours was almost destroyed by Cyclone Bola, which blew away most of the earth. In the south are the Star Keys Islands and The Pyramid, a remote and inaccessible rock covered in albatrosses and mollymawks, their nests and their leavings.

Better known, because of the story of the saving of the black robin, are the slightly bigger islands around Pitt Island – Mangere with its steep summit, which is the highest point in the Chathams, Little Mangere and South East Island, home of a shore-based whaling station in the mid-1800s.

Mangere was farmed from the end of the 19th century but in 1966 Forest and Bird joined with the then Forest Service to buy the island and set it aside as a nature reserve. And it was to Mangere Island that legendary bird conservation officer Don Merton transferred the last seven black robins in the world from their precarious home on Little Mangere and proceeded with his revolutionary cross-fostering programme. There are now around 250 black robins in two populations on Mangere Island and South East Island.

Michael King in his book first published in 1993 called the Chatham Islands *A Land Apart.* I have no doubt that the islanders consider themselves a people apart, with a fierce loyalty to these wild islands that give them their identity and which they hold to their heart.

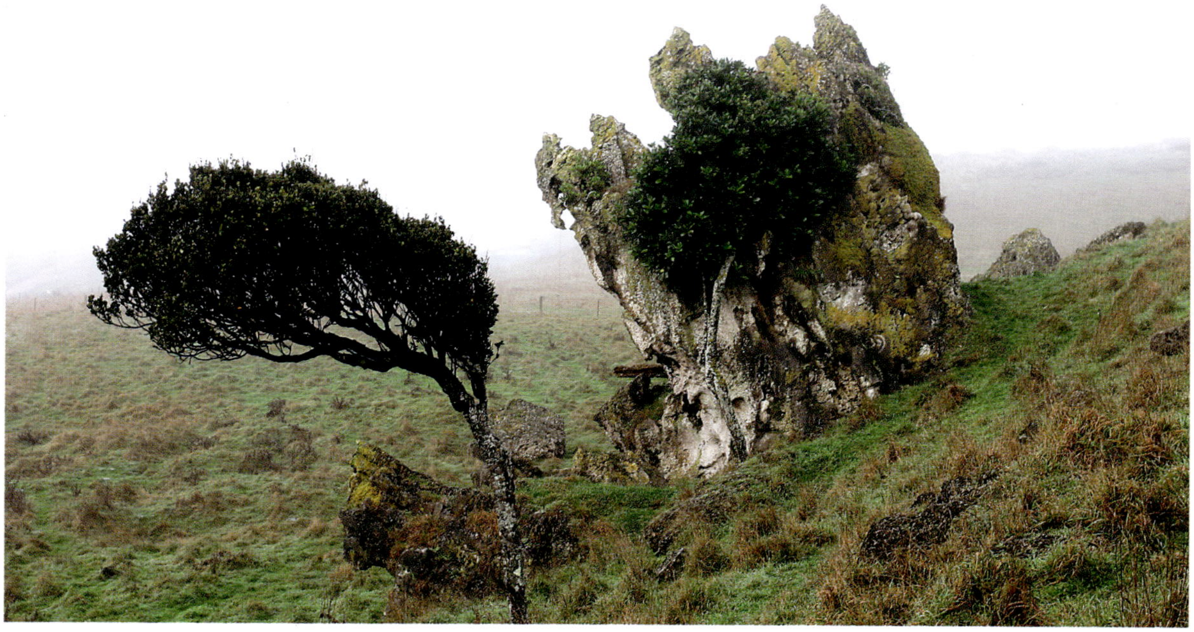

The prevailing south-westerlies sculpt the local trees and shrubs, in this case a Chatham Island akeake.

Bibliography

Begg, A. Charles and Neil C. *Dusky Bay.* Whitcoulls, 1966.

Brown, Derek. *Stephens Island, Ark of the Light.* Derek Brown, no date.

Cemmick, David and Veitch, Dick. *Black Robin Country.* Hodder and Stoughton, 1985.

Dawson, Bee. *Lady Painters.* Viking, 1999.

Dawson, John and Lucas, Rob. *New Zealand Coast and Mountain Plants.* Victoria University Press, 1996.

Ell, Gordon. *Wild Islands.* The Bush Press, 1982.

Hall-Jones, John. *Fiordland Explored.* A H & A W Reed, 1976.

Hall-Jones, John. *Stewart Island Explored.* Craig Printing, 1994.

Hamilton, Mary, Whiting, D'Arcy and Battersby, Philip. *Royal Akarana Yacht Club Coastal Cruising Handbook.* Landsdowne Press, 1980.

Heberley, Heather. *Weather Permitting.* Cape Catley Ltd, 1996.

Holmes, David. *My Seventy Years on the Chatham Islands.* Shoal Bay Press, 1993.

Jones, Clair. *New Zealand's Bay of Islands – The Land and Sea Guide.* Port of Opua Trading Co., no date.

King, Michael and Morrison, Robin. *A Land Apart: The Chatham Islands of New Zealand.* Random House (NZ) Ltd, 1993.

King, Michael. *The Collector – Andreas Reischek, A Biography.* Hodder and Stoughton, 1981.

King, Michael. *The Penguin History of New Zealand.* Penguin Books (NZ) Ltd, 2003.

Lee, Jack. *'I have named it the Bay of Islands…'.* Hodder and Stoughton, 1983.

Maclean, Chris. *Kapiti.* Whitcombe Press, 1999.

Maddock, Shirley. *Islands of the Gulf.* Collins, 1966.

McGill, David. *Island of Secrets.* Steele Roberts and Silver Owl Press, 2001.

Metcalf, Lawrie. *The Cultivation of New Zealand Trees and Shrubs.* Reed Books (NZ) Ltd, 1991 (fully revised edition).

Monin, Paul. *Waiheke Island – a history.* The Dunmore Press Ltd, 1992.

New Zealand Geographic, Numbers 20, 29.

Richards, E.C. *The Chatham Islands: Plants, Birds and People.* Simpson and Williams Ltd, 1952.

Roxburgh, Irvine. *Wanaka Story.* Otago Centennial Historical Publications, 1957.

Scanlan, A.B. *Harbour at the Sugar Loaves.* Taranaki Harbour Board, 1975.

Tait, Peter and Jenny. *White Island, New Zealand's Most Active Volcano.* Godwit, Random House (NZ) Ltd, 2001.

The Story of the Bay of Islands Maritime and Historic Park. The Bay of Islands Maritime and Historic Park, 1989.

Wilson, Nora Creina, *Memories of Mansion House.* Richards Publishing, 1980

Websites

Department of Conservation
www.doc.govt.nz

Dictionary of New Zealand Biography
www.dnzb.govt.nz

Royal New Zealand Forest and Bird Society
www.forest-bird.org.nz

Index

Page numbers in italics indicate a photograph.

PHOTOGRAPHY CREDITS

All photographs taken by Russell McGeorge except:

Andy Belcher/Legend Photography Ltd, p 85 Motiti Island;

GNS Images, p 21 Motuopao Island, p 91 White Island;

Roy Grose/Mike Avis DOC Picton, p 15 tuatara, p 115 giant weta;

Warren Judd, p 20 Three Kings Islands;

Photobank, p 47 Rocky Bay, Waiheke p 88 Matakana Island;

PhotoNewZealand.com, p 61 Port Abercrombie (Chris MacDonald), p 107 Queen Charlotte Sound (Ian Trafford), p 123 Doubtful Sound (Geoff Mason);

Bruce Shanks FPSNZ, ANZIPP, p 58 whitehead, p 68 stitchbird.